T0191720

Texts & Monographs in Symbolic Computation

A Series of the Research Institute for Symbolic Computation, Johannes Kepler University, Linz, Austria

Founding Editor

Bruno Buchberger, Linz, Austria

Series Editor

Peter Paule, Linz, Austria

 MyCopy

Mathematics is a key technology in modern society. Symbolic Computation is on its way to become a key technology in mathematics. "Texts and Monographs in Symbolic Computation" provides a platform devoted to reflect this evolution. In addition to reporting on developments in the field, the focus of the series also includes applications of computer algebra and symbolic methods in other subfields of mathematics and computer science, and, in particular, in the natural sciences. To provide a flexible frame, the series is open to texts of various kind, ranging from research compendia to textbooks for courses.

Indexed by zbMATH.

Clemens G. Raab · Michael F. Singer

Editors

Integration in Finite Terms: Fundamental Sources

 Springer

Editors
Clemens G. Raab
Institute for Algebra
Johannes Kepler University of Linz
Linz, Austria

Michael F. Singer
Department of Mathematics
North Carolina State University
Raleigh, NC, USA

ISSN 0943-853X   ISSN 2197-8409   (electronic)
Texts & Monographs in Symbolic Computation
https://doi.org/10.1007/978-3-030-98767-1

Responsible Editor: Martin Peters
This Springer imprint is published by the registered company Springer Nature Switzerland AG
The registered company address is: Gewerbestrasse 11, 6330 Cham, Switzerland
(www.springernature.com/mycopy)

Preface

The Fundamental Theorem of Calculus tells us that if $f(x)$ is a Riemann integrable function on an interval $[a, b]$ and $F(x)$ is a function with $\frac{dF(x)}{dx} = f(x)$, that is, $F(x)$ is an antiderivative of $f(x)$, then

$$\int_a^b f(x)dx = F(b) - F(a).$$

This theorem motivated the development and teaching of various heuristic techniques for finding antiderivatives in order to evaluate definite integrals. In basic calculus courses, the focus is usually on the so-called elementary functions $(e^x, \sin(x), \log(x), \arctan(x), \sqrt{x}, \ldots)$ and these heuristic techniques yield antiderivatives expressible again in terms of these elementary functions. Two natural questions arise:

1. Do all elementary functions have an elementary antiderivative and, if not, how does one characterize those that do?
2. Is there an algorithm to determine if an elementary function has an elementary antiderivative and, if so, can we find one?

Liouville's Theorem answers the first question and can be used to show that e^{x^2} does not have an elementary antiderivative and Risch's Algorithm gives a positive answer to the second question. These results are now central to a well-developed area devoted to finding antiderivatives and, more generally, solutions of differential equations, expressible in terms of classes of elementary and special functions; the subject of Integration in Finite Terms.

This volume presents four of the foundational texts related to this topic as well as commentaries discussing these texts and their impact. The first is a reprint of Maxwell Rosenlicht's paper *Integration in Finite Terms* [Ros72], presenting, in a very elementary way, a purely algebraic proof of Liouville's Theorem. A commentary by Michael F. Singer gives an outline of the history of Liouville's Theorem and its modern developments. The second text is a reprint of Joseph Fels Ritt's book *Integration in Finite Terms: Liouville's Theory of Elementary Functions* [Rit48]. This 1948 book presents Liouville's (and others') analytic approach to this subject. It is followed by a commentary by Askhold Khovanskii presenting his topological approach to this problem. The third is a revised version of Robert H. Risch's unpublished *On the integration of elementary functions which are built up using algebraic operations* [Ris68] incorporating unpublished results of [Ris69]. This reduced the problem of finding elementary antiderivatives of elementary functions to a problem in arithmetic algebraic geometry: the problem of determining if a point on the jacobian of a curve is of finite order. Risch presented a solution of the latter problem in [Ris70] in the context of deciding the elementary integrability of an algebraic function. Clemens G. Raab's commentary discusses the impact of this paper and further developments. The final text is Barry M. Trager's unpublished thesis *Integration of Algebraic Functions* [Tra84], which provided practical algorithms for performing the integration in finite terms of algebraic functions after the theoretical results of Risch. It is followed by a commentary by Barry Trager, who gives further insight into the methods of his thesis as well as subsequent developments.

Special thanks go to Peter Paule, whose encouragement and support were crucial to beginning this project and seeing it to completion.

References

[Ris68] Robert H. Risch, *On the integration of elementary functions which are built up using algebraic operations*, SDC Document SP-2801/001/00, June 1968.

[Ris69] _____, *Further Results on Elementary Functions*, IBM report RC-2401, 1969.

[Ris70] _____, The solution of the problem of integration in finite terms, *Bull. Amer. Math. Soc.* **76** (1970), 605–608. MR 0269635

[Rit48] Joseph F. Ritt, *Integration in Finite Terms. Liouville's Theory of Elementary Methods*, Columbia University Press, New York, N. Y., 1948. MR 0024949

[Ros72] Maxwell Rosenlicht, Integration in finite terms, *Amer. Math. Monthly* **79** (1972), 963–972. MR 0321914

[Tra84] Barry M. Trager, *Integration of Algebraic Functions*, Ph.D. thesis, Massachusetts Institute of Technology, 1984.

Contents

Contents

Integration in Finite Terms

Maxwell Rosenlicht

Amer. Math. Monthly **79**(9), 963–972 (1972)

1. The question arises in elementary calculus: Can the indefinite integral of an explicitly given function of one variable always be expressed "explicitly" (or "in closed form", or "in finite terms")? Liouville gave the answer one would expect, "No", and he proved in particular that such is not the case with $\int e^{x^2} dx$. Since we have all fallen into the habit of quoting this result and giving neither proof nor reference, it may be worthwhile to actually state it as precisely as possible and give a proof that is as elementary as the subject matter might suggest.

We must define our terms carefully. To begin with, we are not interested in arbitrary functions, but in **elementary functions**, which are functions of one variable

Maxwell Rosenlicht did his Ph.D. work at Harvard, under Oscar Zariski. He was a National Research Fellow at Chicago and Princeton, and held a position at Northwestern Univ. before his present position at the Univ. of California, Berkeley. He has spent visits at the Univ. of Rome, the IHES-Paris, the Univ. of Mexico, Harvard Univ., and Northwestern Univ., and he has had Fulbright and Guggenheim Fellowships. In 1960, he won the American Mathematical Society's Cole Prize in Algebra (with S. Lang). His main research is in algebraic geometry, algebraic groups, and differential algebra, and he is the author of *Introduction to Analysis* (Scott, Foresman, 1968). *Editor.*

built up by using that variable and constants, together with repeated algebraic operations and the taking of exponentials and logarithms. Since we lose no generality by doing so, we shall take all exponentials and logarithms to the base e. We allow ourselves the convenience of the use of complex numbers, for with these the various trigonometric and inverse trigonometric functions turn out to be elementary, as seems reasonable. Thus the integral of a rational function of one real variable is elementary, since it is a linear combination of logarithms, inverse tangents, and rational functions. But we are still deficient in precision, because of the multi-valuedness of algebraic functions and logarithms. The functions we work with must be specific objects, each susceptible of an unambiguous sense. We choose to avoid the difficulties associated with multivaluedness by the simplest method, that of restricting ourselves, in any given discussion, to functions on some specific region (that is, nonempty connected open subset) of the real numbers $\mathbb{R}$ or the complex numbers $\mathbb{C}$, and furthermore considering only meromorphic functions on the region in question, a meromorphic function on a region being a function whose values are complex numbers or the symbol ∞, with the property that sufficiently near any point z_0 of the region the function is given by a convergent Laurent series in $z - z_0$, that is, a convergent power series in $z - z_0$, with the possible addition of a finite number of negative powers. Thus the rational functions of one variable, which form the field $\mathbb{C}(z)$ got by adjoining the identity function z to the field of constant functions $\mathbb{C}$, are all meromorphic on all of $\mathbb{R}$ or $\mathbb{C}$. The exponential of a function f meromorphic on a certain region of $\mathbb{R}$ or $\mathbb{C}$ is a function meromorphic on the subregion obtained by deleting those points where the value of f is ∞ (and then taking a connected component, if we are working in $\mathbb{R}$), while $\log f$ can be taken to be meromorphic on any simply connected subregion where f takes on neither of the values 0 or ∞, by arbitrarily choosing one of its many values at any particular point of the subregion. Furthermore, the implicit function theorem shows that if we are given a polynomial equation with coefficients which are functions meromorphic on a certain region, the leading coefficient not being zero, then there exists a meromorphic solution on a suitable subregion. Thus any complicated expression for an elementary function, compounded of algebraic operations, exponentials and logarithms, has a realization as a meromorphic function on some region. Now the totality of all meromorphic functions on a given region form a field under the usual operations of functional addition and multiplication, and the restriction of all these functions to any given subregion gives an embedding of fields. The derivative of a function meromorphic on a given region is again meromorphic, as is an indefinite integral, if one exists, of the function. Note that the rational functions on a region, that is the restriction of $\mathbb{C}(z)$ to this region, are a field of meromorphic functions on the region that are closed under differentiation, and that if we have any field of meromorphic functions on a region that is closed under differentiation and get a larger field of meromorphic functions on the region by adjoining the exponential or a logarithm of a function in our field, or a solution of a polynomial equation with coefficients in the field, we again get a field of mero-morphic functions on the region that is closed under differentiation. Thus the proper objects of study are seen to be fields of meromorphic functions on given regions in $\mathbb{R}$ or $\mathbb{C}$ which are closed under differentiation. If a function in such a field has an indefinite integral that is expressible "in finite terms," then by restricting all func-tions, if necessary, to a suitable subregion, we see that we have a tower of such fields of meromorphic functions, each larger field being obtained by adjunction of an exponential, or a logarithm, or the solution of an algebraic equation, the tower starting with the original field and culminating in a field containing the indefinite integral. Thus the original loosely worded analytic problem, when formulated as a precise analytic problem, becomes algebraic.

2. Define a **differential field** to be a field F, together with a **derivation** on F, that is, a map of F into itself, usually denoted $a \mapsto a'$, such that $(a + b)' = a' + b'$ and $(ab)' = a'b + ab'$ for all $a, b \in F$. Immediate consequences are that $(a/b)' = (ab' - a'b)/b^2$ if $a, b \in F$, $b \neq 0$, and $(a^n)' = na^{n-1}a'$ for all integers n. Furthermore, $1' = (1^2)' = 2 \cdot 1 \cdot 1'$, so $1' = 0$. Therefore the **constants** of F, that is, all $c \in F$ such that $c' = 0$, are a subfield of F.

If a, b are elements of the differential field F, a being nonzero, let us agree to call a **an exponential of** b, or b **a logarithm of** a, if $b' = a'/a$; this terminology is not unreasonable for our present purposes since the only properties of exponentials and logarithms in which we are interested are their differential properties. We im-mediately get the "logarithmic derivative identity,"

$$\frac{(a_1^{v_1} \cdots a_n^{v_n})'}{a_1^{v_1} \cdots a_n^{v_n}} = v_1 \frac{a_1'}{a_1} + \cdots + v_n \frac{a_n'}{a_n},$$

for $a_1, \cdots, a_n$ nonzero elements of F and $v_1, \cdots, v_n$ integers.

3. There is a standard result on algebraic extensions of differential fields which we shall need later. For completeness we prove it here. The result is that if F is a differential field of characteristic zero and K an algebraic extension field of F, then the derivation on F can be extended to a derivation on K, and this extension is unique. (Thus K has a unique differential field structure extending that of F. We remark that the restriction to characteristic zero is not essential; it suffices to assume that K is separable over F, and the following proof will hold in this more general case.) For the reader who is interested only in the classical function-theoretic case, where the fields in question are fields of meromorphic functions on a region of $\mathbb{R}$ or $\mathbb{C}$, the proof is immediate, the existence proof being a direct consequence of the implicit function theorem, uniqueness following from the ordinary method of computing derivatives of functions given implicitly. To prove the result generally, let X be an indeterminate and define the maps D_0, D_1 of the polynomial ring $F[X]$ into itself by

$$D_0\left(\sum_{i=0}^{n} a_i X^i\right) = \sum_{i=0}^{n} a_i' X^i, \quad D_1\left(\sum_{i=0}^{n} a_i X^i\right) = \sum_{i=0}^{n} i a_i X^{i-1}$$

for $a_0, a_1, \cdots, a_n \in F$. If K has a differential field structure extending that of F, then for any $x \in K$ and any $A(X) \in F[X]$ we have

$$(A(x))' = (D_0 A)(x) + (D_1 A)(x) \cdot x'.$$

If we replace $A(X)$ by the minimal polynomial $f(X)$ of x over F, (that is, the monic irreducible polynomial of which x is a root, indeed a simple root, so that $(D_1 f)(x) \neq 0$), we get $x' = -(D_0 f)(x) / (D_1 f)(x)$. Thus the differential field structure on K that extends that on F is unique, if it exists. We now show that such a structure on K exists. Using the usual field-theoretic arguments, we may assume that K is a finite extension of F, so that we can write $K = F(x)$, for a certain $x \in K$. For some $g(X) \in F[X]$, to be determined later, let the map $D: F[X] \to F[X]$ be defined by

$$DA = D_0 A + g(X) D_1 A,$$

for any $A \in F[X]$. It follows immediately that $D(A + B) = DA + DB$ and $D(AB) = (DA)B + A(DB)$ for all $A, B \in F[X]$, since the analogous identities hold for both D_0 and D_1. Note that $Da = a'$ for all $a \in F$. Now look at the natural surjective ring homomorphism $F[X] \to F[x]$, which is the identity on F and sends X into x. Since $F[x] = F(x) = K$, the map D on $F[X]$ will induce a derivation on K extending that on F if it so happens that D maps the kernel of our ring homomorphism into itself. But the kernel of the homomorphism is the ideal $F[X]f(X)$, where $f(X)$ is the minimal polynomial of x over F. Hence we shall have proved our result once we have shown that D maps $F[X]f(X)$ into itself. The condition for this is simply that D map $f(X)$ into a multiple of itself, that is that Df be any element of $F[X]$ of which x is a root, or that $(Df)(x) = 0$. But this last condition reduces to $(D_0 f)(x) + g(x)(D_1 f)(x) = 0$. Since $(D_1 f)(x) \neq 0$ and $F(x) = F[x]$, a polynomial $g(X) \in F[X]$ can actually be found such that $(Df)(x) = 0$, and this completes the proof of our statement.

4. By a **differential extension field** of a differential field F we mean, of course, a differential field which is an extension field of F whose derivation extends the derivation on F. The following result will be the principal tool for proving the theorem of the next section, and will be used for the verification of our subsequent examples.

LEMMA. *Let F be a differential field, $F(t)$ a differential extension field of F having the same subfield of constants, with t transcendental over F, and with either $t' \in F$ or $t'/t \in F$. If $t' \in F$, then for any polynomial $f(t) \in F[t]$ of positive degree, $(f(t))'$ is a polynomial in $F[t]$ of the same degree as $f(t)$, or degree one less, accrding as the highest coefficient of $f(t)$ is not, or is, a constant. If $t'/t \in F$, then for any nonzero $a \in F$ and any nonzero integer n we have $(at^n)' = ht^n$, for some nonzero*

$h \in F$, and furthermore, for any polynomial $f(t) \in F[t]$ of positive degree, $(f(t))'$ is a polynomial in $F[t]$ of the same degree, and is a multiple of $f(t)$ only if $f(t)$ is a monomial.

We first consider the case $t' = b \in F$. Let the degree of $f(t)$ be $n > 0$, so that $f(t) = a_n t^n + a_{n-1} t^{n-1} + \cdots + a_0$, with $a_0, \cdots, a_n \in F$, $a_n \neq 0$. Then

$$(f(t))' = a_n' t^n + (na_n b + a_{n-1}')t^{n-1} + \cdots.$$

This is clearly a polynomial in $F[t]$, of degree n if a_n is not constant. If a_n is constant and $na_n b + a_{n-1}' = 0$, then $(na_n t + a_{n-1})' = na_n b + a_{n-1}' = 0$, so that $na_n t + a_{n-1}$ is a constant, therefore an element of F, so that $t \in F$, contrary to the assumption that t is transcendental over F. Thus if a_n is constant, $(f(t))'$ has degree $n - 1$.

Now suppose that we are in the case $t'/t = b \in F$. Let $a \in F$, $a \neq 0$, and let n be a nonzero integer. Then

$$(at^n)' = a't^n + nat^{n-1}t' = (a' + nab)t^n.$$

If $a' + nab = 0$, then $(at^n)' = 0$, so that at^n is constant, therefore an element of F, contradicting the transcendence of t over F. Therefore $a' + nab \neq 0$. Finally, let $f(t) \in F[t]$ have positive degree. Clearly $(f(t))'$ has the same degree. If $(f(t))'$ is a multiple of $f(t)$, it must be by a factor in F. Therefore if $f(t)$ is not a monomial, $a_n t^n$ and $a_m t^m$ being two of its different terms, and $(f(t))'$ is a multiple of $f(t)$, we have

$$\frac{a_n' + na_n b}{a_n} = \frac{a_m' + ma_m b}{a_m},$$

so

$$\frac{a_n'}{a_n} + n\frac{t'}{t} = \frac{a_m'}{a_m} + m\frac{t'}{t},$$

or $(a_n t^n / a_m t^m)' = 0$, so that $a_n t^n / a_m t^m \in F$, again contradicting the transcendence of t over F. This completes the proof.

5. Let F be a differential field. Define an **elementary extension of** F to be a differential extension field of F which is obtained by successive adjunctions of elements that are algebraic, or logarithms, or exponentials, that is, a differential extension field of the form $F(t_1, \cdots, t_N)$, where for each $i = 1, \cdots, N$, the element t_i is either algebraic over the field $F(t_1, \cdots, t_{i-1})$, or the logarithm or exponential of an element of $F(t_1, \cdots, t_{i-1})$. Note that each intermediate field $F(t_1, \cdots, t_{i-1})$ is a differential field and an elementary extension of F.

The following result is the abstract generalization of Ostrowski's 1946 generalization of Liouville's 1835 theorem on the subject. A proof of the analytic case may be found in Ritt's classic exposition [4]. Other algebraic proofs, essentially the same as the one given here, may be seen in [2] and [5].

5

THEOREM. *Let F be a differential field of characteristic zero and $\alpha \in F$. If the equation $y' = \alpha$ has a solution in some elementary differential extension field of F having the same subfield of constants, then there are constants $c_1, \cdots, c_n \in F$ and elements $u_1, \cdots, u_n, v \in F$ such that*

$$\alpha = \sum_{i=1}^{n} c_i \frac{u_i'}{u_i} + v'.$$

A number of comments are in order before we proceed with the proof. First, in the case of greatest interest, in which our fields are fields of meromorphic functions on some subregion of $\mathbb{R}$ or $\mathbb{C}$, the condition that F and its elementary extension field have the same constants will be automatically satisfied as long as $\mathbb{C} \subset F$, since any constant meromorphic function is a complex number. In the general case however, the condition that F and its elementary extension field have the same constants, or some related condition, is essential. This can be seen from the example $F = \mathbb{R}(x)$, the field of real rational functions of a real variable, with $x' = 1$ as usual, and $\alpha = 1/(x^2 + 1)$. Clearly $\int (1/(x^2 + 1)) dx$ is an element of an elementary extension field of $\mathbb{R}(x)$, and our claim is that the assumption that we can write $1/(x^2 + 1)$ in the desired form, with $c_1, \cdots, c_n \in \mathbb{R}$ and $u_1, \cdots, u_n, v \in \mathbb{R}(x)$, will lead to a contradiction. For if $x^2 + 1$ occurs v_i times in the expression of u_i as a power product of monic irreducible elements of $\mathbb{R}[X]$, then $u_i'/u_i - 2v_ix/(x^2 + 1)$ is an element of $\mathbb{R}(x)$ without $x^2 + 1$ in its denominator, while $x^2 + 1$, if it occurs in the denominator of v, will occur at least twice in the denominator of v'. Thus $x^2 + 1$ divides the denominator of neither v nor v', implying that $1 - \Sigma 2c_iv_ix$ is divisible by $x^2 + 1$, which is impossible. The final comment is that the theorem has an easy converse: if α can be written as indicated then α has an integral in some elementary extension field of F. This is quite easy to show in the abstract case and is immediate in the classical case where F is a field of meromorphic functions on a subregion of $\mathbb{R}$ or $\mathbb{C}$, as we see by passing to a suitable subregion, where the various $\log u_i$'s can be defined.

Now for the proof of Liouville's theorem. By assumption there is a tower of differential fields

$$F \subset F(t_1) \subset \cdots \subset F(t_1, \cdots, t_N),$$

all with the same subfield of constants, each t_i being algebraic over $F(t_1, \cdots, t_{i-1})$, or the logarithm or exponential of an element of this field, such that there exists an element $y \in F(t_1, \cdots, t_N)$ such that $y' = \alpha$. We shall prove the theorem by induction on N. The case $N = 0$ is trivial, so assume that $N > 0$ and that the theorem holds for $N - 1$. Applying the case $N - 1$ to the fields $F(t_1) \subset F(t_1, \cdots, t_N)$, we deduce that we can write α in the desired form, but with $u_1, \cdots, u_n, v$ in $F(t_1)$. Setting $t_1 = t$, we have t algebraic over F, or the logarithm or exponential of an element of F, and we know that

$$\alpha = \sum_{i=1}^{n} c_i \frac{u_i'}{u_i} + v',$$

with $c_1, \cdots, c_n$ constants of F and $u_1, \cdots, u_n, v \in F(t)$, and it remains to find a similar expression for α, possibly with a different n, but with all of $u_1, \cdots, u_n, v$ in F.

First suppose that t is algebraic over F. Then there are polynomials $U_1, \cdots, U_n$, $V \in F[X]$ such that $U_1(t) = u_1, \cdots, U_n(t) = u_n$, $V(t) = v$. Let the distinct conjugates of t over F in some suitable algebraic closure of $F(t)$ be $\tau_1 (= t), \tau_2, \cdots, \tau_s$. (In case we are dealing with fields of meromorphic functions on a region in $\mathbb{R}$ or $\mathbb{C}$, the functions $\tau_2, \cdots, \tau_s$ can be taken to be meromorphic functions on a suitable sub-region, and it suffices to carry the proof through for functions on the subregion.) Now bear in mind the result of Section 3 on algebraic extensions of differential fields. We have

$$\alpha = \sum_{i=1}^{n} c_i \frac{(U_i(\tau_j))'}{U_i(\tau_j)} + (V(\tau_j))'$$

for $j = 1, \cdots, s$, since this is true for $j = 1$. Application of the operation $(1/s) \sum_{j=1}^{s}$ to both sides of the equation yields

$$\alpha = \sum_{i=1}^{n} \frac{c_i}{s} \frac{(U_i(\tau_1) \cdots U_i(\tau_s))'}{U_i(\tau_1) \cdots U_i(\tau_s)} + \left(\frac{V(\tau_1) + \cdots + V(\tau_s)}{s} \right)'.$$

Since each $U_i(\tau_1) \cdots U_i(\tau_s)$ and $V(\tau_1) + \cdots + V(\tau_s)$ are symmetric polynomials in $\tau_1, \cdots, \tau_s$ with coefficients in F, each of these expressions is actually in F. Hence the last equation is an expression for α of the desired form.

In the remaining cases, where t is the logarithm or exponential of an element of F, we may assume that t is transcendental over F. Then we have

$$\alpha = \sum_{i=1}^{n} c_i \frac{(u_i(t))'}{u_i(t)} + (v(t))',$$

with $u_1(t), \cdots, u_n(t), v(t) \in F(t)$. Each $u_i(t)$ can be written as a power product of a nonzero element of F and various monic irreducible elements of $F[t]$. Hence we may, if necessary, use the logarithmic derivative identity to rewrite $\sum c_i(u_i(t))'/u_i(t)$ in a similar form, but with each $u_i(t)$ either in F or a monic irreducible element of $F[t]$. We therefore assume that $u_1(t), \cdots, u_n(t)$ are distinct, each being an element of F or a monic irreducible element of $F[t]$, and that no c_i is zero. Now look at the partial fraction decomposition of $v(t)$, which expresses $v(t)$ as the sum of an element of $F[t]$ plus various terms of the form $g(t)/(f(t))^r$, where $f(t)$ is a monic irreducible element of $F[t]$, r a positive integer, and $g(t)$ is a nonzero element of $F[t]$ of degree less than that of $f(t)$. Clearly $u_1(t), \cdots, u_n(t), v(t)$ must be of very special form for the right hand side of the last equation to add up to α, which doesn't involve t. To investigate this special form in detail, it now becomes convenient to separate cases. In each case the lemma provides the basic arguments.

First, suppose that t is the logarithm of an element of F, so that $t' = a'/a$, for some $a \in F$. Let $f(t)$ be a monic irreducible element of $F[t]$. Then $(f(t))'$ is also in $F[t]$, and it has degree less than that of $f(t)$, so that $f(t)$ does not divide $(f(t))'$.

7

Thus if $u_i(t) = f(t)$, then the fraction $(u_i(t))'/u_i(t)$ is already in lowest terms, with denominator $f(t)$. If $g(t)/(f(t))^r$ occurs in the partial fraction expression for $v(t)$, with $g(t) \in F[t]$ of degree less than that of $f(t)$ and $r > 0$ and maximal for given $f(t)$, then $(v(t))'$ will consist of various terms having $f(t)$ in the denominator at most r times plus $(g(t)(1/(f(t))^r)')' = -rg(t)(f(t))'/(f(t))^{r+1}$. Since $f(t)$ does not divide $g(t)(f(t))'$, we see that a term with denominator $(f(t))^{r+1}$ actually appears in $(v(t))'$. Thus if $f(t)$ appears as a denominator in the partial fraction expansion of $v(t)$, it will appear in α, which is impossible. Therefore, $f(t)$ does not appear in the denominator of $v(t)$. Therefore $f(t)$ cannot be one of the $u_i(t)$'s either. Since this is true for each monic irreducible $f(t)$, we have each $u_i(t) \in F$ and $v(t) \in F[t]$. Since $(v(t))' \in F$, the lemma implies that $v(t) = ct + d$, with c constant and $d \in F$. Thus

$$\alpha = \sum_{i=1}^{n} c_i \frac{u_i'}{u_i} + c\frac{a'}{a} + d'$$

is an expression for α of the desired form.

Finally, consider the case where t is the exponential of an element of F, say $t'/t = b'$, with $b \in F$. The lemma implies that if $f(t)$ is a monic irreducible element of $F[t]$ other than t itself, then $(f(t))' \in F[t]$ and $f(t)$ does not divide $(f(t))'$. Precisely the same reasoning as above shows that $f(t)$ cannot occur in the denominator of $v(t)$, nor can any $u_i(t)$ equal $f(t)$. Thus $v(t)$ can be written as $v(t) = \sum_j a_j t^j$, where each $a_j \in F$ and j ranges over a finite set of integers, positive, negative, or zero, and each of the quantities $u_1(t), \cdots, u_n(t)$ is in F, with the possible exception that one of these may be t itself. Since each $(u_i(t))'/u_i(t)$ is in F, we have $(v(t))' \in F$, so the lemma implies that $v(t) \in F$. If each $u_i(t)$ is in F, we already have α in the desired form, and are done. If not, only one $u_i(t)$, say $u_1(t)$, is not in F. Then $u_1(t) = t$ and $u_2(t), \cdots, u_n(t) \in F$, so we can write

$$\alpha = c_1 \frac{t'}{t} + \sum_{i=2}^{n} c_i \frac{u_i'}{u_i} + v' = \sum_{i=2}^{n} c_i \frac{u_i'}{u_i} + (c_1 b + v)',$$

with $u_2, \cdots, u_n, c_1 b + v$ all in F. This completes the proof of the theorem.

6. An elementary function is a meromorphic function on some region in $\mathbb{R}$ or $\mathbb{C}$ that is contained in an elementary extension field of the field of rational functions $\mathbb{C}(z)$. We now give some examples of elementary functions with nonelementary indefinite integrals.

As a preliminary comment we note that if $g(z)$ is a non-constant rational function of the complex variable z then e^g is not algebraic over $\mathbb{C}(z)$. This can easily be shown analytically by noting that since $g(z)$ must have at least one pole on the Riemann sphere, e^g will have at least one essential singularity, unlike any algebraic function. Or it can be shown algebraically by looking at the irreducible equation over $\mathbb{C}(z)$ that e^g would otherwise satisfy, say

$$e^{ng} + a_1 e^{(n-1)g} + \cdots + a_n = 0,$$

where $a_1, \cdots, a_n \in \mathbb{C}(z)$, then differentiating this to get

$$ng'e^{ng} + (a_1' + (n-1)a_1 g')e^{(n-1)g} + \cdots + a_n' = 0,$$

which must be proportional to the first equation, so that $ng' = a_n'/a_n$, then noting that a_n'/a_n is either zero or a sum of fractions with constant numerators and linear denominators, whereas ng' can have no linear denominator, so that $g' = 0$, contradicting the assumption that g is nonconstant.

We now want to derive a criterion, due to Liouville, that $\int f(z)e^{g(z)}dz$ be elementary, where $f(z), g(z)$ are given rational functions of z, $f(z)$ being nonzero, and $g(z)$, as above, non-constant. Writing $e^g = t$, we have $t'/t = g'$. Working in the differential field $\mathbb{C}(z, t)$, a pure transcendental extension of $\mathbb{C}(z)$, we see that if $\int fe^g dz$ is elementary, then we can write

$$ft = \sum_{i=1}^{n} c_i \frac{u_i'}{u_i} + v',$$

with $c_1, \cdots, c_n \in \mathbb{C}$ and $u_1, \cdots, u_n, v \in \mathbb{C}(z, t)$. Now let $F = \mathbb{C}(z)$, so that $f, g \in F$ and $u_1, \cdots, u_n, v \in F(t)$. By factoring each u_i as a power product of irreducible elements of $F[t]$ and using logarithmic derivatives, if necessary, we can guarantee that the u_i's which are not in F are distinct monic irreducible elements of $F[t]$. Imagine v expanded into partial fractions with respect to $F[t]$. The lemma implies immediately that the only possible monic irreducible factor of a denominator in v is t, which is also the only possible u_i not in F. Thus v is of the form $\sum b_j t^j$, for j ranging over some set of integers and each $b_j \in F$. Since $\sum c_i u_i'/u_i \in F$, we have $ft = (b_1' + b_1 g')t$. Writing $b_1 = a$, we have $f = a' + ag'$, with $a \in \mathbb{C}(z)$. Conversely, if there is an $a \in \mathbb{C}(z)$ such that $f = a' + ag'$ then one elementary integral of fe^g is ae^g. Thus fe^g has an elementary integral if and only if there is an $a \in \mathbb{C}(z)$ such that $f = a' + ag'$.

For given $f, g \in \mathbb{C}(z)$, the possibility of finding $a \in \mathbb{C}(z)$ such that $f = a' + ag'$ can be decided by considering partial fraction expansions for f, g, and a. For $\int e^{z^2}dz$ we have the equation $1 = a' + 2za$, which is easily seen to have no solution $a \in \mathbb{C}(z)$. For $\int (e^z/z)dz$, we have the equation $1/z = a' + a$, which also has no solution in $\mathbb{C}(z)$. Therefore $\int e^{z^2}dz$ and $\int (e^z/z)dz$ are not elementary. By certain changes of variable we can get other nonelementary integrals. For example, if we replace z by e^z in the second integral we get $\int e^{e^z}dz$ nonelementary, and replacing z by $\log z$ we get $\int (1/\log z)dz$ nonelementary. The integral $\int \log \log z\, dz$ reduces to the previous integral by integration by parts, so it also is nonelementary.

It is slightly more complicated to show that $\int (\sin z/z)dz$ is not elementary. To do this, first change the variable to $\sqrt{-1}\, z$ to slightly simplify the problem to that of showing that $\int ((e^z - e^{-z})/z)dz$ is not elementary. Here again consider the differential field $\mathbb{C}(z, t)$, where $t = e^z$. If our integral is elementary, Liouville's theorem enables us to write

$$\frac{t^2 - 1}{tz} = \sum_{i=1}^{n} c_i \frac{u_i'}{u_i} + v',$$

9

with $c_1, \cdots, c_n \in \mathbb{C}$ and $u_1, \cdots, u_n, v \in \mathbb{C}(z, t)$. Again write $F = \mathbb{C}(z)$, so that $u_1, \cdots, u_n$, $v \in F(t)$, again arrange that the u_i's which are not in F are distinct monic irreducible elements of $F[t]$ and that v is expressed in its partial fraction form, and use the lemma. We again get that the only possible u_i not in F is t, so that $\Sigma c_i u_i'/u_i \in F$, and the only possible monic irreducible factor of a denominator in v is t. Writing $v = \Sigma b_j t^j$, as before, with each $b_j \in F$, we deduce as before that $1/z = b_1' + b_1$, which is impossible. Therefore $\int (\sin z/z) dz$ is not elementary.

7. The question arises whether for any explicitly given elementary function of the complex variable z it can be decided whether or not the function has an elementary integral, and if so, finding it. It is not difficult to see, using the method of the previous section, that this can be done for any function in $\mathbb{C}(z, e^g)$, where g is any nonconstant element of $\mathbb{C}(z)$, but the general question is not so easy. Hardy's book [1] discusses the systematic integration of the kinds of elementary functions that occur in calculus, the main point being that there really *is* a system (contrary to the sometimes expressed opinion that integration in calculus is as much an art as a science), but the book barely broaches the general decision question, which very quickly leads to once intractable questions about points of finite order on abelian varieties over finitely generated ground fields. A solution to this decision problem has recently been announced by Risch [3].

References

1. G. H. Hardy, The Integration of Functions of a Single Variable, 2nd ed., Cambridge Univ. Press, New York, 1916.

2. R. Risch, The problem of integration in finite terms, Trans. Amer. Math. Soc., 139 (1969) 167–189.

3. ———, The solution of the problem of integration in finite terms, Bull. Amer. Math. Soc., 76 (1970) 605–608.

4. J. F. Ritt, Integration in Finite Terms, Columbia Univ. Press, New York, 1948.

5. M. Rosenlicht, Liouville's theorem on functions with elementary integrals, Pacific J. Math., 24 (1968) 153–161.

Comments on Rosenlicht's
Integration in Finite Terms

Contents

In 1968, Maxwell Rosenlicht [Ros68] published the first purely algebraic proof of Liouville's Theorem on Integration in Finite Terms (which we will simply refer to as "Liouville's Theorem") . This paper, together with Robert Risch's paper [Ris69], stimulated renewed interest in both the mathematical and algorithmic aspects of this area. The paper *Integration in Finite Terms* [Ros72] appearing in this volume presents the material of [Ros68] in a simplified form, suitable for an advanced undergraduate. It is a beautiful paper and is the best introduction to the subject. In this commentary, I will review the history of Liouville's Theorem prior to these papers, discuss Rosenlicht's and Risch's contributions and then describe the related results that have appeared

[1] Michael F. Singer
Department of Mathematics, North Carolina State University, Box 8205, Raleigh, NC 27695-8205, USA
e-mail: singer@ncsu.edu
The work of the author was partially supported by a grant from the Simons Foundation (#349357, Michael Singer).

© Springer Nature Switzerland AG 2022
C. G. Raab und M. F. Singer (Hrsg.), *Integration in Finite Terms:
Fundamental Sources*, Texts & Monographs in Symbolic Computation,
https://doi.org/10.1007/978-3-030-98767-1_2

11

subsequently. I will focus mainly on the theoretical aspects since the algorithmic aspects are well described in Clemens Raab's and Barry Trager's commentaries to this volume. Although I will almost exclusively describe results related to elementary functions, I will occasionally mention results related to Liouvillian functions (those built up using exponentials, *arbitrary integrals* and algebraic functions) when this appears in the work of researchers concerned with elementary functions but I will not delve deeper into the problem of solving differential equations in terms of these.

Several books and articles present the history of Liouville's Theorem and related topics. In particular Jesper Lützen's thoughtful and detailed book [Lüt90] thoroughly discusses the contributions of Laplace, Abel, Liouville and others and I have relied heavily on this book in my summaries below. Anyone interested in the subject should go directly to [Lüt90] to get a much more complete picture. Hardy has also given a pleasantly excursive presentation of these contributions in [Har71]. Much of the theoretical and computational aspects of this subject and related results are to be found in Bronstein's book [Bro97]. Finally, the articles of Kasper [Kas80] and Marchisotto and Zakeri [MZ94] give concise histories of some aspects of these results as well.

I will assume the reader is familiar with [Ros72] and will not restate the definitions of differential field, elementary extension, etc.

1 Before Rosenlicht and Risch

Initial glimmerings of phenomena related to Liouville's Theorem can be found in the writings of Fontaine and Condorcet (cf. [Lüt90, pp. 352–357]). A clear statement of the principle underlying Liouville's Theorem was given by Laplace in [Lap12, pp. 4–5]:

> Thus, since the differentiation lets the exponential and the radical quantities subsist and only makes the logarithmic quantities disappear when they are multiplied by constants, one may conclude that the integral of a differential function cannot include any other exponentials and radicals than those already included in this function.[2]

but no rigorous proof seems to have been published by him [Lüt90, p. 358].

1.1 Abel

Abel stated several results related to Liouville's Theorem and stated that he had a general theory of integration of algebraic functions aimed at reducing these objects using algebraic and logarithmic functions (cf. [Lüt90, pp. 358–369], [Rit48, pp. 28–31]). Part of Abel's research has reached modern times as half of the Abel–Jacobi Theorem (see [Gri76, Gri04]) and part has a direct bearing on the subject of this volume. First of all, Abel stated Liouville's Theorem, although any proof that he had now seems lost. He showed that certain elliptic integrals could not be expressed in the form prescribed by this theorem (as did Chebyshev, Zolotarev and others [Lüt90, pp. 367, 415–417]). Concerning Liouville's Theorem, Abel presented an argument

---

[2] I am using the translation that appears in [Lüt90, p. 358].

that if an algebraic function y has an integral of the form $v + \sum_{i=1}^{n} c_i \log v_i$, where v and the u_i are algebraic functions and the c_i are complex numbers, then v and the u_i may be taken to be rational functions of x and y (cf. [Lüt90, pp. 363–364] and [Rit48, pp. 28–31]; note that Littlewood criticized this proof [Lüt90, p. 365] and [Har71, Preface, pp. 38–43]). In Rosenlicht's paper [Ros68] this fact results from taking norms and traces. An even more precise statement of where the c_i, v, and u_i lie can be found in [Ris69, p. 171] or [Bro97, p. 141].

Abel also stated generalizations of Liouville's Theorem. For example, he stated that if one has a relation of the form $F(x, \int y_1 dx, \int y_2 dx, \ldots, \int y_n dx) = 0$, where F is an elementary function of $n + 1$ variables and $y_1, \ldots, y_n$ are algebraic functions, then some constant linear combination of the $\int y_i dx$ is of the form $u + \sum_{i=1}^{m} \log v_i$ with c_i constants and u, v_i algebraic functions. We do not have Abel's proof but a modern statement and proof of this result appears in [PS83, Corollary 2].

Another generalization of Liouville's Theorem stated by Abel is: if the integral of an algebraic function can be expressed in terms of implicitly or explicitly defined elementary functions then it is elementary and so satisfies the conclusion of Liouville's Theorem. Lützen remarks that it is not clear what Abel meant by implicit elementary functions as no proof survives, but Ritt [Rit48, pp. 71–98] and Risch [Ris76] proved precise versions of this result.

1.2 Liouville

Lützen [Lüt90, Chapter IV] and Ritt [Rit48] give detailed analyses of Liouville's papers and methods, so I will only give an overview.

Liouville's first paper on integration in finite terms appeared in 1832 (when he was 23!) and by 1840 he had essentially finished his work on the subject. Liouville began by considering the question of integrating an algebraic function in terms of algebraic functions, obtaining many of Abel's results. Liouville then turned to the problem of expressing integrals of algebraic functions in terms of more general elementary functions. He began by examining when such an integral is of a special form (e.g., when the integral of an expression involving a square root is the sum of a similar expression and logarithms of similar expressions). He then turned to prove what we now know as Liouville's Theorem in the special case when the integrand is an algebraic function and published this result in 1834.

To prove this result, Liouville first gave a classification of elementary functions. Briefly, algebraic functions are called functions of order zero. A function is elementary of order at most n if it is an algebraic function of elementary functions of order at most $n - 1$ and logarithms and exponentials of elementary functions of order at most $n - 1$. The smallest n such that an elementary function is of order at most n is called its order. Since Liouville dealt with actual functions, one needs to take into account branches of these functions and this led to criticisms of imprecision, especially by Ritt (see below). Nonetheless, the concept of order is key to Liouville's proofs. He relies heavily on the following observation (called *Liouville's principle* by Lützen [Lüt90, p. 387] and Ritt [Rit48, p. 16]):

Let f be an elementary function of order n and assume that the number r_n of logarithms and exponentials of order n appearing in its definition is as small as possible. Any algebraic relations among these r_n quantities and elementary functions of order at most $n-1$ must hold when any quantity is substituted for any of these r_n quantities.

Liouville then proceeds as follows. Assume that one expresses the integral of an algebraic function as an elementary function of order n with r_n as above minimal. Differentiating one then has a relation as in the statement of Liouville's principle. Liouville then uses this principle to show that exponentials cannot appear and logarithms must only appear linearly in the expression of the integral. To get a feeling of Liouville's techniques (following closely Lützen's exposition [Lüt90, p. 388]), assume $n = 1$ and

$$\int y \, dx = \phi(x, \theta),$$

where y is algebraic, ϕ is an elementary function of order n and $\theta = \log u$ is a logarithm of order n. Differentiating, we have

$$y = \frac{\partial \phi(x, \theta)}{\partial x} + \frac{\partial \phi(x, \theta)}{\partial \theta} \left(\frac{du}{dx} / u \right).$$

From Liouville's principle, this expression remains unchanged when we replace θ with $\theta + c$ and equating the two expressions for y, we get

$$\frac{\partial \phi(x, \theta + c)}{\partial x} + \frac{\partial \phi(x, \theta + c)}{\partial \theta} \left(\frac{du}{dx} / u \right) = \frac{\partial \phi(x, \theta)}{\partial x} + \frac{\partial \phi(x, \theta)}{\partial \theta} \left(\frac{du}{dx} / u \right).$$

Therefore

$$\phi(x, \theta + c) = \phi(x, \theta) + \text{constant}$$

so

$$\frac{\partial \phi(x, \theta)}{\partial \theta} = A,$$

where A is a constant. Liouville's principle implies we can replace θ with a variable ζ in this equation. Doing this and solving the resulting differential equation we have

$$\phi(x, \zeta) = A\zeta + \phi(x, \zeta_0) - A\zeta_0,$$

where ζ_0 is some arbitrary value of ζ. We then have

$$\int y \, dx = A\theta + \phi(x, \zeta_0) - A\zeta_0,$$

so $\theta = \log u$ appears linearly. Arguing analogously, Liouville showed that for $n = 1$ and $\theta = e^u$, ϕ will be independent of θ. When $n > 1$, an induction and similar use of Liouville's principle yields the result.

In 1835, Liouville published the general version of his theorem, allowing y to be an algebraic function of $u_1(x), \ldots, u_n(x)$, where each u_i satisfies an equation of the form $\frac{du_i}{dx} = p_i(u_1, \ldots, u_n)$, p_i an algebraic function (in modern terms, y belongs to a differential of finite transcendence degree over the constants; a precursor of

Ostrowski's approach [Ost46b]). In this paper he also proved the result concerning the elementary integrability of fe^g proven in Rosenlicht's paper. In 1837 he published a paper continuing his research into the structure of elementary functions. He was able to show that e^x cannot be expressed in terms of logarithms and algebraic functions and $\log x$ cannot be expressed in terms of exponential and algebraic functions and that Kepler's equation $x = y - a \sin y$, $a \in \mathbb{C}$, has no elementary solution y (a modern proof is given in [Ros69]). Furthermore, Lützen cites some unpublished 1840 notes where Liouville claims to show that if the integral of an algebraic function is among a set of functions that are defined implicitly by elementary functions of several variables, then it must be elementary. Lützen outlines Liouville's argument but concludes that it does not represent a proof in the modern sense. Such a result has been proven by Risch [Ris76] and in a slightly weaker form in [PS83].

Liouville also considered solving differential equations in terms of (what we now call) *Liouvillian functions*, that is, functions expressible in terms of exponentials, arbitrary integrals and algebraic functions (cf. [Lüt90, pp. 401–411]). His main result is that if a second-order linear differential equation $y'' - Py = 0$, P a polynomial, has a Liouvillian solution then the associated Riccati equation $u' + u^2 = P$ has an algebraic solution. He was able to use this to show for which parameters the Bessel equation has Liouvillian solutions. Questions of this nature can now be handled using differential Galois theory [vdPS03] and I will not describe this further.

1.3 The Russian School

As already noted, Chebyshev and Zolotarev considered the question of elementary integrability of elliptic integrals and this, together with work of Ostrogradsky and Dolbnia, is discussed in [Lüt90, pp. 414–418].

Mordukhai-Boltovskoi produced a significant body of work around the problem of integration in finite terms. Most of his papers are not easily available. The book [MB10] is available online and the library at the Research Institute for Symbolic Computation, Johannes Kepler University Linz has photocopies (due to Robert Risch) of several of Mordukhai-Boltovskoi's articles. In addition [MB09] is a translation of a much cited article and [DLS98] also mentions the work of Mordukhai-Boltovskoi and related works. Regrettably, I do not read Russian and so most of what I know of this author is what is explained in [Rit48].

In [MB13] (cf. [Rit48, p. 52]) Mordukhai-Boltovskoi describes methods for deciding if elementary functions have elementary integrals, although no general algorithm is given. In [MB10] (cf. [Rit48, p. 76]) Mordukhai-Boltovskoi describes methods for solving linear differential equations in terms of Liouvillian functions. These are only briefly mentioned in [Rit48] but Ritt does go into greater detail concerning two other works of Mordukhai-Boltovskoi in [Rit48, Ch. VII].

Ritt first presents a result of [MB09] concerning elementary first integrals of first-order differential equations $y' = f(x, y)$, f an algebraic function. Mordukhai-Boltovskoi showed, using analytic techniques in the spirit of Liouville, that if there exists an elementary function $g(x, y)$ of two variables such that $g(x, y)$ is constant on solutions of this equation (i.e., g is a first integral) then there is a first integral of the form $\phi_0(x, y) + \sum_{i=1}^{m} c_i \log \phi_i(x, y)$, where the c_i and ϕ_i are algebraic functions of two variables. This result is given a modern, purely algebraic proof in [PS83]. A result dealing with Liouvillian first integrals appears in [Sin92]. There is a very large literature concerning elementary and Liouvillian first integrals and this deserves its own survey, but that will not be given here.

The second result of Mordukhai-Boltovskoi presented by Ritt is from [MB37] and concerns explicit elementary solutions of first-order differential equations $F(x, y, y') = 0$, F a polynomial. Mordukhai-Boltovskoi showed, again using analytic techniques, that if such a differential equation has a nonalgebraic elementary solution, then it has a one-parameter family of solutions of the form

$$y = G\left(x, \sum_{i=1}^{m} c_i \log(\phi_i(x)) + c\right) \text{ or } y = G(x, e^{\phi_0(x) + \sum_{i=1}^{m} c_i \log(\phi_i(x)) + c}),$$

where G is an algebraic function of two variables, c is the parameter, c_i are fixed constants and the ϕ_i are algebraic functions. This follows from a result characterizing differential subfields of elementary extensions presented and proven algebraically in [Sin75],[RS77] and [Ris79] (a result anticipating these results can be found in [Koe87]). A result characterizing differential subfields of Liouvillian extensions is given in [Sri20] (see also [Sri17, Sri18]).

1.4 Hardy

Hardy's book [Har71, pp. 38–44] contains a proof of Abel's Theorem on the algebraic integrability of algebraic functions. In addition, he describes techniques for integrating various kinds of elementary functions and outlines a method to determine if the integral of an algebraic function is algebraic. A large part of the book is devoted to integrals of algebraic functions and how the theory of algebraic curves informs this. Hardy also gives an exposition of Liouville's theorem (without proof) and some of its consequences. He also mentions two problems [Har71, p. 7]: *(i) if f(x) is an elementary function, how can one determine whether its integral is also an elementary function?* and *(ii) if the integral is an elementary function, how can we find it?*. He then makes the regrettable prognostication *Complete answers to these questions have not and probably never will be given.*

1.5 Ostrowski

In [Ost46b], Ostrowski introduced what he called a *corps liouvillien* (and what is now called a differential field) into the study of integration in finite terms. He showed that Liouville's Theorem holds for *functions* $f(x)$ lying in such a field and $\int f(x) dx$ lying in an elementary extension of this field. Outside of this formalism, Ostrowski's approach is still analytic and follows Liouville's general approach, making several simplifications and clarifications. Ostrowski also published [Ost46a] in which he showed if the integrals of elements of a differential field are algebraically dependent over this field then a constant linear combination of these will lie in the field. This, together with a statement concerning algebraic dependence among exponentials, is now known as the Kolchin–Ostrowski Theorem [Kol68] (cf. [Ax71, Theorem 4], [Ros76, Corollary])).

1.6 Ritt

In [Rit48], Ritt describes the work of Abel, Liouville, Mordukhai-Boltovskoi and his own contributions. A main feature of his presentation is his emphasis on dealing with the function-theoretic properties of elementary functions. As he says in his introduction.

> There are, however, certain questions connected with the many-valued character of the elementary functions which could be pressed back behind the symbols in Liouville's time but which have since learned to assert their rights. Such matters are mulled over in the first chapter. The mulling is inescapable. It might be great fun to talk just as if the elementary functions were one-valued. I might even sound convincing to some readers; I certainly could not fool the functions. [Rit48, p. vi]

The first three chapters of [Rit48] focus on giving rigorous foundations to the functional properties of elementary function and presenting Liouville's Theorem taking this (and Ostrowski's work) into account. Askold Khovanskii discusses some of these issues and also generalizes this approach in his appreciation of Ritt and Ritt's book in his article in this volume.

In Chapter IV, Ritt shows that Kepler's Equation (mentioned above) has no elementary solution and briefly alludes to two of his papers, [Rit25] and [Rit29]. In [Rit25], Ritt shows: *if a function $F(z)$ and its inverse are both elementary, then there exist n functions $\phi_1(z), \ldots, \phi_n(z)$ where each $\phi_i(z)$ with odd index is algebraic, and each $\phi_i(z)$ with even index is either e^x or $\log x$ such that $F(x) = \phi_n \phi_{n-1} \ldots \phi_2 \phi_1(x)$ each $\phi_i(z)$, $(i < n)$, being substituted for z in $\phi_{i+1}(z)$.* To prove this, Ritt gives a careful analysis (very much in the spirit of Liouville) as to how the order of a composite of two elementary functions depends on the orders of each of these functions. A modern algebraic proof of this result is given in [Ris79].

The paper [Rit29] briefly alluded to in [Rit48, p. 59] describes the interaction of analytic and algebraic properties of finite exponential sums $c_1 e^{a_1 x} + \cdots + c_r e^{a_r x}$, where the c_i and the a_i are complex numbers. Using properties of Dirichlet series, Ritt shows that if w is a solution of a polynomial equation $a_n w^n + \cdots + a_1 w + a_0 = 0$, where the a_i are finite exponential sums and w is analytic in a sector of angle greater than π, then w is a finite exponential sum. He also shows that if the quotient of two finite exponential sums is analytic in such a sector, then this quotient equals a finite exponential sum. In an earlier paper [Rit27a], Ritt gave a factorization theorem for finite exponential sums. Finite exponential sums form a $\mathbb{C}$-algebra and one can talk about division, irreducibility, etc in this ring. Ritt showed that any finite exponential sum can be written uniquely as a product of irreducible finite exponential sums and simple finite exponential sums, where these latter objects have the property that their exponents are integer multiples of a fixed complex number. A recent proof and generalization can be found in [EvdP97]. Other properties of finite exponential sums can be found in [HRS89]. Ritt's Factorization Theorem also relates to the model theory of fields with exponentiation (cf. [Mac16, p. 921]).

In Chapter V, Ritt develops properties of fractional power series and uses these to prove, in Chapter VI, Liouville's result that if the equation $y' + y^2 = P(x)$ has a Liouvillian solution then it has an algebraic solution. In particular, if a second-order homogeneous linear differential equation with algebraic function coefficients has a Liouvillian solution, it will have a solution y such that y'/y is algebraic. This result (in fact for a homogeneous linear differential equation of arbitrary order) can now be derived using differential Galois theory. A proof of this and related results using power series techniques (in the guise of valuation theory) can be found in [Ros73] and [Sin76].

In Chapter VII, Ritt presents the results of Mordukhai-Boltovskoi mentioned above and in Chapter VIII he presents the results from his papers [Rit23] and [Rit27b]. In [Rit23], Ritt proves that if the integral of an elementary function satisfies an elementary relation, then the integral is actually elementary. As mentioned above, Risch proved algebraically a generalization of this result in [Ris76]. In [Rit27b], Ritt proved that if a solution of a second-order homogeneous linear differential equation satisfies a Liouvillian relation then all solutions of the linear equation are Liouvillian. This latter result is proven algebraically in [Sin92]. Ritt's proofs of these results depend on analytic considerations and proceed using an induction on the elementary order or Liouvillian order of the functions used in building the elementary or Liouvillian relations.

2 The Fundamental Papers of Rosenlicht and Risch

2.1 Rosenlicht

In [Ros68], Rosenlicht gave the first purely algebraic statement and proof of Liouville's Theorem. The proof relies on little more than a knowledge of partial fraction decomposition and a few simple facts from Galois theory. This is reproduced in [Ros72], the paper presented in this volume, with an even simpler presentation and additional comments. A proof of Abel's Theorem concerning algebraic integrals is embedded in the proof of Liouville's Theorem (see the argument on lines 2–15 on page 969). Using an induction argument, the proof is then reduced to showing that if K is a differential field, $\alpha \in K$ and

$$\alpha = \sum_{i=1}^{n} c_i \frac{u_i'}{u_i} + v', \tag{1}$$

where the c_i are constants and $u_i, v \in K(t)$, t transcendental over K and t an exponential or a logarithm of an element in K, then one can find a similar expression with $u_i, v \in K$. The underlying idea is to compare the partial fraction decompositions of the u_i'/u_i and v' and to understand what cancelations occur in order for the right-hand side of (1) to be an element of K. Rosenlicht explains this clearly and I have nothing to add. Besides proving Liouville's Theorem, Rosenlicht reproves another result of Liouville:

If f and g are algebraic functions, then $\int f e^g dx$ is an elementary function if and only if there is an $a \in \mathbb{C}(x)$ such that $f = a' + ag'$. Using this he can easily show that $\int e^{x^2} dx$ is not elementary. Rosenlicht's other papers concerning elementary functions will be discussed in Section 3.2.

2.2 Risch

Risch published his first contribution to the algorithmic aspects of Liouville's Theorem in [Ris69]. Risch's algorithm will be more fully discussed in Clemens Raab's commentary in this volume but I want to discuss his more theoretical contributions here. In [Ris69], Risch gave a refinement of Liouville's Theorem. In Rosenlicht's version, it was assumed that the constants were algebraically closed and that the elementary tower containing the integral had no new constants. Risch did away with these two restrictions and showed that if the integral of an element α lies in an elementary extension of a differential field k containing α then there are $v \in k$, $c_i \in \overline{C}$ (the algebraic closure of the constants C of k), $v_i \in \overline{C}k$ such that $\alpha = v' + \sum_{i=1}^{n} c_i v_i' / v_i$ and every automorphism of $\overline{C}k$ over k permutes the terms of the sum. Further results concerning the precise subfield of $\overline{C}$ needed in the above refinement are due to Lazard, Riboo, Rothstein and Trager and are described in [Bro97, Chapter 5.6]. A finer description of the u, v_i that can appear is also given in [Bro07].

The preprint [Ris67] (contemporaneous with the preprint version of [Ris69]) contains a treatment of elementary integrals of real functions. Risch showed that if an element α in a real differential field k has an elementary integral then $\alpha = w_0' + \sum_{i=1}^{s} c_i w_i' / w_i + \sum_{i=s+1}^{n} c_i w_i' / (w_i^2 + 1)$, where $w_0 \in k$, the c_i are in the real closure $\tilde{C}$ of the constant field C (with respect to some fixed order on C) of k and $w_1, \ldots, w_n \in \tilde{C}k$, that is, the integral of α can be expressed as an element of k and a sum of logarithms and arctangents of elements of $\tilde{C}k$. Risch also discusses related algorithmic questions.

3 The Aftermath

Two interrelated themes have dominated the topic of expressing integrals in finite terms after the papers of Rosenlicht and Risch: generalizations of Liouville's Theorem and understanding the algebraic relations that can occur among elementary functions and other special functions.

3.1 Generalizations of Liouville's Theorem

A multivariate generalization of Liouville's Theorem was presented by Caviness and Rothstein in [CR75]. Its proof is very much in the spirit of [Ros68]. Another proof of this result was presented by Rosenlicht in [Ros76, Theorem 3] based on his reworking of techniques introduced by Ax [Ax71] (more about this in Section 3.2).

In [Ris76], Risch presented a geometric approach to Liouville's Theorem and in the process showed that if the integral of a function is among a set of functions that are defined implicitly by elementary functions of several variables, then it must be elementary. As mentioned above, this generalizes Liouville's and Ritt's results of a similar nature.

An early generalization of Liouville's Theorem appears in [MZ79] (see also [ND79]). The authors consider the problem of expressing the antiderivative of an elementary function in terms of elementary functions and special functions defined by indefinite integrals, such as the Spence function or error function. The natural conjecture to make is that if an integral can be expressed in this way, the special functions must appear linearly. The authors show that this conjecture is false in general. They also show that if the special functions involved are integrals of elements in the field of definition of the integrand and functional composition is not allowed, then this conjecture is true.

The paper [SSC85] begins to consider what happens when one expresses integrals in terms of compositions of algebraic functions, elementary functions, and special functions. The authors define a class of special functions, the so-called $\mathcal{EL}$-elementary functions, which include the error function

$$\operatorname{erf}(x) = \frac{2}{\sqrt{\pi}} \int_0^x e^{-t^2} dt$$

and the logarithmic integral

$$\operatorname{li}(x) = \int_0^x \frac{1}{\log t} dt$$

but does not include the dilogarithm

$$\ell_2(x) = -\int_0^x \frac{\log(1-t)}{t} dt$$

or the exponential integral

$$\operatorname{Ei}(x) = -\int_x^\infty \frac{e^{-t}}{t} dt$$

(the form of these functions is slightly different in [SSC85] but differ from these by additive or multiplicative constants and so the results still hold). The authors give a Liouville-type Theorem for integration in terms of these special functions. Rather than state the general result, I will restrict myself to the question of integration in terms of error functions. One then defines an *erf-elementary extension of a differential field* k as a differential field K such that there is a tower $k = K_0 \subset \ldots \subset K_n = E$ where each $K_i = K_{i-1}(\theta_i)$ and θ_i is either algebraic over K_{i-1}, or there exist $u_i, v_i \in K_{i-1}$ such that $\theta_i' = u_i'/u_i$, $\theta_i'/\theta_i = u_i'$, or $\theta_i' = u_i'v_i$ where $v_i'/v_i = -2u_iu_i'$. One says that an element $\alpha \in k$ is *erf-elementary integrable* if there is an element y in an erf-elementary extension of k such that $y' = \alpha$. The main result of [SSC85] is that, assuming the constants C of k are algebraically closed, then such an α has an erf-elementary integral if and only if there are constants $a_i, b_i \in C$, $w_i \in k$ and u_i, v_i algebraic over k such that

$$\alpha = w'_0 + \sum_{i=1}^{n} a_i \frac{w'_i}{w_i} + \sum_{i=1}^{m} b_i u'_i v_i,$$

where $v'_i / v_i = -2u_i u'_i$ and u_i^2, v_i^2 and $u'_i v_i$ are in F. In other words, if α is erf-elementary integrable then its integral can be expressed as a constant linear combination of an element in k, logarithms of elements in k and error functions of elements algebraic over k. The authors give an example to show that one cannot strengthen the conclusion to conclude that the u_i, v_i are actually in k. They also give an example to show that such a result guaranteeing that the logarithms appear linearly in the integral does not hold when one integrates in terms of the dilogarithm. They present a procedure to decide if an element in a purely exponential extension of $C(x)$ has an erf-elementary integral and find this if it does. The procedure proceeds by induction on the defining tower of the integrand and an interesting aspect of this is that one must rewrite this tower (if necessary) so that the elements at each stage are selected so as to be independent in a certain way from the previous ones.

A similar result for logarithmic integrals can be deduced from the general result of [SSC85] and Cherry gave in [Che85] a procedure to determine if an element of a transcendental elementary extension of $C(x)$, $x' = 1$, C algebraically closed, can be integrated in terms of logarithmic integrals. In [Che86] the procedure of [SSC85] is extended to apply to a more general class of elementary functions. In [Kno92, Kno93], Knowles gives a procedure dealing with an even further extended class of elementary functions that allows one to decide if they are erf-elementary integrable. Cherry [Che89] also gave a procedure to determine if fe^g, $f, g \in C(x)$, can be expressed in terms of a class of special functions called the special incomplete gamma functions. This class of special functions includes the exponential integral, the error function, the sine and cosine integrals, and the Fresnel integrals.

The results of [SSC85] do not apply to integration in terms of the exponential integral $\mathrm{Ei}(x)$ and elementary functions. In [LL02] the authors sharpen the results of [Ros76] and extend the results of [SSC85] to include these functions as well as the exponential integrals and special cases of incomplete gamma functions. The conclusion is, as in [SSC85], if an integral can be expressed in terms of these functions, then they will appear in the integral in constant linear combinations composed with algebraic functions. In [Heb15], the author considers integration in terms of elementary functions, exponential integrals and general incomplete gamma functions and derives a similar result together with more precise structural information aimed at efficient algorithms and implementations which have been included in the CAS computer algebra system (see also [KS19]). In [Heb21], the author extends Liouville's Theorem to include integration in terms of elementary functions and elliptic integrals.

The results of [SSC85] also do not apply to integration in terms of dilogarithms and elementary functions. This case of dilogarithms was taken up by Baddoura [Bad94, Bad11]. A key element in these results is a characterization of the algebraic identities among the dilogarithm and elementary functions. More formally, a *transcendental-dilogarithmic-elementary-extension* of a differential field k is a field E for which there exists a tower of fields $k = K_0 \subset \ldots \subset K_n$ where for each $i > 0$, $K_i = K_{i-1}(\theta_i)$, θ_i transcendental over K_{i-1} and there exist $u_i, v_i \in K_{i-1}$ such that either $\theta_i' = u_i'/u_i$, $\theta_i' = u_i'\theta_i$, or $\theta_i' = -v_i/u_i$, where $v_i' = -u_i'/(1-u_i)$, that is θ_i is a logarithm, exponential or dilogarithm of an element in K_{i-1}. Baddoura shows that if k is a Liouvillian extension of an algebraically closed field of constants C and $\alpha \in k$ and there exists an element y in a transcendental-dilogarithmic-elementary-extension such that $y' = \alpha$ then there exist $c_0, \ldots, c_m, v_0, \ldots, v_m, w_1, \ldots, w_n \in k$ and constants $d_1, \ldots, d_n \in C$ such that in a suitable extension of k we have

$$\int \alpha = v_0 + \sum_{i=1}^{m} c_i \log(v_i) + \sum_{i=1}^{n} d_i \left(\ell_2(w_i) + \frac{1}{2} \log(w_i) \log(1 - w_i)\right).$$

Notice that, although the dilogarithms appear linearly, we have introduced products of logarithms into the expression. Reworkings, refinements, and extensions of these results appear in [KS19, KS21] and [Heb18].

The dilogarithm is one of a sequence of functions referred to as polylogarithms $\ell_n(x)$, which are defined inductively for $n > 2$ as

$$\ell_n(x) = \int_0^x \frac{\ell_{n-1}(t)}{t} dt.$$

In [Bad94], Baddoura considers integration in terms of polylogarithms and elementary functions and made a conjecture concerning the form that such an integral must take. In [Bad11], an argument is presented to verify this conjecture in a special case.

A problem that arises when one deals with integration in terms of nonelementary functions is the following: *Given $y_1, \ldots, y_n$ in a differential field K determine the set of constants $c_1, \ldots, c_n$ such that $c_1 y_1 + \cdots + c_n y_n$ has an integral elementary over K.* Risch solved[3] the related question of determining the set of constants $c_1, \ldots, c_n$ such that $c_1 y_1 + \cdots + c_n y_n$ has an integral in K, when K is an elementary extension of $C(x), C$ algebraically closed and $x' = 1$. He showed how one can construct a system of linear equations over C such that the c_i satisfy these equations if and only if this expression has an integral in K. In [Bro97, Chapter 7], the author considers a similar question concerning determining if $c_1 y_1 + \cdots + c_n y_n$ has an integral in K when K is a transcendental Liouvillian extension of the constants (or even a more general extension), as well as related problems. When K is a purely transcendental Liouvillian extension of an algebraically closed field of constants, Mack ([Mac76]; see also [SSC85, Appendix] and [Raa12, Chapters 3 and 4]) showed that one could effectively find a system of linear equations with constant coefficients such $c_1, \ldots, c_n$ such that $c_1 y_1 + \cdots + c_n y_n$ has an integral elementary over K if and only if the c_i satisfy this system This leads to the general question of describing the set of

---

[3] Setting $f = 0$ in Theorem 1(b) of *On the Integration of Elementary Functions which are Built Up Using Algebraic Operations* in this volume yields this result. When K is a purely transcendental elementary extension of $C(x)$, this already appears in [Ris69].

parameters for which a parameterized expression has an elementary integral. The first case of this question is *Let $f(x,t)$ be an element of an algebraic extension of $C(x,t)$ where C is algebraically closed, $t' = 0$ and $x' = 1$. Describe the set $\{c \in C \mid f(x,c)$ has an integral elementary over $C(x)\}$.* In [Dav81, p. 90] Davenport asserted that if $f(x,t)$ is not generically integrable in elementary terms then there are only finitely many values of $c \in C$ such that $f(x,c)$ has an elementary integral over $C(x)$. Recently, Masser and Zannier [MZ20] (see [Zan14] for earlier results) have shown that this assertion is false (see [Mas17] for an elementary introduction to their results). For example, the function

$$\frac{x}{(x^2 - t^2)\sqrt{x^3 - x}}$$

is not generically integrable but is integrable for infinitely many algebraic values of t. Furthermore, in [MZ20] they can characterize those algebraic functions which satisfy Davenport's assertion. From the paper of Risch and the thesis and commentary of Trager in this volume, one sees that integration of algebraic functions is intimately connected with the question of whether integer multiples of certain divisors on a curve are principal or not, that is, whether points on the associated Jacobian are torsion or not. In [MZ20] the authors study how points on a Jacobian become torsion under specialization in the parameterized case and this allows them to deduce their results on elementary integrability in this case.

Another generalization of Liouville's Theorem appears in [DS86]. One can consider Liouville's Theorem as describing elementary solutions of first-order inhomogeneous linear differential equations of the form $y' = a$. In [DS86, Theorem 2], the authors show that if an nth order linear differential equation $L(y) = b$, with coefficients in a differential field k, has a solution elementary over k, then it has a solution of the form $P(\log u_1, \ldots, \log u_n)$, where P is a polynomial with coefficients algebraic over k whose degree is at most equal to n, and the u_i are algebraic over k. Furthermore, if P_n is the homogeneous term of P of degree n and L has no order zero term, then the coefficients of P_n are constant. References to other papers and other results concerning elementary and Liouvillian solutions of linear differential equations also appear in [DS86].

Finally, as mentioned in Section 1.3, the result of Mordukhai-Boltovskoi on elementary first integrals has given rise to a large field aimed at understanding rational, algebraic, elementary and Liouvillian first integrals, a field too large to be surveyed here.

3.2 Structure Theorems

In algorithmic considerations, it is useful to know when two expressions represent the same object. For elementary functions this question was considered in [Cav70] and the references given in this paper (see also [Eps79]). Key to this is an understanding the algebraic relations that can occur among a set of elementary functions. The structure theorems developed by Epstein, Caviness, Risch, Rosenlicht, and Rothstein address this issue. These results now follow from results of Rosenlicht based on ideas of Ax and I will begin by describing these.

A fundamental conjecture in transcendental number theory is Schanuel's Conjecture: *Let* $\alpha_1, \ldots, \alpha_n$ *be* $\mathbb{Q}$*-linearly independent complex numbers. Then* tr.deg$_{\mathbb{Q}}\mathbb{Q}(\alpha_1, \ldots, \alpha_n, e^{\alpha_1}, \ldots, e^{\alpha_n}) \geq n$. This conjecture implies many of the known transcendence facts (the transcendence of π, Lindemann's Theorem, ...). In [Ax71], Ax proved function-theoretic and differential algebraic analogues of this conjecture. Among other results, he gave a new proof of the Kolchin–Ostrowski Theorem [Kol68, Ost46a]: *Let* $k \subset K$ *be differential fields with commuting derivations* $\{D\}_{D \in \Delta}$ *of characteristic* 0 *with the same field of constants* $C = \cap_{D \in \Delta} \mathrm{Ker} D$ *and let* $y_1, \ldots, y_n, z_1, \ldots, z_n \in K^*$ *satisfy* $Dy_i, Dz_i/z_i \in k$ *for* $1 \leq i \leq n$ *and* $D \in \Delta$. *If the* y_i *are C-linearly independent modulo* k *and no non-trivial power product of the* z_i *is in* k, *then* $y_1, \ldots, y_n, z_1, \ldots, z_n$ *are algebraically independent over* k. Ax's approach to this problem is to linearize the property of algebraic dependence. This is done by considering the module of differentials $\Omega_{K/k}$ (see [Ros76] or [Bro97, Chapter 9.1] for an exposition of this object aimed at its use in these kinds of questions). Ax's proof technique for his general results depends on the fact that if $k \subset K$ are fields then $u_1, \ldots, u_m \in K$ are algebraically dependent over k if and only if their differentials $du_1, \ldots, du_m$ are k-linearly dependent in $\Omega_{K/k}$. In addition, following earlier results of Johnson [Joh69a, Joh69b], one can put a differential structure on $\Omega_{K/k}$ to take into account the differential relations among elements in a differential field K. Finally residue calculations allow Ax to restrict the kind of algebraic relations that can occur. The earlier work of Kolchin is based on his differential Galois theory and general criteria for solutions of differential equations to be algebraically dependent. He can additionally deduce a statement about dependence among integrals, exponentials and Weierstrass functions as well as criteria for Bessel functions. Algebraic independence statements for Weierstrass functions, using refinements of Ax's techniques, are also shown in [BK77].

In [Ros76], Rosenlicht gives a simplified presentation of some of the basic results of [Ax71]. He proves results concerning algebraic dependence among elementary and Liouvillian functions. From these he not only deduces the multivariate generalization of Liouville's Theorem mentioned above but can use these new techniques to recover and generalize work from his earlier paper [Ros69]. In this latter paper Rosenlicht uses valuation theory and the associated power series techniques to show that if k is a differential field and $y_1, \ldots, y_n, z_1, \ldots, z_n$ are elements of a Liouvillian extension of k such that $y_i'/y_i = z_i', i = 1, \ldots, n$ and that $k(y_1, \ldots, y_n, z_1, \ldots, z_n)$ is algebraic over each of $k(y_1, \ldots, y_n)$ and $k(z_1, \ldots, z_n)$ then $y_1, \ldots, y_n, z_1, \ldots, z_n$ are all algebraic over k. Using this he can show that $y = e^{y/x}$ and Kepler's equation have no Liouvillian solutions. Rosenlicht's results were also used in [BCDJ08] to show that the Lambert W-function is not Liouvillian.

The results of [Ros76] can be used to prove the structure theorems of Epstein–Caviness [EC79], Risch [Ris79] and Rothstein–Caviness [RC79]. These structure theorems describe the relations that occur between an exponential or integral in an elementary or Liouvillian extension and the elements used to build the tower defining such an extension. For example, the Risch Structure Theorem is formulated in the following way. Let K be an elementary extension of $k = C(x)$, $x' = 1$, with the same field of constants C. We may write $K = k(t_1, \ldots, t_n)$ where, for each i, t_i is algebraic over $K_{i-1} = k(t_1, \ldots, t_{i-1})$, or there exists a $u_i \in K_{i-1}$ such that $t_i' = u_i' t_i$ or $t_i' = u_i'/u_i$. Let $E_{K/k} = \{i \in \{1, \ldots, n\} \mid t_i \text{ transcendental over } K_{i-1} \text{ and } t_i' = u_i' t_i, u_i \in K_{i-1}\}$ and $L_{K/k} = \{i \in \{1, \ldots, n\} \mid t_i \text{ transcendental over } K_{i-1} \text{ and } t_i' = u_i'/u_i, u_i \in K_{i-1}\}$. The Risch Structure Theorem states that for $u, v \in K$, if $v' = u'/u$ then there are $r_i \in \mathbb{Q}$ such that

$$v + \sum_{i \in L_{K/k}} r_i t_i + \sum_{i \in E_{K/k}} r_i u_i \in C,$$

where $t_i' = u_i' t_i$ for $i \in E_{K/k}$. The results of Epstein–Caviness give a more precise special version of this latter result and the result of Rothstein–Caviness generalizes Risch's result to include Liouvillian extensions. The importance of these results lies in the fact that they lead to algorithms to determine the structure of elementary/Liouvillian towers as one builds these towers. An excellent exposition of the various structure theorems is given in [Bro97, Chapter 9]. Related questions are considered in [CDL18], where the authors define a canonical decomposition of an element in an extension field generated by an integral. This is useful in determining if the integral of an element already lies in this field.

Structure theorems also play a key role in the ongoing work concerning a parallel approach to algorithms determining if an elementary function has an elementary integral. The original algorithm of Risch proceeded in a recursive manner, working down the elementary tower that defines the given elementary function. Let us restrict to functions defined by elementary towers of the form $K = C(x, \theta_1, \ldots, \theta_n)$, where $x' = 1$ and, for each i, θ_i is transcendental over $K_{i-1} = C(x, \theta_1, \ldots, \theta_{i-1})$ and either $\theta_i' = u_i'/u_i$ or $\theta_i' = u_i' \theta_i$ for some $u_i \in K_{i-1}$. The parallel approaches to determining if an element $y \in K$ has an elementary integral proceed by determining, all at once, the rational functions of $x, \theta_1, \ldots, \theta_n$ $v, u_1, \ldots, u_m$ that can occur in the expression

$$y = v' + \sum_{j=1}^{m} c_j \frac{u_j'}{u_j}$$

predicted by Liouville's Theorem. This approach is clearly described in [Bro07] where references to previous work is given. At present the parallel approach still contains heuristic elements but results concerning degree bounds and other information about the possible v and u_i are given in this latter work.

References

[Ax71] James Ax, On Schanuel's conjectures, *Ann. of Math.* (2) **93** (1971), 252–268. MR 0277482

[Bad94] Jamil Baddoura, A Conjecture on Integration in Finite Terms with Elementary Functions and Polylogarithms, *Proceedings of the International Symposium on Symbolic and Algebraic Computation*, ACM, 1994, pp. 158–162.

[Bad11]   _____ , A note on symbolic integration with polylogarithms, *Mediterr. J. Math.* **8** (2011), no. 2, 229–241. MR 2802326

[BCDJ08] Manuel Bronstein, Robert M. Corless, James H. Davenport, and David J. Jeffrey, Algebraic properties of the Lambert *W* function from a result of Rosenlicht and of Liouville, *Integral Transforms Spec. Funct.* **19** (2008), no. 9-10, 709–712. MR 2454730

[BK77]   W. Dale Brownawell and Kenneth K. Kubota, The algebraic independence of Weierstrass functions and some related numbers, *Acta Arith.* **33** (1977), no. 2, 111–149. MR 0444582

[Bro97]   Manuel Bronstein, *Symbolic integration. I Transcendental functions*, Algorithms and Computation in Mathematics, vol. 1, Springer-Verlag, Berlin, 1997, With a foreword by B. F. Caviness. MR 1430096

[Bro07]   _____ , Structure theorems for parallel integration, *J. Symbolic Comput.* **42** (2007), no. 7, 757–769, Paper completed by Manuel Kauers. MR 2348061

[Cav70]   Bob F. Caviness, On canonical forms and simplification, *J. Assoc. Comput. Mach.* **17** (1970), 385–396. MR 0281386

[CDL18]   Shaoshi Chen, Hao Du, and Ziming Li, Additive decompositions in primitive extensions, *ISSAC'18—Proceedings of the 2018 ACM International Symposium on Symbolic and Algebraic Computation*, ACM, New York, 2018, pp. 135–142. MR 3840374

[Che85]   Guy W. Cherry, Integration in finite terms with special functions: the error function, *J. Symbolic Comput.* **1** (1985), no. 3, 283–302. MR 849037

[Che86]   _____ , Integration in finite terms with special functions: the logarithmic integral, *SIAM J. Comput.* **15** (1986), no. 1, 1–21. MR 822189

[Che89]   _____ , An analysis of the rational exponential integral, *SIAM J. Comput.* **18** (1989), no. 5, 893–905. MR 1015263

[CR75]   Bob F. Caviness and Michael Rothstein, A Liouville theorem on integration in finite terms for line integrals, *Comm. Algebra* **3** (1975), no. 9, 781–795. MR 0384764

[Dav81]   James H. Davenport, *On the integration of algebraic functions*, Lecture Notes in Computer Science, vol. 102, Springer-Verlag, Berlin-New York, 1981. MR 617377

[DLS98]   Viacheslav Alekseevich Dobrovol'skii, Natalia Vasilevna Lokot', and Jean-Marie Strelcyn, Mikhail Nikolaevich Lagutinskii (1871–1915): un mathématicien méconnu, *Historia Math.* **25** (1998), no. 3, 245–264. MR 1649949

[DS86]   James H. Davenport and Michael F. Singer, Elementary and Liouvillian solutions of linear differential equations, *J. Symbolic Comput.* **2** (1986), no. 3, 237–260. MR 860538

[EC79]   Harvey I. Epstein and Bob F. Caviness, A structure theorem for the elementary functions and its application to the identity problem, *Internat. J. Comput. Inform. Sci.* **8** (1979), no. 1, 9–37. MR 523633

[Eps79]   Harvey I. Epstein, A natural structure theorem for complex fields, *SIAM J. Comput.* **8** (1979), no. 3, 320–325. MR 539250

[EvdP97] Graham R. Everest and Alf J. van der Poorten, Factorisation in the ring of exponential polynomials, *Proc. Amer. Math. Soc.* **125** (1997), no. 5, 1293–1298. MR 1401740

[Gri76] Phillip A. Griffiths, Variations on a theorem of Abel, *Invent. Math.* **35** (1976), 321–390. MR 0435074

[Gri04] _____, The legacy of Abel in algebraic geometry, *The legacy of Niels Henrik Abel*, Springer, Berlin, 2004, pp. 179–205. MR 2077573

[Har71] Goeffry H. Hardy, *The integration of functions of a single variable*, Hafner Publishing Co., New York, 1971, Reprint of the second edition, 1916, Cambridge Tracts in Mathematics and Mathematical Physics, No. 2. MR 0349924

[Heb15] Waldemar Hebisch, Integration in terms of exponential integrals and incomplete gamma functions, *ACM Commun. Comput. Algebra* **49** (2015), no. 3, 98–100. MR 3434590

[Heb18] _____, Integration in terms of polylogarithm, Preprint. arXiv:1810.05865, 2018.

[Heb21] _____, Symbolic integration in the spirit of Liouville, Abel and Lie, Preprint. arXiv:2104.06226, 2021.

[HRS89] C. Ward Henson, Lee A. Rubel, and Michael F. Singer, Algebraic properties of the ring of general exponential polynomials, *Complex Variables Theory Appl.* **13** (1989), no. 1-2, 1–20. MR 1029352

[Joh69a] Joseph Johnson, Differential dimension polynomials and a fundamental theorem on differential modules, *Amer. J. Math.* **91** (1969), 239–248. MR 0238822

[Joh69b] _____, Kähler differentials and differential algebra, *Ann. of Math.* (2) **89** (1969), 92–98. MR 0238823

[Kas80] Toni Kasper, Integration in finite terms: the Liouville theory, *Math. Mag.* **53** (1980), no. 4, 195–201. MR 600228

[Kno92] Paul H. Knowles, Integration of a class of transcendental Liouvillian functions with error-functions. I, *J. Symbolic Comput.* **13** (1992), no. 5, 525–543. MR 1170095

[Kno93] _____, Integration of a class of transcendental Liouvillian functions with error-functions. II, *J. Symbolic Comput.* **16** (1993), no. 3, 227–241. MR 1259671

[Koe87] Leo Koenigsberger, Bermekungen zu Liouville's Classificirung der Transcendenten, *Math. Ann.* **28** (1887), 482–492.

[Kol68] Ellis R. Kolchin, Algebraic groups and algebraic dependence, *Amer. J. Math.* **90** (1968), 1151–1164. MR 0240106

[KS19] Yashpreet Kaur and Varadharaj R. Srinivasan, Integration in finite terms with dilogarithmic integrals, logarithmic integrals and error functions, *J. Symbolic Comput.* **94** (2019), 210–233. MR 3945065

[KS21] _____, Integration in finite terms: dilogarithmic integrals, *Appl. Algebra Engrg. Comm. Comput.* (2021).

[Lap12] Pierre-Simon Laplace, *Théorie analytique de probabilité*, 3eme ed., Ve Courcier, Paris, 1812.

[LL02]   Utsanee Leerawat and Vichian Laohakosol, A generalization of Liouville's theorem on integration in finite terms, *J. Korean Math. Soc.* **39** (2002), no. 1, 13–30. MR 1872579

[Lüt90]   Jesper Lützen, *Joseph Liouville 1809–1882: master of pure and applied mathematics*, Studies in the History of Mathematics and Physical Sciences, vol. 15, Springer-Verlag, New York, 1990. MR 1066463

[Mac76]   Carolla Mack, *Integration of affine forms over elementary functions*, Tech. report, Computer Science Department, University of Utah, 1976, VCP-39.

[Mac16]   Angus Macintyre, Turing meets Schanuel, *Ann. Pure Appl. Logic* **167** (2016), no. 10, 901–938. MR 3522649

[Mas17]   David Masser, Integration in elementary terms, *Newsletter London Math. Soc.* **473** (2017), 30–36.

[MB09]   Dimitry Mordukhai-Boltovski, Researches on the integration in finite terms of differential equations of the first order, *Communications de la Société Mathématique de Kharkov* **X** (1906–1909), 34–64, 231–269 (Russian), Translation of pp. 34–64, B. Korenblum and M.J. Prelle, SIGSAM Bulletin, Vol.15, No. 2, May 1981, pp. 20–32.

[MB10]   ———, *On the integration in finite terms of linear differential equations (in Russian)* available at http://eqworld.ipmnet.ru/ru/library/mathematics/ode.htm., Warsaw, 1910.

[MB13]   ———, *On the integration of transcendental functions (in Russian)*, Warsaw, 1913.

[MB37]   ———, Sur la résolution des équations différentielles du premier ordre en forme finie, *Rendiconti del circolo matematico di Palermo* (1937), 49–72.

[MZ79]   Joel Moses and Richard Zippel, An extension of Liouville's theorem, *Symbolic and algebraic computation (EUROSAM '79, Internat. Sympos., Marseille, 1979)*, Lecture Notes in Comput. Sci., vol. 72, Springer, Berlin-New York, 1979, pp. 426–430. MR 575703

[MZ94]   Elana A. Marchisotto and Gholam-Ali Zakeri, An Invitation to Integration in Finite Terms, *Coll. Math. J.* **25** (1994), no. 4, 295–308.

[MZ20]   David Masser and Umberto Zannier, Torsion points, Pell's equation, and integration in elementary terms, *Acta Math.* **225** (2020), no. 2, 227–313. MR 4205408

[ND79]   Arthur C. Norman and James H. Davenport, Symbolic integration—the dust settles?, *Symbolic and algebraic computation (EUROSAM '79, Internat. Sympos., Marseille, 1979)*, Lecture Notes in Comput. Sci., vol. 72, Springer, Berlin-New York, 1979, pp. 398–407. MR 575700

[Ost46a]   Alexandre Ostrowski, Sur les relations algébriques entre les intégrales indéfinies, *Acta Math.* **78** (1946), 315–318. MR 0016764

[Ost46b]   ———, Sur l'intégrabilité élémentaire de quelques classes d'expressions, *Comment. Math. Helv.* **18** (1946), 283–308. MR 0016763

[PS83]   Myra J. Prelle and Michael F. Singer, Elementary first integrals of differential equations, *Trans. Amer. Math. Soc.* **279** (1983), no. 1, 215–229. MR 704611

[Raa12] Clemens G. Raab, *Definite Integration in Differential Fields*, Ph.D. thesis, Johannes Kepler Univ. Linz, Austria, 2012, http://www.risc.jku.at/publications/download/risc_4583/PhD_CGR.pdf.

[RC79] Michael Rothstein and Bob F. Caviness, A structure theorem for exponential and primitive functions, *SIAM J. Comput.* **8** (1979), no. 3, 357–367. MR 539254

[Ris67] Robert H. Risch, *On Real Elementary Functions*, SDC Document SP-2801/001/00, May 1967.

[Ris69] _____, The problem of integration in finite terms, *Trans. Amer. Math. Soc.* **139** (1969), 167–189. MR 0237477

[Ris76] _____, Implicitly elementary integrals, *Proc. Amer. Math. Soc.* **57** (1976), no. 1, 1–7. MR 0409427

[Ris79] _____, Algebraic properties of the elementary functions of analysis, *Amer. J. Math.* **101** (1979), no. 4, 743–759. MR 536040

[Rit23] Joseph F. Ritt, On the integrals of elementary functions, *Trans. Amer. Math. Soc.* **25** (1923), no. 2, 211–222. MR 1501240

[Rit25] _____, Elementary functions and their inverses, *Trans. Amer. Math. Soc.* **27** (1925), no. 1, 68–90. MR 1501299

[Rit27a] _____, A factorization theory for functions $\sum_{i=1}^{n} a_i e^{\alpha_i x}$, *Trans. Amer. Math. Soc.* **29** (1927), no. 3, 584–596. MR 1501406

[Rit27b] _____, On the integration in finite terms of linear differential equations of the second order, *Bull. Amer. Math. Soc.* **33** (1927), no. 1, 51–57. MR 1561321

[Rit29] _____, Algebraic combinations of exponentials, *Trans. Amer. Math. Soc.* **31** (1929), no. 4, 654–679. MR 1501505

[Rit48] _____, *Integration in Finite Terms. Liouville's Theory of Elementary Methods*, Columbia University Press, New York, N. Y., 1948. MR 0024949

[Ros68] Maxwell Rosenlicht, Liouville's theorem on functions with elementary integrals, *Pacific J. Math.* **24** (1968), 153–161. MR 0223346

[Ros69] _____, On the explicit solvability of certain transcendental equations, *Inst. Hautes Études Sci. Publ. Math.* (1969), no. 36, 15–22. MR 0258808

[Ros72] _____, Integration in finite terms, *Amer. Math. Monthly* **79** (1972), 963–972. MR 0321914

[Ros73] _____, An analogue of l'Hospital's rule, *Proc. Amer. Math. Soc.* **37** (1973), 369–373. MR 0318117

[Ros76] _____, On Liouville's theory of elementary functions, *Pacific J. Math.* **65** (1976), no. 2, 485–492. MR 0447199

[RS77] Maxwell Rosenlicht and Michael F. Singer, On elementary, generalized elementary, and Liouvillian extension fields, *Contributions to algebra (collection of papers dedicated to Ellis Kolchin)* (1977), 329–342. MR 0466093

[Sin75] Michael F. Singer, Elementary solutions of differential equations, *Pacific. J. Math.* **59** (1975), no. 2, 535–547. MR 0389874

[Sin76] _____, Solutions of linear differential equations in function fields of one variable, *Proc. Amer. Math. Soc.* **54** (1976), 69–72. MR 0387260

[Sin92] _____, Liouvillian first integrals of differential equations, *Trans. Amer. Math. Soc.* **333** (1992), no. 2, 673–688. MR 1062869

[Sri17] Varadharaj R. Srinivasan, Liouvillian solutions of first order nonlinear differential equations, *J. Pure Appl. Algebra* **221** (2017), no. 2, 411–421. MR 3545269

[Sri18] _____, Corrigendum to "Liouvillian solutions of first order nonlinear differential equations" [J. Pure Appl. Algebra **221** (2) (2017) 411–421], *J. Pure Appl. Algebra* **222** (2018), no. 6, 1372–1374. MR 3754430

[Sri20] _____, Differential subfields of Liouvillian extensions, *J. Algebra* **550** (2020), 358–378. MR 4058220

[SSC85] Michael F. Singer, B. David Saunders, and Bob F. Caviness, An extension of Liouville's theorem on integration in finite terms, *SIAM J. Comput.* **14** (1985), no. 4, 966–990. MR 807895

[vdPS03] Marius van der Put and Michael F. Singer, *Galois theory of linear differential equations*, Grundlehren der Mathematischen Wissenschaften [Fundamental Principles of Mathematical Sciences], vol. 328, Springer-Verlag, Berlin, 2003. MR 1960772

[Zan14] Umberto Zannier, Elementary integration of differentials in families and conjectures of Pink, *Proceedings of the International Congress of Mathematicians—Seoul 2014. Vol. II*, Kyung Moon Sa, Seoul, 2014, pp. 531–555. MR 3728626

Joseph Fels Ritt

Columbia University Press, New York (1948)

INTEGRATION
IN FINITE TERMS

Liouville's Theory
of Elementary Methods

Joseph Fels Ritt

Davies Professor of Mathematics
Columbia University

New York

Columbia University Press

1 9 4 8

© Springer Nature Switzerland AG 2022
C. G. Raab und M. F. Singer (Hrsg.), *Integration in Finite Terms:*
Fundamental Sources, Texts & Monographs in Symbolic Computation,
https://doi.org/10.1007/978-3-030-98767-1_3

31

PREFACE

During the period between 1833 and 1841, J. Liouville presented a theory of integration in finite terms. He determined the form which the integral of an algebraic function must have when the integral can be expressed with the operations of elementary mathematical analysis, carried out a finite number of times. He showed that the elliptic integrals of the first and second kinds have no elementary expressions. He proved that certain simple differential equations cannot be solved by elementary procedures. His papers contain other remarkable applications of his theory.

The questions treated by Liouville are questions which occur to every strong undergraduate student of mathematics. Nevertheless Liouville's work never received very wide attention. It has always been something which everyone would like very much to know about but which very few undertake to study.

During the nineteenth century, extremely little was done in direct continuation of Liouville's work.* About forty years ago, the Russian mathematician Mordukhai-Boltovskoi began to write on Liouville's theory and contributed extensively to it. In particular, he published a book on the integration of transcendental functions and one on the integration in finite terms of linear differential equations. Through his influence, the subject seems to have been more widely studied in Russia than elsewhere. The present writer published some work on these questions between 1923 and 1927. At the present time, Ostrowski is writing on the subject.

This monograph gives an account of Liouville's work and of some of that of his few followers. On the basis of what has already been said, a glance through the chapters, or even over the table of contents, will give a sufficient idea of the topics covered.

I should like, however, to say something in regard to the treatment given here of Liouville's work. Liouville's methods are ingenious and beautiful. From the formal standpoint, they are entirely sound. There are, however, certain questions connected with the

* To be sure, there appeared the Picard-Vessiot theory of linear differential equations, which furnishes, for such equations, results analogous to those of Galois for algebraic equations. Recent work of E. R. Kolchin has brought rigor and simplicity to the Picard-Vessiot theory.

One should perhaps mention also the remarkable work of Bruns on the algebraic solutions of the equations of celestial mechanics. (*Acta Mathematica*, Vol. XI.)

many-valued character of the elementary functions which could be
pressed back behind the symbols in Liouville's time but which have
since learned to assert their rights. Such matters are mulled over
in the first chapter. The mulling is inescapable. It might be great
fun to talk just as if the elementary functions were one-valued. I
might even sound convincing to some readers; I certainly could not
fool the functions. However, if one is chiefly interested in formal
ideas, one may read the "First Survey of the Elementary Functions"
in Chapter I and then pass to the summary at the end of that chapter. It ought not to be hard, after that, to follow the formal
processes of Chapter II.

As regards the theory of functions, I have assumed, in Chapter I,
an acquaintance with the simpler facts concerning analytic continuation. Riemann surfaces of algebraic functions are mentioned in
Chapter I; their simpler properties permit, in Chapter II, the
swift liquidation of questions on the integrability of special algebraic functions. In algebra, I use in Chapter I the discriminant
of a polynomial and the resultant of a pair of polynomials. For the
rest, special material of algebra or of analysis is developed wherever it is needed.

I should like, in conclusion, to say something concerning Joseph
Liouville (1809-1882). He originated the notion of derivative of
fractional order. The first examples of transcendental numbers were
due to him; his work on this question was the starting point of the
modern researches of Thue and Siegel. He was one of the founders of
the theory of boundary value problems. As Abel had also done, earlier, he solved a problem involving an integral equation; this was
decades before the general theory of such equations came into being.
His work on doubly periodic meromorphic functions was precursive to
Weierstrass's theory of the elliptic functions, just as his work in
the theory of elimination anticipated, to some extent, ideas of
Kronecker. He presented a method for treating classes of Diophantine equations which has been developed extensively, in recent
times, by E. T. Bell. In geometry, he determined the gronp of con-
formal transformations in three dimensions. He was the founder of
the *Journal de mathématiques pures et appliquées.* Considering his
achievements, one may question whether he has been adequately ap-
preciated, even by the mathematicians of his own country. It is
surprising, for instance, that his collected works were never published.

If this monograph should promote the study of what is probably one of the most interesting portions of Liouville's work, the writer will feel amply rewarded.

J. F. Ritt

Columbia University
July, 1947

CONTENTS

Contents

Chapter I

ELEMENTARY FUNCTIONS OF ONE VARIABLE

FIRST SURVEY OF THE ELEMENTARY FUNCTIONS

1. The functions which we shall study in the present chapter are
essentially those which make up the functional world of a student
of the integral calculus. Such a student, if not familiar with the
concept of algebraic function in its most general form, knows the
polynomials and fractional rational functions, has seen functions
involving radicals, and can imagine quite well the most general
algebraic function which can be expressed in terms of radicals. He
knows e^x, log x, sin x, cos x, and the inverses of the latter two
functions. After compounding functions of the foregoing types in
various ways to produce combinations like e^x or log (sec x + tan x),
he is in possession of an extensive class of functions, each constructed with a finite number of operations. A typical example
would be

$$\tan [e^{x^2} - \log_x (1 + x^4)] + [x^x + \log \arc \sin x]^4.$$

The expression x^x is to be interpreted, of course, as $e^{x \log x}$. As
to the logarithm to the base x, it is nothing more than the natural
logarithm divided by log x.

The use of only a finite number of operations needs particular
emphasis. One meets infinite series in a calculus course. The representation of functions by means of such series is a question for-
eign to our present study. We shall be concerned only with what can
be obtained from the basic functions with calculations involving
only a finite number of operations.

2. An inspection of our functions will permit us to describe them
more closely and to classify them. We shall look twice at our material, first casually, just to see how matters stand, then fully and
squarely, with no turning away from hard realities.

We notice first that, if complex numbers are employed, the trigonometric functions and their inverses become redundant. For instance

$$\sin x = \frac{e^{ix} - e^{-ix}}{2i}$$

and

$$\arc \sin x = \frac{1}{i} \log (ix + \sqrt{1 - x^2}).$$

39

Thus the functions of elementary analysis are constructed out of the variable x by repeated use of the following operations:

(a) Algebraic operations performed on one or more expressions

(b) The taking of exponentials

(c) The taking of logarithms.

Just what is to be understood by an algebraic operation will be explained fully later.

Now let us explain how the elementary functions will be classified. The variable x will be called a *monomial of order zero*, and any algebraic function of x will be called a *function of order zero*. The exponential or the logarithm of a nonconstant algebraic function will be called a *monomial of the first order*. An algebraic combination of x and of monomials of the first order will, if it is not an algebraic function, be called a *function of the first order*. The exponential or the logarithm of a function of the first order, will, if it is not a function of order zero or a function of order unity, be called a *monomial of the second order*. For instance, as one will see in Chapter IV, the exponential of e^x, and log (log x) are monomials of the second order. An algebraic combination of monomials of orders 0, 1, 2 which is not a function of one of the orders 0, 1, will be called a *function of the second order*. The classification continues in this way.

All of this needs a closer examination, and to that we now turn.

ALGEBRAIC FUNCTIONS

3. Let us recall the notion of algebraic function. A function u of x is algebraic if it is defined by an irreducible relation

(1) $$\alpha_0(x)\, u^p + \alpha_1(x)\, u^{p-1} + \ldots + \alpha_p(x) = 0$$

with $p > 0$ where the α are polynomials in x with any complex numbers for coefficients, α_0 not being identically zero. In saying that (1) is irreducible, we shall mean that its first member is not the product of two nonconstant polynomials in u and x.

We shall find ourselves at times employing the Riemann surface of u, which is a surface of p sheets. For our immediate purposes, it is more important to consider u as a *monogenic analytic function* (m.a.f.) in the sense of Weierstrass, that is, as the totality of power series which can be obtained from some given power series by analytic continuation. Each power series is called an element of the m.a.f. An element which is a series of powers of $x - x_0$ may be represented by $P(x-x_0)$.

Let $P(x-x_0)$ be any element of our algebraic function u. Let C be any curve with x_0 for first point. We may consider C to be given by a relation

(2) $$x = \varphi(\lambda)$$

where $\varphi(\lambda)$ is a continuous, complex-valued function of the real variable λ on the closed interval $(0, 1)$, with $\varphi(0) = x_0$.

If $P(x-x_0)$ cannot be continued along the entire curve C, it must be that C has on it one or more points which are places at which u has poles or branch points. If that is the case, we can, by bending C slightly so as to avoid the singularities of u, which are finite in number, obtain a new curve, along all of which u can be continued. This simple step of bending a curve so as to permit the continuation of an element leads to a general idea which will help us considerably in the study of the elementary functions.

FLUENT FUNCTIONS

4. An analytic function of x will be said to be *fluent* if every element $P(x-x_0)$ of the function has the following property. *For every curve (2) with $\varphi(0) = x_0$, and for every $\varepsilon > 0$, there can be found a curve*

(3) $$x = \varphi_1(\lambda) \qquad (0 \leq \lambda \leq 1)$$

with $\varphi_1(0) = x_0$, such that

$$|\varphi_1(\lambda) - \varphi(\lambda)| < \varepsilon$$

for $0 \leq \lambda \leq 1$ and such that $P(x-x_0)$ can be continued along the entire curve (3). Broadly speaking, an element of the function, if it cannot be continued along a given path, can be continued along some path in every neighborhood of the given one.

As we have indicated, algebraic functions are fluent. To be formal, one replaces (2) by a curve consisting of short straight segments and then uses arcs of small circles to avoid the finite number of singular points which may lie on the segments.

EXPONENTIALS OF ALGEBRAIC FUNCTIONS

5. Let v be a nonconstant algebraic function. We are interested in the nature of e^v. We consider any element of v and take its exponential. This furnishes an analytic element, the totality of whose continuations constitutes e^v. This process furnishes a unique m.a.f., for if P_1 and P_2 are elements of v, the path which continues P_1 into P_2 continues the exponential of P_1 into that of P_2. The

Riemann surface of e^v is that of v. Our definition of e^v gives e^v a value at each finite point of the surface which is not a branch point and at which v has no pole. It is possible to go further. At a branch point of v which is not a pole, corresponding to a value x_0 of x, we have merely to define e^v as the exponential of the value of v. We secure an expansion of e^v for the branch point, in fractional powers of $x - x_0$, by taking the exponential of the expansion of v. If v has a pole at a point of the surface with $x = x_0$, we secure for e^v an expansion containing an infinite number of negative powers of $x - x_0$, fractional powers appearing in the expansion if we are dealing with a branch point. If x_0, in what precedes, is ∞, we obtain for e^v an expansion in integral or fractional powers of $1/x$.

Two distinct elements $P_1(x-x_0)$ and $P_2(x-x_0)$, formed for the same point x_0, produce two distinct elements Q_1 and Q_2 of e^v. If Q_1 were identical with Q_2, we would have

$$P_2 = P_1 + 2k\pi i$$

with k an integer distinct from 0. The path which continues P_1 into P_2 would thus continue P_2 into $P_3 = P_2 + 2k\pi i$. The same path would continue P_3 into a new element. Continuing, we would find an infinite number of elements for v at x_0, a contradiction of the algebraic nature of v.

In passing, let us notice that if v is not a constant, e^v is not algebraic. If it were, e^x, resulting from e^v by an algebraic substitution for x, would be algebraic.* The nonalgebraic character of e^x follows from the fact that it assumes the same value for an infinite number of values of x, something which cannot happen for an algebraic function, as can be seen from (1), in which a value assigned to u leads to an algebraic equation for x.

That e^v is fluent follows immediately from the fact that v is.

LOGARITHMS OF ALGEBRAIC FUNCTIONS

6. We now examine log v, where v is algebraic and not constant. It will turn out that there may be more than one m.a.f., but always a finite number of such functions, which it is proper to call log v.

Let $P(x-x_0)$ be an element of v, for some x_0, which is not zero at x_0; that is, the series P contains a constant term distinct from 0. We can form an infinite number of elements log P, any

---

* Such questions of substitution will receive formal treatment later.

two differing by an integral multiple of $2\pi i$. Let Q be any of these elements. We obtain from Q, by continuation, an m.a.f. which we shall call log v.

An example will show that distinct elements Q may lead to distinct functions log v. Let $v = x^2$ and let x_0 be any complex number distinct from 0. Let a Q be selected. Let x describe a closed path, beginning and ending at x_0, which does not pass through the origin. As x travels over this path, its amplitude changes by a multiple of 2π, so that the amplitude of x^2 changes by an even multiple of 2π. As Q is continued along the path, its coefficient of i, which is one of the amplitudes of x^2, changes by an even multiple of 2π. Thus there is no path which continues Q into $Q + 2\pi i$. It is easy to see that there are just two functions log x^2, the doubles of log x and log (-x).

In the general case, there is no difficulty in showing that there is more than one log v only when v is of the form u^r with r an integer greater than unity and u a function which is one-valued on the surface of v. If r is taken as large as possible, there are just r functions log v.

All elements $P(x-x_0)$ lead to the same functions log v, since a Q secured from any element P_1 can be continued, along a suitable path, into a Q for any other element P_2.

Thus, given v, we must, in referring to log v, specify one of a set of functions, any two of which differ by a constant. It is easy to see that each log v is fluent; given a path (2) we can bend it slightly to avoid the singularities and zeros of v.

ALGEBRAIC COMBINATIONS

7. Let $\mathcal{E}_1, \ldots, \mathcal{E}_r$ be monogenic analytic functions, not necessarily distinct from one another. Let x_0 be any point at which each $\mathcal{E}_i$ has at least one analytic branch and let $P_i(x-x_0)$, $i = 1, \ldots, r$, be an element of $\mathcal{E}_i$. Let u(x) be an m.a.f. for which some x_0 exists as above at which u has an element which satisfies an equation

(4) $\alpha_0 u^p + \ldots + \alpha_p = 0$

where each α is a polynomial, with constant coefficients, in the $\smallsmile$ $P_i(x-x_0)$, and where α_0 is not identically zero. We shall call u(x) an *algebraic combination* of the $\mathcal{E}_i$.

Thus the integral function

$$u = 2 \sum_{n=0}^{\infty} \frac{x^n}{(2n)!}$$

which may be written $e^{\sqrt{x}} + e^{-\sqrt{x}}$ may be regarded as an algebraic combination of the two identical monomials $e^{\sqrt{x}}$, $e^{\sqrt{x}}$. We may use any $x_0 \neq 0$ and take P_1 and P_2 as the two elements of $e^{\sqrt{x}}$ at x_0. One need not fear that the definition of algebraic combination makes the symbol $e^{\sqrt{x}} + e^{-\sqrt{x}}$ ambiguous. Our definition of algebraic combination does not make it necessary to break with established notation and to create a symbolism which will describe everything involved in making algebraic combinations.

FUNCTIONS OF FIRST ORDER

8. If v is a nonconstant algebraic function, each of the monogenic analytic functions e^v and $\log v$ will be called a *complete monomial of order 1*. The variable x will be called a complete monomial of order 0. By a *monomial* we shall mean a branch of a complete monomial which is analytic in some area.

An m.a.f. u(x) will be called a *function of order 1* if it is not algebraic and if it is an algebraic combination of complete monomials of orders 0 and 1. Frequently we shall use the term *function of order 1* to represent a branch of a function of order 1.

Let u be a function of order 1. We are going to prove that u is fluent. We start by considering the *field* $\mathcal{D}$ of x and the P_i which figure in the equation (4) defining u. $\mathcal{L}$ is the totality of rational combinations, with complex coefficients, of x and the P_i. The functions in $\mathcal{D}$ are all meromorphic in some circle with center at x_0. We may assume the first member of (4) to be irreducible in $\mathcal{D}$. If it is not, we replace it by that one of its irreducible factors which vanishes for the given element of u. Then the discriminant D of (4), which is a polynomial in the coefficients in (4) and is a function analytic at x_0, cannot vanish identically. If it did, the first member of (4) would have a factor of positive degree in common with its derivative with respect to u; this is because the resultant of the first member of (4) and its derivative equals $a_0 D$ or its negative.

Now let us see what will assure the possibility of continuing u along a curve C which starts at x_0. Suppose that each P_i in (4) can be continued along C. This means that a_0 and the discriminant can be continued along C. Suppose now that neither a_0 nor the discriminant vanishes at any point of C. Then (4), considered an algebraic equation for a quantity u, will provide p distinct values for u at each point of C. These values will furnish p distinct functions, each analytic along C. One of these functions will

44

effect the continuation along C of the given element of u at x_0.
If C does not satisfy the foregoing conditions, we first bend it
slightly so as to have a curve C' along which each P can be continued. This is possible because each complete monomial is fluent.
It may be that α_0 or the discriminant has zeros along C'. Let us
consider, for instance, α_0. The continuation of α_0 along C' is ef-
fected by the use of a finite set of elements. Each of these elements represents an analytic function, the zeros of which are
naturally isolated. Thus a slight bending of C' will give us a
curve along which α_0 is not zero, except perhaps at x_0, which is
the only point being held fast. Now if α_0 is zero at x_0, we can
start operations from a point on C close to x_0 at which α_0 is not
zero. Similar remarks apply to the discriminant.

Thus a function of the first order is fluent.

9. Now let $\mathfrak{U}$ be any area in the complex plane and suppose that
we can continue the above-mentioned element of u with center at x_0
into and all over $\mathfrak{U}$, so that u has a branch which is uniform and
analytic throughout $\mathfrak{U}$. Let C be some curve along which u can be
continued from x_0 into $\mathfrak{U}$. Bending C slightly if necessary, we may
suppose that each P can be continued along C. Thus, in every area $\mathfrak{U}$
in which u is analytic, there is an area $\mathfrak{U}_1$ in which the complete
monomials ξ of which the P are elements have analytic branches
$\theta_i(x)$, i = 1, ..., r and in which there is satisfied an equation

(5) $$\beta_0 u^P + \ldots + \beta_p = 0$$

with each β_i the same polynomial in x and the $\theta(x)$ which α_i is in
x and the P. We may take $\mathfrak{U}_1$ so that each $\theta_i(x)$ is an exponential
or a logarithm of a branch, analytic in $\mathfrak{U}_1$, of the algebraic function of which ξ_i is an exponential or a logarithm.

Suppose that in (4) we replace each P_i by a variable x_i'. Replacing u by a variable v, we form an equation

(6) $$\alpha_\sigma(x, x_i') v^P + \ldots + \alpha_p(x, x_i') = 0.$$

Let C be a curve, joining x_0 to a point $\underline{a}$, along which u and the P
can be continued. C continues each P_i into a function $\theta_i(x)$ analytic in a neighborhood of $\underline{a}$. Suppose that α_0 in (4) and the discriminant of (4), both continued along C, are not zero at a. Then, for
x = a, $x_i' = \theta_i(a)$, i = 1, ..., r, the first coefficient in (6) and
the discriminant of (6) do not vanish. This means that for the same
values of x and the x', and for v equal to u(a), the derivative
with respect to v of the first member of (6) does not vanish. Using
the implicit function theorem on (6), we can solve for v, obtaining

an algebraic function of x and the x' which is analytic when those
variables remain in the neighborhood of $x = a$, $x'_i = \theta_i(a)$ and which,
when each x'_i is replaced by $\theta_i(x)$, reduces, in a neighborhood of a,
to the continuation of u(x) along C.

10. On the basis of §9, the structure of a function u of order 1
may be described as follows. Given any area in which some branch of
u is analytic, there exist

(0) a point $\underline{a}$ interior to the area, a $\rho > 0$ and a $\rho_1 > \rho$;

(I) r algebraic functions of x, each analytic for $|x - a| < \rho_1$;

(I') r monomials $\theta_1 \ldots, \theta_r$, each an exponential or a logarithm
   of one of the functions in (I), each analytic for $|x - a| < \rho$
   and such that $|\theta_i(x) - \theta_i(a)| < \rho_1$ for $|x - a| < \rho$, $(i = 1,$
   $\ldots, r)$;

(II) an algebraic function of the variables $x, x'_1, \ldots, x'_r$ which
   is analytic for $|x - a| < \rho_1$, $|x'_i - \theta_i(a)| < \rho_1$, $(i = 1, \ldots, r)$
   and which reduces to the given branch of u for $|x - a| < \rho$ if
   each x'_i is replaced by $\theta_i(x)$.*

Furthermore, the integer r, the algebraic equations satisfied
by the functions in (I) and that in (II), and the exponential or
logarithmic characters of the θ_i are independent of the area in
which u is considered and of the branch of u.

The first member of (6) is irreducible as a polynomial in v, x,
and the x'. This does not mean that when the x' are replaced by
their monomials we obtain an equation (4) which determines a single
analytic function u. For instance,

$$u^2 - e^x = 0$$

gives two distinct analytic functions. There is no point of difficulty here. A function u of the first order is a definite function,
for which a scheme of construction can be given as above. There may
be other functions whose schemes of construction employ the same
material which appear in the scheme for u.

FUNCTIONS OF ANY ORDER

11. We can now define, by induction, functions of any order n.
The exponential or a logarithm of a function of order n-1 will be
called a *complete monomial of order n*, provided that it is not
among the functions of orders 0, 1, ..., n-1. A branch of a complete monomial, analytic in some area, will be called a monomial.

* One may ask why, in arranging our material, we do not first choose ρ_1 and
then a sufficiently small ρ, without requiring that $\rho < \rho_1$. The reason is that
the algebraic function in (II) may depend effectively on x.

By a *function of order n*, we shall mean an algebraic combination
of monomials of orders 0, 1, $\ldots$, n which is not a function of one
of the orders 0, 1, $\ldots$, n-1. A branch of such an m.a.f. will also
called, at times, a function of order n.

As above, we may assume that the discriminant of the equation
like (4) which describes the algebraic combination does not vanish
identically.

The existence of functions of all orders will be seen in Chapter
IV. One sees by a quick induction that a function of any order n is
fluent.

The structural scheme of a function of order n is built up by induction, with the help of the notion of fluency. Given a function u
of order n, analytic in some area, there exist

(0) a point $\underline{a}$ interior to the area, a $\rho > 0$ and a $\rho_1 > \rho$;

(I) r_1 algebraic functions of x, each analytic for $|x - \underline{a}| < \rho_1$;

(I') r_1 monomials θ_1', $\ldots$, θ_r', each either an exponential or a
   logarithm of one of the functions in (I), each analytic for
   $|x - \underline{a}| < \rho$ and such that $|\theta_i'(x) - \theta_i'(\underline{a})| < \rho_1$, for $|x - \underline{a}| < \rho$,
   $(i = 1, \ldots, r_1)$;

(II) r_2 algebraic functions of x and of r_1 other variables
   x_1', $\ldots$, x_{r_1}', each analytic for $|x - \underline{a}| < \rho_1$, $|x_i' - \theta_i'(\underline{a})| < \rho_1$;

(II') r_2 monomials θ_1'', $\ldots$, θ_{r_2}'', each an exponential or a logarithm of one of the functions of order 1 to which the algebraic
   functions in (II) reduce when each x_i' is replaced by θ_i'; each
   θ_i'' is analytic for $|x - \underline{a}| < \rho$ and $|\theta_i''(x) - \theta_i''(\underline{a})| < \rho_1$ for
   $|x - \underline{a}| < \rho$;

   . . . .

(N + 1) an algebraic function of x; $\ldots$; $x_1^{(n)}$, $\ldots$, $x_{r_n}^{(n)}$, analytic
   for $|x - \underline{a}| < \rho_1$, $\ldots$, $|x_1^{(n)} - \theta_1^{(n)}(\underline{a})| < \rho_1$, which reduces to
   the given branch of u when each variable is replaced by the mo-
   nomial which corresponds to it.*

Furthermore, the integers r_i, the algebraic equations satisfied
by the functions in (I), (II), $\ldots$, (N + 1), and the character of
the θ as exponentials or logarithms are independent of the area in
which u is considered and of the branch of u.

A function u of order n, described as above at a point $\underline{a}$, will be
said to be of *regular structure* at $\underline{a}$.

We see that an accented x may be used in forming a monomial of
higher order than that to which it corresponds, and may be used

---

* Of course, x itself is not actually replaced.

again by itself. Such would be the case, for instance, in the
scheme of structure of

$$(\log \log x) \log x.$$

We have chosen a symbolism which allows this, for the purposes of
our work on integration. To hint at what is involved here, let us
consider the function

$$u = (\log \log x)^2$$

in which log x is used only to build a higher monomial. In the derivative of u, the monomial log x appears by itself, as well as in
log log x.

A monomial of order n will frequently be described as an *n-monomial* and indeed as an *n-exponential* or an *n-logarithm*, according
to its character.

The functions to which orders are assigned by the preceding definitions will be called *elementary functions of x*.

As Bieberbach has pointed out,* the fluency of the elementary
functions makes visible immediately.the nonelementary character of
a function which has a curve made up of singularities, for instance,
a natural boundary. Thus many of the higher automorphic functions,
functions which satisfy simple algebraic differential equations,
are not elementary. The integrals of algebraic functions and the
elliptic functions cannot be treated in this way.

ALGEBRAIC COMBINATIONS OF ELEMENTARY FUNCTIONS

12. We shall now inquire into the nature of algebraic combinations of elementary functions. To get an idea of what is in this
question, let us consider the process of forming a function u of
order 2. One would use an equation (4) in which the a are polynomials in elements of complete monomials of orders 0, 1, 2. It is
natural to ask what would happen if, with greater apparent generality, one permitted the a to be polynomials in complete monomials
of order 2 and in elements of *any functions* of orders 0, 1. The
answer is that nothing new would be obtained; the use of monomials
is sufficient. This will be seen after we have obtained a set of
theorems that parallel those theorems in the theory of algebraic
numbers which show that the system of algebraic numbers is closed
with respect to algebraic operations.

In what follows, all functions will be analytic in some area and
n will be a given positive integer.

* *International Congress of Mathematicians* (Zurich, 1932), I, 164.

LEMMA: *Let φ be an analytic function. Let there exist, for some* $k \geq 1$, k *analytic functions* $\xi_1, \cdots, \xi_k$, *not all zero, such that*

(7)
$$\varphi \xi_1 = a_{11} \xi_1 + \ldots + a_{1k} \xi_k$$
$$\vdots$$
$$\varphi \xi_k = a_{k1} \xi_1 + \ldots + a_{kk} \xi_k$$

where the a_{ij} *are polynomials in monomials of the first n orders. Then φ is elementary and its order does not exceed n.*

The equations (7) are a system of homogeneous linear equations in the ξ which have a solution with some ξ not zero. The determinant

$$\begin{vmatrix} \varphi - a_{11}, & -a_{12}, & \cdots, & -a_{1k} \\ & & & \\ -a_{k1}, & -a_{k2}, & \cdots, & \varphi - a_{kk} \end{vmatrix}$$

must vanish. This gives an equation

$$\varphi^k + c_1 \varphi^{k-1} + \ldots + c_k = 0$$

where the c are polynomials in monomials of orders $0, \ldots, n$. Thus, φ is elementary, and its order does not exceed n.

THEOREM: *Let α and β be elementary functions of orders not exceeding n. Then* $\alpha + \beta$, $\alpha - \beta$ *and αβ are elementary and their orders do not exceed n.*

We know that α and β satisfy equations

$$\alpha^p + a_1 \alpha^{p-1} + \ldots + a_p = 0$$
$$\beta^q + b_1 \beta^{q-1} + \ldots + b_q = 0$$

with coefficients which are polynomials in monomials of orders $0, \ldots, n$. We let $k = pq$ and represent the k functions $\alpha^r \beta^s$, $r = 0, 1, \ldots, p-1$, $s = 0, 1, \ldots, q-1$, taken in any order, by $\xi_1, \ldots, \xi_k$. We consider any ξ_i, call it simply ξ, and write

$$\xi = \alpha^P \beta^Q, \qquad 0 \leq P < p, \qquad 0 \leq Q < q.$$

Consider $\alpha\xi$. If $P < p - 1$, $\alpha\xi$ is some ξ_j. If $P = p - 1$, we have

$$\alpha\xi = - (a_1 \alpha^{p-1} + \ldots + a_p) \beta^Q$$

so that

(8)
$$\alpha\xi = c_1 \xi_1 + \ldots + c_k \xi_k$$

where the c are polynomials in monomials of orders $0, \ldots, n$. We have similarly, with d of obvious character,

(9)
$$\beta\xi = d_1 \xi_1 + \ldots + d_k \xi_k.$$

Then

$$(\alpha + \beta)\, \xi = (c_1 + d_1)\, \xi_1 + \ldots + (c_k + d_k)\, \xi_k.$$

Now unity is one of the ξ_i so that the ξ_i are not all zero. It follows from the lemma above that $\alpha + \beta$ is elementary and of order not greater than n. We proceed similarly for $\alpha - \beta$. To treat $\alpha\beta$, we use (8) and write

$$\alpha\,\beta\,\xi = c_1\,\beta\,\xi_1 + \ldots + c_k\,\beta\,\xi_k$$

so that, by (9), $\alpha\beta\xi$ is linear in the ξ_i.

It follows from the theorem just proved that if a function α is a a polynomial, with constant coefficients, in elementary functions of orders not exceeding n, then α is elementary and of order not more than n.

THEOREM: *Let an analytic function* μ *satisfy an equation*

(10) $$\mu^t + \alpha\,\mu^{t-1} + \beta\,\mu^{t-2} + \ldots + \lambda = 0$$

where $\alpha, \beta, \ldots, \lambda$ *are elementary and of orders not exceeding* n. *Then* μ *is elementary and its order is not more than* n.

Let $\alpha, \beta, \ldots, \lambda$ be given by equations with coefficients rational in monomials,

$$\alpha^p + a_1\,\alpha^{p-1} + \ldots + a_p = 0,$$
$$\beta^q + b_1\,\alpha^{q-1} + \ldots + b_q = 0,$$

$$\cdot \quad \cdot \quad \cdot \quad \cdot \quad \cdot$$

$$\lambda^r + l_1\,\lambda^{r-1} + \ldots + l_r = 0.$$

We let $k = pq \ldots rt$, and consider the k functions

$$\alpha^P\,\beta^Q \ldots \lambda^R\,\mu^T, \quad 0 \leq P < p, \ldots, 0 \leq T < t,$$

which functions are not all zero, since unity is one of them. We represent these functions, in some order, by $\xi_1, \ldots, \xi_k$. For each ξ,

$$\mu\xi = \alpha^P\,\beta^Q \ldots \lambda^R\,\mu^{T+1}.$$

Thus, if $T < t - 1$, $\mu\xi$ is some ξ_1. If $\mu = T - 1$, we use (10) and find that

(11) $$\mu\xi = -\alpha^{P+1}\,\beta^Q \ldots \lambda^R\,\mu^T - \ldots - \alpha^P\,\beta^{Q'} \ldots \lambda^{R+1}$$

Consider the first power product in the second member of (11). It is a ξ_1 if $P < p - 1$. If $P = p - 1$, the usual reduction gives, for the power product, a linear combination of the ξ_1. Thus, $\mu\xi$ is a linear combination of the ξ_1 with coefficients as desired, and our theorem is proved.

LEMMAS ON ALGEBRAIC FUNCTIONS

13. Let $y = f(x_1, \ldots, x_r)$ be algebraic in its r variables and analytic at $(a_1, \ldots, a_r)$. Let $\varphi_1, \ldots, \varphi_s$, where $s < r$, be algebraic functions of $x_{s+1}, \ldots x_r$, analytic at $(a_{s+1}, \ldots, a_r)$, with

$$\varphi_i (a_{s+1}, \ldots, a_r) = a_i, \quad i = 1, \ldots, s.$$

When each x_i, $i = 1, \ldots, s$, is replaced in f by φ_i, f becomes a function $g(x_{s+1}, \ldots, x_s)$, analytic at $(a_{s+1}, \ldots, a_r)$. We wish to see that g is algebraic in all its variables.

Let the irreducible equation which defines f be

$$\alpha_0 y^p + \ldots + \alpha_p = 0.$$

The substitution of the several φ for $x_1, \ldots, x_s$ may annul every α. However, it is clearly legitimate to assume that $s = 1$ and to make the substitutions in succession. The replacement of x_1 by φ_1 cannot annul every α identically in $x_2, \ldots, x_r$. Otherwise the α would have a common factor. The algebraic character of g can then be shown by the methods of §12. In §14, we shall use the case of $s = 1$. In II, 4, we shall use an $s > 1$. The φ_i will then be constants.

Taking another situation, let us suppose that f, given as above, is zero at $(a_1, \ldots, a_r)$, while $\partial f / \partial x_1$ is not zero there. The equation $f = 0$ defines x_1 as a function g of $x_2, \ldots, x_r$, analytic at $(a_2, \ldots, a_r)$, and equal to zero there. We wish to see that g is algebraic. In the equation defining f, the term α_p cannot be free of x_1. Otherwise, as $\alpha_p = 0$ when $f = 0$, we could not take $x_2, \ldots, x_r$ as independent variables and determine x_1 so as to make f zero. Thus g, which satisfies $\alpha_p = 0$, is algebraic.

LIOUVILLE'S PRINCIPLE

14. The order of an elementary function is a perfectly definite integer. However, the structural scheme of a function of order $n > 0$, as presented in §11, is not at all unique. For instance, e^{x^2} and $e^{-x} e^{x^2 + x}$ are the same function of order 1.

Among all representations of a given function u of order n, there are certain ones which employ a least number of n-monomials; in any such representation, the r_n in $(N + 1)$ of §11 is not greater than the r_n of any other representation of u. Representations which have this feature of economy will be employed throughout our work on integration, and we shall now establish for them an important principle.

Under the assumption that r_n is a minimum, we shall show that *no algebraic relation can exist among the r_n monomials of order* n *in* u *and monomials of order less than* n. To make this statement explicit, we refer to §11 and use the point a there mentioned. Let $\xi_1, \ldots, \xi_p$ be monomials of orders less than n. Suppose that a function

$$(12) \qquad f(x_1^{(n)}, \ldots, x_{r_n}^{(n)}; \quad y_1, \ldots, y_p),$$

algebraic in all its variables and analytic for $x_1^{(n)} = \theta_1^{(n)}(a)$, $y_1 = \xi_1(a)$, vanishes in the neighborhood of a when each $x_1^{(n)}$ is replaced by its $\theta_1^{(n)}$ and each y_1 by ξ_1. We shall show that f *vanishes for all values of the* $x^{(n)}$ *close to the* $\theta^{(n)}(a)$, *if only each* y_1 *is replaced by* ξ_1.

The principle which we have just formulated underlies all of Liouville's work on the elementary functions. We shall call it *Liouville's principle*. The proof amounts to nothing more than solving for one of the n-monomials in any nonidentical relation which may exist. When that monomial is replaced in u by the expression found for it, u acquires an expression involving too few n-monomials. A question connected with the implicit theorem accounts for the details which follow.

Suppose that (12) is not an identity in the $x^{(n)}$ for $y_j = \xi_j$, $j = 1, \ldots, p$. We can find a point b, as close as one pleases to a, such that

$$(13) \qquad f(x_1^{(n)}, \ldots, x_{r_n}^{(n)}; \xi_1(b), \ldots, \xi_p(b))$$

does not vanish identically in the $x^{(n)}$.

Consider the partial derivatives of f, of all orders, cross-derivatives included, with respect to the $x^{(n)}$. Not all of them can vanish for $y_1 = \xi_1(b)$, $x_1^{(n)} = \theta_1^{(n)}(b)$. If they did, the function in (13) would be constant when each $x_1^{(n)}$ varies in the neighborhood of $\theta_1^{(n)}(b)$. The constant would be zero, since (13) vanishes for $x_1^{(n)} = \theta_1^{(n)}(b)$; this would contradict what we know of (13).

Working now in a neighborhood 𝔓 of x = a, let us suppose that all of the above-mentioned derivatives of f up to and including those of order j^* vanish throughout 𝔐 when the variables in (12) are replaced by their monomials, but that some derivative of order $j + 1$ does not vanish at x = b, where b is in 𝔐. That the required j and b exist follows from what precedes. To fix our ideas, suppose that

$$G(x_1^{(n)}, \ldots, y_p)$$

---

* In this, we consider f to be its own derivative of order zero.

is a partial derivative which vanishes over $\mathfrak{M}$ but that the derivative of G with respect to $x_1^{(n)}$ does not vanish at b.

By §13, the equation G = 0 determines $x_1^{(n)}$ as an algebraic function of $x_2^{(n)} \ldots, y_p$, analytic at $(\theta_2^{(n)}$ (b), $\ldots, \xi_p(b))$ which reduces to $\theta_1^{(n)}$ for the familiar replacements. If we substitute this algebraic function for $x_1^{(n)}$ in (N + 1) of §11, and have regard to §13, we find a contradiction of the assumption that r_n is a minimum.

One must bear well in mind that the y in (12) must be replaced by their monomials before we get a function which is identically zero. For instance, the function

$$F = (3 y_1 - y_2) x_1'',$$

which vanishes for every x when y_1 is replaced by e^x, y_2 by $e^{x + \log 3}$ and x_1'' by log log x, is certainly not zero identically in $y_1, y_2 \ x_1''$. It vanishes in x and x_1'' when y_1 and y_2 are replaced by their monomials.

DIFFERENTIATION

15. Let u be a function of order 1, of regular structure (§11) at some point $\underline{a}$. We wish to form du/dx for the neighborhood of $\underline{a}$. Let $f(x_1', \ldots, x_r'; x)$ be the algebraic function appearing in (II). The derivative of u can be written

$$\sum_{i=1}^{r} \frac{\partial f(\theta_1, \ldots, \theta_r; x)}{\partial x_i'} \frac{d\theta_i}{dx} + \frac{\partial f(\theta_1, \ldots, \theta_r; x)}{\partial x}$$

If θ_i is an exponential, e^{v_i}, its derivative is $\theta_i v_i'$. If θ_i is a logarithm, log v_i, its derivative is v_i'/v_i. In the logarithmic case, v_i is not zero at $\underline{a}$, since θ_i is analytic at $\underline{a}$.

Let $\theta_1, \ldots, \theta_t$ be exponentials and the remaining θ logarithms. Let

$$g = \sum_{i=1}^{t} \frac{\partial f(x_1', \ldots, x_r'; x)}{\partial x_i'} x_i' v_i' + \sum_{i=t+1}^{r} \frac{\partial f}{\partial x_i'} \frac{v_i'}{v_i} + \frac{\partial f}{\partial x}.$$

Then g is algebraic in x and the x'. If we take ρ_1 of §11 sufficiently small, g will be analytic for $|x - \underline{a}| < \rho_1, \ldots, |x_r' - \theta_r(\underline{a})| < \rho_1$. If now we take ρ correspondingly small, so as to limit the variation of the monomials, we see that g reduces to the derivative of u for $|x - a| < \rho$ when each variable is replaced by its monomial.

It is now easy to treat a function u of any order n. Of the algebraic functions introduced in (I), $\ldots$, (N) of §11, there are some which are used for forming logarithmic monomials. As each monomial is analytic at $\underline{a}$, such an algebraic function is not zero when x = a and each accented x is its $\theta(a)$; the function is therefore distinct from zero if the x are close to these values. If now ρ_1 is taken sufficiently small and if ρ is taken correspondingly small, so as to limit the variation of the monomials, we may assume that none of the algebraic functions which give logarithmic monomials vanish when x differs from a, and each accented x from its $\theta(a)$, by a quantity less than ρ_1 in modulus.

This understood, the formulas for the differentiation of composite functions show that *if u is a function of order n, of regular structure at a point a, there exists an algebraic function of the x, analytic for* $|x - a| < \rho_1$, ..., $|x_{r_n}^{(n)} - \theta_{r_n}^{(n)}| < \rho_1$ *which reduces to the derivative of u for* $|x - a| < \rho$ *when each variable is replaced by the monomial which corresponds to it.*

Thus if u is elementary and of order n, its derivative is elementary and of order not exceeding n.

16. The matter which we shall now examine takes care of itself quite well in Chapter II. Still it strikes one of the deeper notes of Liouville's theory and may well be lifted out from among other details.

Let u of order n be derived from

$$g(x_1^{(n)}, ..., x).$$

Suppose first that $\theta_1^{(n)}$ is an exponential, e^v. The algebraic function which yields du/dx can be written

(14) $$\frac{\partial g}{\partial x_1^{(n)}} x_1^{(n)} \varphi + \text{other terms.}$$

In (14), φ is an algebraic function of $x_1^{(n-1)}$, ..., x which yields the derivative of v. The "other terms" are derivatives of g with respect to $x_2^{(n)}$, ..., x times algebraic functions which yield the derivatives of $\theta_2^{(n)}$, ..., x.

If, in g, we replace $x_1^{(n)}$ by $\mu \, \theta_1^{(n)}$, where μ is a constant close to unity, and the other variables by the monomials which correspond to them, we secure a function of x, analytic at a, which we shall call u_μ. It is clear that du_μ/dx is obtained from (14) by replacing $x_1^{(n)}$ by $\mu \, \theta_1^{(n)}$ and the other variables by their monomials. If we consider u_μ as a function of x and μ, it is seen to be analytic for x = a, $\mu = 1$.

Now let $\theta_1^{(n)}$ be a logarithm. We secure du/dx from an algebraic function

(15) $$\frac{\partial g}{\partial x_1^{(n)}} \varphi + \text{other terms,}$$

where φ, algebraic in $x_i^{(n-1)}$, ..., x, reduces to the derivative of
$\theta_i^{(n)}$. Let μ be a constant close to zero and let u_μ be the function,
analytic at x = a, which is obtained from g on replacing $x_i^{(n)}$ by
$\theta_i^{(n)} + \mu$ and the other variables by their monomials. The same substitutions performed in (15) will produce the derivative of u_μ.
Furthermore u_μ, as a function of x and μ, is analytic at x = a,
$\mu = 0$.

SUMMARY

17. In §§ 3-16, we made a close examination of the elementary
functions. In formal work, it suffices to bear a few facts in mind.
In the structure of a function u of order n, there appear certain
monomials of orders 0, 1, ..., n. The function u is algebraic in
these monomials, although some of the monomials may be used only
for building monomials of higher order and may not appear effectively by themselves in the final expression for u. The derivative
of u is algebraic in monomials which appear in the structure of u.
If an expression for u is taken which employs as few as possible
n-monomials, any algebraic relation among these n-monomials and
any monomials of orders less than n, must hold identically in the
monomials of order n. The reason is that a nonidentical relation
would furnish an expression for one of the n-monomials which would
permit us to write u with fewer such monomials. If θ is an n-exponential in u, and if θ is replaced in u by $\mu\theta$, where μ is a constant, we secure a function whose derivative is obtained from the
derivative of u by the same substitution. If θ is an n-logarithm
in u, and if θ is replaced in u by $\theta + \mu$, we secure a function
whose derivative is obtained from that of u by the same substitution.

We are now prepared to take up questions of integration in finite
terms.

Chapter II

ALGEBRAIC FUNCTIONS WITH ELEMENTARY INTEGRALS

LIOUVILLE'S FIRST THEOREM

1. The meaning of the expression *integrable in finite terms* depends on the material with which one is working. An elementary function is said to be integrable in finite terms if its integral is elementary.

The first problem considered by Liouville in the field of integration in finite terms deals with the integration of algebraic functions.* We consider an algebraic function $y(x)$ defined by an irreducible relation

(1) $$\alpha_0(x) y^m + \alpha_1(x) y^{m-1} + \ldots + \alpha_m(x) = 0.$$

In dealing with the integral of y, we shall need only the barest knowledge of the nature, as a function, of the integral. Any element $P(x - x_0)$ of y can be integrated into an element Q. The m.a.f. obtained from Q can be called *an integral of* y and can be written $\int y\, dx$. Q is determined to within an additive constant. By using all constants we obtain every integral of y, that is, every function whose elements have derivatives which are elements of y.

Most of the time we can work on an even simpler basis. Let us consider some branch of y which is analytic in some simply connected area $\mathfrak{A}$. Then $\int y\, dx$, which is defined to within an additive constant, is analytic throughout $\mathfrak{A}$. If now we ask whether y is integrable in finite terms, our question is a perfectly clear one. We are asking whether there is an elementary function $u(x)$, of regular structure** at a point $\underline{a}$ in $\mathfrak{A}$, whose derivative coincides with y for a neighborhood of $\underline{a}$. Any further questions which may arise can be taken care of by considerations of analytic continuation.

2. We now state Liouville's first theorem on integration.

THEOREM: *Let* $y(x)$ *be an algebraic function whose integral is elementary. Then*

(2) $$\int y\, dx = v_0(x) + c_1 \log v_1(x) + \ldots + c_r \log v_r(x)$$

where r *is a positive integer, each* $v(x)$ *an algebraic function, and each* c *a constant.*

---

* Liouville [4]. A list of references is given at the end of the monograph.
** 1, 11 (Chapter I, §1). When no chapter is mentioned in a reference, the chapter is that in which one is reading.

Of course, the integral may be algebraic. This case may be considered to be covered by our statement if we allow all c to be zero.

An expression like the second member of (2) has an algebraic derivative. As one approaches the problem, it appears to be a foregone conclusion that the only possibility for an elementary integral is that described in (2). If the answer contained an exponential, the exponential ought to survive after differentiation. Similarly one cannot imagine a logarithm disappearing unless it enters linearly. Considerations of this vague type had led Laplace, before Liouville's time, to conjecture the theorem established by Liouville, and Liouville claimed for his method of proof the merit of following these intuitive ideas.

The proof of Liouville's theorem will be conducted as follows. Letting u be an integral of y, we shall suppose that u is elementary but not algebraic. Let n > 0 be the order of u. We shall suppose that we have for u an expression of the type described in I, 14, which involves as few as possible n-monomials, as in I, 11. We shall show first that none of the n-monomials is an exponential. It will then be shown that u has an expression like the second member of (2) with each log v_i an n-logarithm and v_0 a function of order less than n. The proof will be completed by showing that n = 1.

3. As has just been stated, we understand r_n to be small as possible. We work at a point a at which u has regular structure. Let the algebraic function of $(N + 1)$ of I, 11, from which u is obtained be represented by

$$g(x_i^{(n)}, \ldots, x).$$

Let every variable except $x_i^{(n)}$ be replaced by its monomial. We obtain a function of $x_i^{(n)}$ and x which we shall represent by $f(x_i^{(n)}, x)$. This latter function produces u when $x_i^{(n)}$ is replaced by $\theta_i^{(n)}$. On this basis, letting θ represent $\theta_i^{(n)}$, we write

(3) $$u = f(\theta, x).$$

We now obtain the derivative of u, having regard to I, 15. On the basis of (3), we write

(4) $$\frac{du}{dx} = f_\theta(\theta, x)\frac{d\theta}{dx} + f_x(\theta, x).$$

The expression f_θ is obtained by replacing $x_i^{(n)}$ by θ in $\partial f(x_i^{(n)}, x)/\partial x_i^{(n)}$. The description of f_x is similar. We notice also that the second member of (4) may be considered to be obtained by replacements from (14) or (15) of Chapter I.

Let us suppose now that θ is an exponential and let $\theta = e^w$ with w of order $n-1$. Observing that the derivative of u is y, we obtain from (4)

$$(5) \qquad\qquad y = f_\theta(\theta, x)\, \theta\, \frac{dw}{dx} + f_x(\theta, x).$$

Now (5) is an algebraic relation among the monomials in u, of the type studied in I, 14. By Liouville's principle (5) must hold identically in θ. This means, on the one hand, that if, in the algebraic function out of which the second member of (5) is obtained, we replace all variables except $x_i^{(n)}$ by their monomials, we will get an expression in $x_i^{(n)}$ and x which is identical with $y(x)$ for every $x_i^{(n)}$. It means again, perhaps more simply, that the relation

$$y = f_{x_i^{(n)}}(x_i^{(n)}, x)\, x_i^{(n)} \frac{dw}{dx} + f_x(x_i^{(n)}, x)$$

is an identity in $x_i^{(n)}$ and x.

In particular, we may replace θ in (5) by any function of x which is analytic at the point $\underline{a}$ and close in value to $\theta(a)$ at $\underline{a}$. We shall replace θ by $\mu\theta$ where μ is a constant close to unity. We may thus write, for x close to $\underline{a}$,

$$(6) \qquad\qquad y = f_{\mu\theta}(\mu\theta, x)\, \mu\theta\, \frac{dw}{dx} + f_x(\mu\theta, x).$$

where $f_{\mu\theta}$ is the result of replacing $x_i^{(n)}$ by $\mu\theta$ in $f_{x_i^{(n)}}$.

By 1, 15, and even on the basis of its own structure, the second member of (6) is the derivative of $f(\mu\theta, x)$. By (5) and (6), $f(\theta, x)$ and $f(\mu\theta, x)$ have the same derivative and thus differ by a constant. This constant depends on μ. We write

$$(7) \qquad\qquad f(\mu\theta, x) = f(\theta, x) + \beta(\mu).$$

As $f(\mu\theta, x)$ is analytic in μ and x for $\mu = 1$, $x = \underline{a}$, the function $\beta(\mu)$ must be analytic for $\mu = 1$.

We differentiate (7) with respect to μ and find

$$(8) \qquad\qquad f_{\mu\theta}(\mu\theta, x)\, \theta = \beta'(\mu),$$

the accent indicating differentiation. We put $\mu = 1$ and represent $\beta'(1)$ by c, obtaining the relation

$$(9) \qquad\qquad f_\theta(\theta, x)\, \theta = c.$$

Again we apply Liouville's principle; (9) holds identically in θ. We replace θ by an independent variable z and have, for the neighborhood of $x = a$, $z = \theta(a)$,

$$(10) \qquad\qquad z\, f_z(z, x) = 0.$$

Now (10) is a partial differential equation for $f(z, x)$. It gives

$$f_z(z, x) = \frac{c}{z}$$

so that

(11) $$f(z, x) = c \log z + \gamma(x),$$

where γ is analytic at $x = a$. We notice that, as $\theta(a) \neq 0$, $\log z$ is analytic for $z = \theta(a)$.

To determine $\gamma(x)$, we replace z in (11) by any value z_0 close to $\theta(a)$. We have

$$f(z_0, x) = c \log z_0 + \gamma(x).$$

Hence, identically in z and x,

(12) $$f(z, x) = c \log z - c \log z_0 + f(z_0, x).$$

In particular, we may replace z by θ in (12) for the neighborhood of $x = a$. Thus

(13) $$u = f(\theta, x) = c \log \theta - c \log z_0 + f(z_0, x).$$

In (13), $\log \theta$ is of order n-1 while $f(z_0, x)$, which by I, 13, is an elementary function of regular structure at $\underline{a}$, involves fewer than r_n monomials of order n. This contradicts the fact that the expression (3) of u is as economical as possible in n-monomials. We have thus proved that θ is not an exponential.

4. We know now that the $\theta_i^{(n)}$ are logarithms. We shall prove that u is a function of order less than n plus terms of the form $c_i \theta_i^{(n)}$ with constant c. We use (4). Let $\theta = \log w$ with w of order n-1. Then

(14) $$y = f_\theta(\theta, x) \frac{w'}{w} + f_x(\theta, x).$$

As w and w' are of order less than n, (14) is an identity in θ.

We replace θ by $\theta + \mu$ where μ is a constant close to zero. Then

(15) $$y = f_{\theta+\mu}(\theta+\mu, x) \frac{w'}{w} + f_x(\theta+\mu, x).$$

The second member of (15) is the derivative of $f(\theta+\mu, x)$. Then

(16) $$f(\theta+\mu, x) = f(\theta, x) + \beta(\mu)$$

where β is analytic for $\mu = 0$. Differentiation of (16) with respect to μ gives

$$f_{\theta+\mu}(\theta+\mu, x) = \beta'(\mu).$$

Putting $\mu = 0$ and writing $\beta'(0) = c_1$, we have

(17) $$f_\theta(\theta, x) = c_1.$$

Now (17) is an identity in θ. We replace θ by a variable z and write

$$f_z(z, x) = c_1.$$

Then

$$f(z, x) = c_1 z + \gamma(x).$$

Determining γ by a special value z_0 of z, we have

$$f(z, x) = c_1 z - c_1 z_0 + f(z_0, x).$$

Hence

(18) $u = f(\theta, x) = c_1 \theta - c_1 z_0 + f(z_0, x).$

In $g(x_1^{(n)}, \ldots, x)$ of §3, let the variables corresponding to monomials of order less than n be replaced by their monomials. There results a function

$$h(x_1^{(n)}, \ldots, x_r^{(n)}; x)$$

where r represents r_n. We have

(19) $u = h(\theta_1^{(n)}, \ldots, \theta_r^{(n)}; x).$

We equate the second members of (18) and (19), remembering that $\theta = \theta_1^{(n)}$. The resulting equation holds identically in the $\theta^{(n)}$. This means that

$$\frac{\partial h}{\partial x_1^{(n)}} = c_1.$$

There exist similarly, for $i = 2, \ldots, r$, constants c_i such that

$$\frac{\partial h}{\partial x_i^{(n)}} = c_i.$$

Thus

(20) $h = c_1 x_1^{(n)} + \ldots + c_r x_r^{(n)} + v(x)$

with $v(x)$ analytic for $x = a$. To determine $v(x)$, we replace the $x^{(n)}$ in (20) by constants. We find, referring to I, 13, that $v(x)$ is elementary, of regular structure at $\underline{a}$ and of order less than n. We have, by (20),

$$u = v(x) + c_1 \theta_1^{(n)} + \ldots + c_r \theta_r^{(n)}$$

as was to be proved.

5. We now prove that $n = 1$. Let us assume that $n > 1$. We know that u has an expression

(21) $c_1 \log v_1 + \ldots + c_r \log v_r + v$

with each $\log v_i$ of order n and v of order less than n.* There will

---

* It should be emphasized that r is the least number of n-monomials in terms of which u can be expressed.

be many expressions (21) for u. These will all have the same r, but different v_1, v, and c_i. In each expression, a certain number of (n-1)-monomials are used in building the r + 1 functions v_1, ..., v_r, v found in that expression. We consider those expressions which use as few as possible (n-1)-monomials.

From the expressions just described, we select one in which as few as possible (n-1)-monomials are used in building the r functions v_1, ..., v_r. We understand (21) to be the expression just selected. Let θ be one of the (n-1)-monomials appearing in the v_1. We write

(22) $u = c_1 \log v_1(\theta, x) + ... + o_r \log v_r(\theta, x) + v(\theta, x)$.

We have

(23) $y = \sum_{j=1}^{r} c_j \dfrac{v_{j\theta}(\theta, x) \theta' + v_{jx}(\theta, x)}{v_j(\theta, x)} + v_\theta(\theta, x) \theta' + v_x(\theta, x)$.

where literal subscripts indicate partial differentiation.

We show first that θ cannot be an exponential. Suppose that $\theta = e^w$ with w of order n-2. Then

(24) $y = \sum_{j=1}^{r} c_j \dfrac{v_{j\theta}(\theta, x) \theta w' + v_{jx}(\theta, x)}{v_j(\theta, x)} + v_\theta(\theta, x) \theta w' + v_x(\theta, x)$.

In (24) we have an algebraic relation among the (n-1)-monomials in u and monomials of order less than n-1. The relation (24) must be an identity in θ. If it were not, we could, with all the formality of I, 14, solve for θ in (24) and, by a substitution, express u with fewer (n-1)-monomials than are found in (22).

We replace θ by μθ in (24). The second member becomes the derivative of

$$\sum_{j=1}^{r} c_j \log v_j(\mu\theta, x) + v(\mu\theta, x).$$

Hence

$$\sum_{j=1}^{r} c_j \log v_j(\mu\theta, x) + v(\mu\theta, x) = u(x) + \beta(\mu)$$

for every μ close to unity and for every x close to the point a at which our functions are being studied. We differentiate with respect to μ and put μ = 1. Then

(25) $\sum_{j=1}^{r} c_j \dfrac{v_{j\theta}(\theta, x) \theta}{v_j(\theta, x)} + v_\theta(\theta, x) \theta = \beta'(1) = c$.

Again, (25) is an identity in θ. Thus, for a variable z,

$$\sum_{j=1}^{r} c_j \dfrac{v_{jz}(z, x)}{v_j(z, x)} + v_z(z, x) = \dfrac{c}{z}.$$

Then

$$\sum_{j=1}^{r} c_j \log v_j(z, x) + v(z, x) = c \log z + \gamma(x)$$
$$= c \log z - c \log z_0 + \sum_{j=1}^{r} c_j \log v_j(z_0, x) + v(z_0, x)$$

where z_0 is close to $\theta(a)$. For $z = \theta(x)$ we find

$$(26) \quad u = c \log \theta - c \log z_0 + \sum_{j=1}^{r} c_j \log v_j(z_0, x) + v(z_0, x).$$

In the second member of (26), each $\log v_j$ must be of order n; otherwise u would either be of order less than n or be expressible with less than r n-monomials. Now $\log \theta = w$, so that (26) gives an expression of type (21) for u involving fewer (n-1)-monomials than appear in the expression selected for u. Thus, θ cannot be an exponential.

We now assume that θ is a logarithm, Let $\theta = \log w$. Then (23) gives

$$(27) \quad y = \sum \frac{c_j}{v_j(\theta, x)} [v_{j\theta}(\theta, x) \frac{w'}{w} + v_{jx}(\theta, x)] + v_\theta(\theta, x) \frac{w'}{w} + v_x(\theta, x),$$

an equation which must hold identically in θ. We replace θ by $\theta + \mu$ and find that

$$(28) \quad \sum c_j \log v_j(\theta+\mu, x) + v(\theta+\mu, x) = u(x) + \beta(\mu)$$

with $\beta(\mu)$ analytic for $\mu = 0$. We differentiate (28) with respect to μ and put $\mu = 0$. Then

$$\sum c_j \frac{v_{j\theta}(\theta, x)}{v_j(\theta, x)} + v_\theta(\theta, x) = \beta'(0) = c.$$

As this holds identically in θ, we write

$$\sum c_j \frac{v_{jz}(z, x)}{v_j(z, x)} + v_z(z, x) = c.$$

Integrating, we have

$$\sum c_j \log v_j(z, x) + v(z, x) = cz + \gamma(x)$$
$$= cz - cz_0 + \sum c_j \log v_j(z_0, x) + v(z_0, x).$$

For $z = \theta(x)$,

$$(29) \quad u = c\theta - cz_0 + \sum c_j \log v_j(z_0, x) + v(z_0, x).$$

In (29), we have an expression in which the (n-1)-monomials are among those in (22). The n-logarithms in (29) employ fewer (n-1)-monomials than appear in (22). This final contradiction renders

untenable the assumption that n > 1, and the proof of Liouville's
theorem is completed.

6. Let us try to describe Liouville's method of proof, as represented, in particular, by the work of §§ 3 and 4. We have a function
u of order n whose structure is to be examined. We use an expression for u which involves a least number of n-monomials. Let θ be
one of these. When we differentiate, we secure a relation which
holds identically in θ. We replace θ by μθ if θ is an exponential
and by θ + μ if θ is a logarithm. We find that when one of these
substitutions is made in u, that function is increased by a function of μ. The relation thus secured is differentiated with respect
to μ; μ is put equal to unity in the exponential case and to zero
in the logarithmic case. We apply Liouville's principle again, replacing θ by a variable z. We have now a partial differential equation, which we integrate, determining the arbitrary function of x
which appears by using a special value of z. Replacing z by θ, we
have an expression for u which permits us to draw conclusions as to
the structure of that function.

In what way did Liouville, working on these questions at the age
of about twenty-three, assemble the ideas which underlie his method? It is a simple theory that he found inspiration in Abel's investigation of the unsolvability of the quintic equation.* Abel
gives a classification of expressions involving radicals which resembles Liouville's arrangement of the elementary functions. He
selects, for a given function, the expression most economical in
radicals and is then able to replace certain radicals by arbitrary
quantities.

7. Let us take another look at Liouville's procedure. If the integral u is elementary and if proper economy in monomials is observed, each exponential in u may be replaced by a constant times itself and each logarithm by a constant plus itself; the replacements
will leave u an integral. Now the integrals of y are a one-parameter family. The monomials which enter into u must not create two
or more arbitrary constants. Here we have a viewpoint from which
one can conjecture Liouville's theorem and the answers to other
questions on integration in finite terms. This idea is due to
Koenigsberger [2]. It underlies the proofs in Mordukhai-Boltovskoi's
paper [15].

* Abel [1]. The criticisms which have been made of Abel's reasoning appear to
be unfounded.

ABEL'S THEOREM

8. It was shown by Abel, about a decade before Liouville proved the theorem of §2, that when the integral of an algebraic function $y(x)$ is expressible as in (2), the algebraic functions v may be taken so as to be rational in x and y, with constant coefficients. In the language of the theory of algebraic functions, we may arrange so that the v are rational on the Riemann surface of y.

To secure a perspicuous proof of Abel's theorem, we shall permit ourselves to use a few of the simpler qualitative properties of algebraic functions.*

As we understand the relation (2), on the basis of our derivation of it, y and the v are analytic in some definite area $\mathfrak{u}$. Now y is defined by (1) and the v_i by similar equations. We have been working with definite branches of y and the v. We shall suppose, shrinking $\mathfrak{u}$ if necessary, that every branch of y, and every branch of each v, is analytic in $\mathfrak{u}$. We shall suppose that (2) has been established for branches

$$(30) \qquad y_1, v_{01}, \ldots, v_{r1}$$

of y and the v.

Let $\mathbf{a}$ be a point in $\mathfrak{u}$ and let C be a curve which starts and ends at $\mathbf{a}$, avoiding the singular points of y and the v. When we continue y_1 and the v_{11} along C, the set (30) is replaced by some set of branches, perhaps (30) itself, of y and the v. We shall employ only curves C which replace y_1 by itself. The number of such curves is infinite, but there are only a finite number of possibilities for the sets by which (30) can be replaced. Let the distinct sets which it is possible to secure be

$$
\begin{aligned}
& y_1, v_{01}, \ldots, v_{r1} \\
& y_1, v_{02}, \ldots, v_{r2} \\
(31) \qquad & \qquad \cdot \quad \cdot \quad \cdot \\
& y_1, v_{ot}, \ldots, v_{rt}
\end{aligned}
$$

For instance, in the second column of (31), $v_{01}, \ldots, v_{ot}$ are various branches of v_0. They need not be distinct branches; it is only the sets as a whole which are distinct. Let C_1 be a curve which converts the first set into $y_1, v_{01}, \ldots, v_{r1}$.

Now let

---

* In Chapter III a more general question will be treated by algebraic methods.

(32) $w_0 = v_{01} + v_{02} + \ldots + v_{0t}.$

We are going to prove that w_0 is unchanged when it is continued
along a curve C, starting and ending at $\underline{a}$, which returns y_1 to itself. For this we prove that C permutes the sets of (31) among themselves. To continue the ith set along C is to continue the first
set along C_i and then along C. Thus every set goes into some set
when it is continued along C. Two distinct sets cannot go into the
same set, since the continuation along C is a reversible process.

On this basis, when w_0 is continued along C, the second subscripts of the v_{0i} in (32) are permuted among themselves. This
shows that w_0 is unchanged.

By 1, 12, w_0 is a branch of an algebraic function of x. We are
going to show that this algebraic function is rational in y and x.
Consider a curve D_1 which starts at $\underline{a}$ and ends at a point b, avoiding singular points of y and v_0. Along it, y_1 and w_0 can be continued. Let $P(x - b)$ be the element secured for y_1 at b and $Q(x - b)$
the element secured for w_0. If a second path D_2 continues y_1 from
$\underline{a}$ into $P(x - b)$, then D_2 must continue w_0 into $Q(x - b)$. Otherwise,
using D_1, and D_2 reversed, we would have a path which leaves y_1
unchanged while changing w_0. Thus the continuation of w_0 furnishes
an algebraic function which is one-valued on the Riemann surface of
y and is therefore a rational combination of y and x.

Let us explain the point just made in greater detail. Let the
branches of y, analytic in $\mathfrak{U}$, be $y_1, \ldots, y_m$. We denote w_0 by w_{01}.
All paths which continue y_1 into y_i continue w_{01}, as was seen above,
into a single definite function analytic in $\mathfrak{U}$, which we denote by
w_{0i}. Now let functions A be determined by equations

$$w_{01} = A_0 + A_1 y_1 + \ldots + A_{m-1} y_1^{m-1}$$

$$w_{02} = A_0 + A_1 y_2 + \ldots + A_{m-1} y_2^{m-1}$$

$$\cdot \quad \cdot \quad \cdot \quad \cdot \quad \cdot \quad \cdot \quad \cdot$$

$$w_{0m} = A_0 + A_1 y_m + \quad + A_{m-1} y_m^{m-1}.$$

The determinant D of this system is not zero, for it is a Vander-
monde determinant and the y_i are distinct. we have

$$A_j = D^{-1} D_j$$

where D_j is a determinant with the w_{0i} in one column. By I, 12, A_j
is an algebraic function. We wish to see that A_j is unchanged when
continued along any path which starts and ends at $\underline{a}$. For this we

notice that such a continuation permutes the rows of D among them-selves and performs the same permutation on D_j. Thus the continua-tion either leaves D and D_j unchanged or replaces them both by their negatives, so that A_j is unchanged. An algebraic function which is unchanged in this manner can have but one branch and so is rational. To sum up,

$$w_0 = A_0 + A_1 y + \ldots + A_{m-1} y^{m-1}$$

with each A a rational function of x.

For $i > 0$, we use a product rather than a sum. If $w_i = v_{i1} v_{i2} \ldots v_{it}$, $i = 1, \ldots, r$, w_i is a rational combination of y and x.

We return now to (2). Let us show that, for $i = 1, \ldots, t$, we have a relation

(33) $d_i + \int y \, dx = v_{0i} + c_1 \log v_{1i} + \ldots + c_r \log v_{ri}$

where the d are constants. We work in the neighborhood of a point $\underline{a}$ in $\mathfrak{U}$ at which no v_{ji} with $j > 0$ is zero. Each $\log v_{ji}$ will have an element at $\underline{a}$, secured in a definite manner.

For $i = 1$, $d_1 = 0$ and (33) is merely (2). For any other i, we use a curve C_i which continues the first set of (31) into y_1, v_{o1}, $\ldots$, v_{ri}, avoiding singularities and values of x at which one or more v_{ji} have zero values. Then each $\log v_{j1}$ is continued into a definite function $\log v_{ji}$. Since y_1 returns to itself, its integral is changed through the addition of a constant, which we call d_i.

Summing the equations (33) for $i = 1, \ldots, t$, we find

(34) $h + t \int y \, dx = w_0 + c_1 \log w_1 + \ldots + c_r \log w_r$

where h is the sum of the d. Thus

(35) $\int y \, dx = \dfrac{w_0 - h}{t} + \dfrac{c_1}{t} \log w_1 + \ldots + \dfrac{c_r}{t} \log w_r.$

In the second member of (35), the algebraic functions are all ra-tional in y and x. Equation (35) has been established for the neigh-borhood of $\underline{a}$. It is preserved under analytic continuation. Perhaps the simplest way to look at the situation is as follows. In (2), y is a function with a definite Riemann surface. It has been shown that the v may be selected so as to have the same surface. Each logarithmic derivative v_i'/v_i is a function rational on this surface, and we have

(36) $y = v_0' + c_1 \dfrac{v_1'}{v_1} + \ldots + c_r \dfrac{v_r'}{v_r}.$

Every idea contained in (2) is contained in the simpler relation (36).

ALGEBRAIC FUNCTIONS WITH ALGEBRAIC INTEGRALS

9. As a special case of Abel's theorem, we have the result that *if y is an algebraic function of x and if the integral of y is algebraic, the integral is rational in y and x.*

For this special case we shall present a separate proof, almost entirely algebraic.

Let u, the integral, be analytic in $\mathfrak{U}$. Let the irreducible equation satisfied by u be

$$(37) \qquad\qquad B_0 u^p + \ldots + B_p = 0,$$

the B being polynomials in x with $B_0 \neq 0$. The existence of (37) is enough to show that u satisfies in $\mathfrak{U}$ various equations of the form

$$(38) \qquad\qquad D_0 u^q + \ldots + D_q = 0$$

with each D a polynomial in y and x with constant coefficients, D_0 not vanishing identically in x in $\mathfrak{U}$. We have in (37) a trivial instance of (38). From among all equations of type (38) satisfied by u, we select one which is of a least degree in u. We shall suppose that (38) is such an equation of least degree and proceed to prove that $q = 1$.

Suppose that $q > 1$. We write (38)

$$(39) \qquad\qquad u^q + \beta_1(x) u^{q-1} + \ldots + \beta_q(x) = 0$$

with each β rational in y and x. Differentiating (39), we have

$$(40) \quad u^{q-1} \left[q y + \frac{d\beta_1}{dx}\right] + u^{q-2} \left[(q-1) y \beta_1 + \frac{d\beta_2}{dx}\right] + \ldots + \left[y \beta_{q-1} + \frac{d\beta_q}{dx}\right] = 0.$$

Now

$$\frac{d\beta_i}{dx} = \frac{\partial\beta_i}{\partial x} + \frac{\partial\beta_i}{\partial y} \frac{dy}{dx}$$

The derivative of y is found from (1) to be rational in y and x. Thus the coefficients in (40) are rational in y and x. If those coefficients were not zero identically in x, we would, clearing fractions, have an equation like (38) for u, of degree less than q. Thus, throughout $\mathfrak{U}$,

$$q y + \frac{d\beta_1}{dx} = 0$$

and

(41) $$u = \int y \, dx = \frac{1}{q} \beta_1(x, y) + c.$$

Now (41) is an equation (38) for u with $q = 1$. We have thus a contradiction of the assumption that $q > 1$. Then $q = 1$ and (38) expresses u rationally in y and x.

As to the nature of the proof, the selection of an equation (38) of least degree is a device which replaces a reduction algorithm. We could instead start by differentiating (37) after dividing by B_0 and replace dy/dx by its expression in terms of y and x. We would either secure an equation (38) of degree less than p or else obtain an expression for the integral as in (41). In the former case, the process would be repeated until a rational expression for u is secured.

From the standpoint of the theory of abelian integrals, the result on algebraic integrals is a very obvious one. If the integral of y is algebraic, it can have no periods, either cyclic or polar. It is thus uniform on the surface of y and so is rational in y and x.

10. Liouville [3] furnished a method for determining whether an algebraic function has an algebraic integral and for obtaining the integral when it is algebraic. Without following the matter through to the very end, let us see what is involved in it.

If $y(x)$, defined by (1), has an algebraic integral, we have, by §8,

$$\int y \, dx = A_0 + A_1 y + \ldots + A_{m-1} y^{m-1}$$

with each A a rational function of x. Then

(42) $$y = \frac{d A_0}{dx} + \sum_{i=1}^{m-1} \left(y^i \frac{d A_i}{dx} + i A_i y^{i-1} \frac{dy}{dx} \right).$$

By (1) each $y^{i-1} dy/dx$ is a rational combination of x and y. Every such rational combination is a linear combination of $1, y, \ldots, y^{m-1}$ with rational functions of x for coefficients. Let such linear combinations be substituted for each $y^{i-1} dy/dx$ in (42). Then (42) becomes

(43) $$y = \sum_{i=0}^{m-1} y^i \left(\frac{d A_i}{dx} + \beta_i \right)$$

with each β a linear combination of $A_1, \ldots, A_{m-1}$ with rational functions of x for coefficients. The expression of a function as a linear combination of $1, \ldots, y^{m-1}$ is unique; two such expressions for a single function would furnish an equation for y of degree

less than m. Thus we may equate coefficients of like powers of y in
both members of (43). This furnishes for the A a system of linear
differential equations of the first order, with known rational functions of x for coefficients. Our problem is to determine whether
this system has a solution with each A rational. Liouville carries
this question through. We shall not go into the details since, from
the standpoint of the theory of linear differential equations as it
exists at present, the question is an essentially routine one. In
III, 11 we work out a simple problem of this type.

No general method exists for determining whether an algebraic
function whose integral is not algebraic can be integrated with
logarithms. This problem, which depends on delicate questions in
the theory of numbers, appears to be a difficult one.

RESIDUES AND INTEGRATION

11. The methods which were at Liouville's disposal for the examination of special algebraic functions were of a somewhat laborious
type. Great simplicity can be secured by the use of the expansions
of an algebraic function at places on its Riemann surface. We shall
recall the facts.

Let $y(x)$ be algebraic and let $u(x)$ be an algebraic function, not
identically zero, which is uniform on the surface of y; u thus is
rational in x and y. At a point on the surface of y which is neither a branch point nor a point at ∞, u either is analytic or has
a pole; it will have a Taylor development at the point in the first
case and a Laurent development in the second. Now consider a branch
point, corresponding to a finite value x_0 of x. There u has an expansion

$$(44) \qquad a_0(x - x_0)^{p/r} + a_1(x - x_0)^{(p+1)/r} + \ldots$$

where p is some integer and r the number of sheets which circulate
at the point. We suppose that $a_0 \neq 0$. If $p > 0$, u is said to have
a zero of order p at the point, and if $p < 0$, u is said to have
there a pole of order $-p$. At each point on the surface at ∞, u has
an expansion

$$(45) \qquad a_0 x^{p/r} + a_1 x^{(p-1)/r} + \ldots$$

with $a_0 \neq 0$. If the point is not a branch point, $r = 1$. There is a
pole of order p if $p > 0$ and a zero of order $-p$ if $p < 0$.

Consider a finite point on the surface of y at which u has a pole.
We write the expansion of u at this point in the form (44), with

r = 1 if the point is not a branch point. The product by r of the coefficient of $(x - x_0)^{-1}$ in (44) is called the *residue* of u at the pole. At a point at ∞, we call the product by -r of the coefficient of x^{-1} in the expansion (45) of u, the residue of u at the point. Thus there are two types of places at which we speak of residues, the poles at finite points and all points at ∞.

We now consider u'/u, the logarithmic derivative of u, which is uniform on the surface of y. We examine a finite point on the surface, at which u has an expansion (44). If p = 0, the expansion of u'/u, found formally from (44), will contain no negative powers if r = 1. If r > 1, the expansion may contain negative powers, but the exponents of $x - x_0$ will all exceed -1. Thus u'/u may have a pole at the point if r > 1, but, if so, the residue is zero. If $p \neq 0$, the expansion of u'/u starts with

$$r^{-1} p (x - x_0)^{-1}$$

so that u'/u has a pole with p for residue. For a point at ∞, we find from (45) that the residue of u'/u is -p.

Consider now u'. None of its expansions can contain a term of exponent -1. Thus the residues of u' are all zero.

12. We return now to Liouville's theorem of §2, supposing that the integral of y is elementary but not algebraic. A set of r complex numbers $c_1, \ldots, c_r$ will be called *independent* if there does not exist a set of rational numbers $q_1, \ldots, q_r$, not all zero, such that

$$q_1 c_1 + \ldots + q_r c_r = 0.$$

Let an expression (2) of the integral of y be given which contains a minimum number of logarithms. We shall prove that $c_1, \ldots, c_r$ are independent. Suppose, for instance, that

$$c_1 = s_2 c_2 + \ldots + s_r c_r$$

with rational s. We can write the second member of (2) as

$$v_0 + c_2 \log v_1^{s_2} v_2 + \ldots + c_r \log v_1^{s_r} v_r.$$

In this expression, the functions of which logarithms are taken are algebraic. Thus r is not a minimum. This proves the independence of the c.

When we have an expression for $\int y\, dx$ which contains a least number of logarithms, we can, as in §8, replace it by an expression with the same number of logarithms and with each algebraic function rational in y and x.

We shall use, for $\int y \, dx$, such an expression as has just been described. We refer to (2). No v_i with $i > 0$ is a constant. An algebraic function which is not a constant has at least one zero and at least one pole. Suppose, for instance, that v_1 has a zero at a finite point on the surface of y, for which $x = x_0$. At this point v_1'/v_1 will have a residue distinct from zero. Let the coefficient of $(x - x_0)^{-1}$ in the expansion of v_1'/v_1, at the point under consideration be q_i, $i = 1, \ldots, r$. The q are all rational and $q_1 \neq 0$. As the residues of v_0' are all zero, we see from (2) that $q_1 \, c_1 + \ldots + q_r \, c_r$ is not zero so that y has a nonzero residue at the point. Similarly, if v_1 has a zero at a point at ∞, y has a nonzero residue at the point.[*]

Thus, *if $y(x)$ is an algebraic function whose integral is elementary but not algebraic, there must exist points on the surface of y at which y has residues distinct from zero.*

ELLIPTIC INTEGRALS

13. We shall now demonstrate the nonelementary character of Legendre's elliptic integrals of the first and second kinds,

$$I_1 = \int \frac{dx}{\sqrt{(1 - x^2)(1 - k^2 x^2)}}, \quad I_2 = \int \frac{x^2 \, dx}{\sqrt{(1 - x^2)(1 - k^2 x^2)}},$$

where k is a constant with $k^2 \neq 0, 1$. We must show first that there are no nonzero residues and then that the integrals are not algebraic.

For those versed in the theory of abelian integrals, the nonelementary character of I_1 follows immediately from §12. An integral of the first kind, on a surface of any positive genus, has an infinite number of branches and has no logarithmic branch points. Thus no integral of the first kind can be elementary. An integral of the second kind cannot be elementary if it is not algebraic. We prefer, however, to treat the question by more elementary methods.

Consider I_1. Let

$$y = \sqrt{(1 - x^2)(1 - k^2 x^2)}.$$

The poles of $1/y$ are at $\pm 1, \pm k^{-1}$. Writing

$$(46) \qquad \frac{1}{y} = \frac{(x - 1)^{-1/2}}{\sqrt{(-x - 1)(1 - k^2 x^2)}},$$

we see that the function under the radical in (46) is not zero at $x = 1$, so that the reciprocal of its square root is analytic for

$x = 1$. Thus the residue of $1/y$ at $x = 1$ is zero. The same is true at the other three branch points. At ∞, the two branches of $1/y$ are uniform, and their developments start with $\pm\, k^{-1} x^{-2}$. Thus the residues at ∞ are zero.*

It remains to be shown that I_1 is not algebraic. Suppose that it is. By §§ 8 and 9

$$(47) \qquad I_1 = A_0 + A_1\, y$$

where A_0 and A_1 are rational functions of x. We differentiate (47). Then

$$\frac{1}{y} = \frac{dA_0}{dx} + y\,\frac{dA_1}{dx} + A_1\,\frac{dy}{dx} = \frac{dA_0}{dx} + y\,\frac{dA_1}{dx} + \frac{A_1}{2y}\,\frac{dy^2}{dx}.$$

Thus

$$(48) \qquad y\,\frac{dA_0}{dx} = 1 - y^2\,\frac{dA_1}{dx} - \frac{A_1}{2}\,\frac{dy^2}{dx}.$$

The second member of (48) is a rational function of x. The first member is irrational unless dA_0/dx is zero. It follows that A_0 is a constant.

By (47), $1/y$ is the derivative of $A_1\, y$, which is a function uniform on the surface of y. Then $A_1\, y$ cannot have a pole at a finite point x_0. If it did, $1/y$ would have an expansion for $x = x_0$ beginning with a term $a_0(x - x_0)^{-p}$ with $p > 1$. The only places where $1/y$ can show negative exponents are the branch points, where the exponents are $-1/2$. If $A_1\, y$ had a pole at ∞, $1/y$ would have an expansion at ∞ beginning with a nonnegative power of x. Thus (47) cannot hold and I_1 is not elementary.

We examine I_2. The integrand, x^2/y, has zero residues at ± 1, $\pm k$. To find the expansions of x^2/y at ∞, we write

$$(49) \qquad \frac{x^2}{y} = [k^2 - \frac{1 + k^2}{x^2} + \frac{1}{x^4}]^{-\frac{1}{2}} = \pm\,(\frac{1}{k} + \frac{1 + k^2}{2\,k^3}\,\frac{1}{x^2} + \dots).$$

Thus the residues at ∞ are 0. We have thus to show that I_2 is not algebraic. Writing $I_2 = A_0 + A_1\, y$, we prove as above that A_0 is a constant. Then x^2/y is the derivative of $A_1\, y$. As above, $A_1\, y$ can have no pole at a finite place. This means that A_1 has no poles in the finite complex plane; if it did, such poles could not be removed by multiplication by y. Thus A_1 is a polynomial. If A_1 is of degree m, the expansions of $A_1\, y$ at ∞ begin with terms in x^{m+2}; the

---

* Note that this does not happen when $k = 0$. That is what allows the integral of $(1 - x^2)^{-\frac{1}{2}}$ to be elementary.

expansions of the derivative of A_1 y begin with x^{m+1}. As $m + 1 > 0$, we see from (49) that the derivative of A_1 y cannot be x^2/y.

The subject of elliptic integrals of the third kind is a more delicate one. Such an integral is sometimes a logarithm of a rational combination of x and y. Abel, and later Chebyshev and Zolotareff wrote noteworthy papers on this question.*

CHEBYSHEV'S INTEGRAL

14. Chebyshev considered the integral

(50) $$u = \int x^p(1 - x)^q \, dx$$

where each of p and q is rational and not zero. He proved that, for the integral to be elementary, it is necessary and sufficient that at least one of p, q and p + q be an integer.

We treat first the question of sufficiency. Let $p = r/t$ and $q = s/t$ where r, s, t are relatively prime integers with t positive. If q is an integer, we put $x = v^t$, and u becomes the integral of a rational function of v. The integral of a rational function can always be obtained in finite terms by the method of partial fractions. When p is an integer, we replace $1 - x$ by v^t. When p + q is integral, we write

$$u = \int x^{p+q} \left(\frac{1 - x}{x}\right)^q \, dx$$

and replace $(1 - x)/x$ by v^t, again securing a rational integrand.

To treat the sufficiency question, we start by showing that the integrand in (50), which we denote by y, is a function of t branches. We have

(51) $$y^t - x^r (1 - x)^s = 0.$$

We consider a branch y_1 of y analytic in a small simply connected area $\mathfrak{A}$ which contains neither 0 nor 1. Let a be a point in $\mathfrak{A}$. Let C_1 be a simple closed curve, passing through a, with 0 interior to C_1 and 1 exterior to it. Let C_2 be a simple closed curve, passing through a, with 1 interior to C_2 and 0 exterior to it. If we move around C_1, y_1 is replaced, according to (51), by $e^{2\pi i r/t} y_1$. A passage around C_2 gives $e^{2\pi i s/t} y_1$. As r, s, t are relatively prime, we can determine integers g, h, k such that

(52) $$gr + hs + kt = 1.$$

* Appel et Goursat, *Fonctions algébriques* (Paris, 1929), I, 354.

Let C be a curve consisting of g turns around C_1 and h turns around C_2. Then C replaces y_1 by

$$e^{2\pi i(gr+hs)/t}\, y_1 = e^{2\pi i(1-kt)/t}\, y_1 = e^{2\pi i/t}\, y_1.$$

Thus t-1 successive turns around C give t-1 new branches of y in addition to y_1. By (51) there are no more than t branches. Thus y has precisely t branches.

Suppose now that none of p, q, p+q is an integer.

We shall prove that the residues of y are all zero. Residues can exist only for x = 0, 1, ∞. For x = 0, the expansions of $(1 - x)^{s/t}$ are all Taylor expansions. Because r/t is fractional, all expansions of y will contain only fractional powers. Thus, when p < 0, the residues of y at 0 are 0. Similarly, at x = 1, there can be only zero residues. For x = ∞, we write

$$y = x^{(r+s)/t}\, (-1 + \frac{1}{x})^{s/t}$$

Again there are only fractional powers and the residues are 0.

We have now to show that u is not algebraic. Let

(53) $u = A_0 + A_1\, y +\ldots+ A_{t-1}\, y^{t-1}.$

We have

(54) $\dfrac{d}{dx} A_i\, y^i = y^i\, (\dfrac{dA_i}{dx} + \dfrac{iA_i}{y} \dfrac{dy}{dx}).$

Now

(55) $\dfrac{1}{y} \dfrac{dy}{dx} = \dfrac{1}{ty^t} \dfrac{dy^t}{dx}$

and the second member of (55) is a rational function of x. Thus, differentiating (53) we have

(56) $y = B_0 + B_1\, y +\ldots+ B_{t-1}\, y^{t-1}$

where, for each i, B_i is given by the expression in parentheses in (54) and is a rational function of x. Equation (56) must be an identity, else y would satisfy an equation of degree less than t. Thus $y = B_1\, y$ and, adding a constant to u if necessary, we may write

(57) $u = A_1\, y.$

Suppose now that $A_1 = P/Q$ with P and Q relatively prime polynomials. We shall show first that Q is a constant. Suppose that this is not so. Then Q has a zero at some finite point $\underline{a}$. If $\underline{a}$ is neither

0 nor 1, y is not zero at $\underline{a}$, so that the second member of (57) will
have poles on the surface of y for x = a. Then the derivative of $A_1 y$
which is y, must have poles at $\underline{a}$. This cannot be, so that a = 0 or
a = 1. Let a = 0. The expansions of n at 0 begin with fractional
exponents less than r/t, and we cannot get a first exponent r/t for
y when we differentiate u. Thus a $\neq$ 0. Similarly, we cannot have
a = 1, so that Q is a constant.

Then A_1 is a polynomial. If A_1 is of degree m, the expansions of
u at ∞ start with terms of exponent $m + (r + s) t^{-1}$, which is fractional. To get y from u by differentiation, we must have m = 1. Let
then

(58) $u = (a + b x) y.$

Unless a = 0, the expansions of the second member of (58) at x = 0
will start with the exponent r/t and we get too low a first power
for y when we differentiate (57). Similarly, a + b x must vanish
for x = 1. These demands are too great. The necessity is proved.

Chapter III

INTEGRATION OF TRANSCENDENTAL FUNCTIONS

LIOUVILLE'S GENERAL THEOREM

1. When we say that a function $u(x)$ is algebraic *in* functions $y_1(x)$, ..., $y_p(x)$, we shall mean that the functions mentioned are analytic at some point and satisfy an equation

$$\alpha_0 u^m + ... + \alpha_m = 0$$

where the α are polynomials in the y with constant coefficients.

Liouville [5] gave the following generalization of his theorem of Chapter II. Let $y_1(x)$, ..., $y_p(x)$ be functions satisfying differential equations

(1) $$\frac{dy_i}{dx} = f_i(y_1, ..., y_p), \quad i = 1, ..., p$$

where the f are algebraic functions of the y. Let $u(x)$ be algebraic in the $y(x)$ and let the integral of u be elementary in the $y(x)$, that is, obtainable from the y in a finite number of steps by performing algebraic operations and taking exponentials and logarithms. Then

(2) $$\int \cdot u \, dx = v_0 + c_1 \log v_1 + ... + c_r \log v_r$$

where the v are algebraic in the y. If the f are rational, the v can be taken as rational in the y. It follows, for instance, that if w is an elementary function and if the integral of w is elementary, the integral has an expression like the second member of (2) with v which are algebraic in the monomials appearing in w.

A neat treatment of these questions is furnished by Ostrowski's recent generalization [17] of Liouville's broader theorem to functions which are algebraic over differential fields. We take up this question.

DIFFERENTIAL FIELDS

2. Let $\mathfrak{A}$ be an area in the plane of the complex variable x. Let $\mathfrak{F}$ be a set of functions each of which is meromorphic in $\mathfrak{A}$ and at least one of which is not identically zero. We shall call $\mathfrak{F}$ a *differential field* if $\mathfrak{F}$ is closed with respect to rational operations and to differentiation. If f and g are functions in $\mathfrak{F}$, $f + g$, $f - g$

and f g are in ℑ. If g is not identically zero, f/g is in ℑ. Again, if f is any function in ℑ, the derivative of f is in ℑ. Thus is made explicit the meaning of the two types of closure.

For instance, the set of all rational functions of x is a differential field. Here 𝔘 is the entire plane.*

FUNCTIONS ELEMENTARY OVER A FIELD

3. In what follows, we deal with a differential field ℑ which contains all complex constants.

A function u will be said to be *algebraic over* ℑ, or of *order zero over* ℑ, if it satisfies an equation

$$(3) \qquad \alpha_0 \, u^p + \cdots + \alpha_p = 0$$

with the α in ℑ and α_0 not zero. To be precise, (3) is satisfied by some element $P(x - x_0)$ of u, and u is taken as the totality of elements obtained by continuing P from x_0 along curves lying in 𝔘. Thus, if 𝔘 is not the entire plane, u may be not a complete m.a.f. but only part of such a configuration. This, of course, is an annoyance, but it is not fatal to our theory. In specific applications, our confinement to an area is not a serious matter.

By a *complete monomial of order* 1, *over* ℑ, we shall mean a function which is not of order 0 over ℑ, and which is the exponential or a logarithm of a function of order 0 over ℑ. Again we use what may be a portion of an m.a.f. A *monomial* of order 1 is a branch of a complete monomial analytic in some area in 𝔘. The definitions continue as in Chapter I. We obtain the functions *elementary over* ℑ. They have the property of fluency in the area 𝔘.

A function of order n over ℑ is described by a scheme like that of I, 11, with a slight modification. In (I), we use r_1 functions of order 0 over ℑ. In (II), we use r_2 functions of x and x_1', ..., x_{r_1}', algebraic in the x_i' over ℑ. Such a function satisfies an equation whose coefficients are polynomials in the x' with coefficients in ℑ. Similar remarks apply to (III), ..., (N + 1).

The differentiation of a function elementary over ℑ is discussed as in I, 15. Because ℑ is closed with respect to differentiation, the derivative of a function of x which is algebraic over ℑ is

---

* No generality would be gained by allowing the functions in ℑ to have isolated essential singularities as well as poles. If f (x) has an isolated essential singularity at a, there is, by Picard's theorem, a rational value c which is assumed by f (x) in every neighborhood of a. Then 1/(f (x)-c) has a pole in every neighborhood of a.

algebraic over $\Im$. If a function f of several $x_j^{(1)}$ and x is algebraic in the $x_j^{(1)}$ over $\Im$, each of its partial derivatives is algebraic in the $x_j^{(1)}$ over $\Im$; this is seen with the help of the
algebraic equation satisfied by f. Thus the theorem of I, 15, on
the differentiation of a function of order n carries over to the
present case. One uses a function which is algebraic in the accented letters over $\Im$.

OSTROWSKI'S GENERALIZATION

4. We may now state the Liouville-Ostrowski theorem, which uses
a field $\Im$ containing all complex constants.

THEOREM: *Let* y(x) *be algebraic over* $\Im$. *Let the integral of* y(x)
be elementary over $\Im$. *Then*

(4) $$\int y \, dx = v_0 + c_1 \log v_1 + \ldots + c_r \log v_r$$

where the v *are algebraic over* $\Im$. *Furthermore, the* v *can be taken
so as to be rational in* y *and in functions of* $\Im$.

Let u, the integral of y, be of order n > 0 over $\Im$, and let it
be expressed with a least number of n-monomials. Let the function,
algebraic over $\Im$, associated with u, be

$$g(x_1^{(n)}, \ldots, x)$$

Let each accented variable except $x_1^{(n)}$ be replaced by its monomial.
There results a function $f(x_1^{(n)}, x)$. Let θ represent $\theta_1^{(n)}$. We have

$$u = f(\theta, x).$$

Then

(5) $$y = f_\theta(\theta, x) \, \theta' + f_x(\theta, x).$$

The partial derivatives in (5) are obtained by replacements from
functions algebraic in the accented letters over $\Im$. We now follow
with no essential change §§ 3, 4, 5 of Chapter II. We secure the
relation (4). In § 6 we treat the rationality question.

OSTROWSKI'S METHOD OF FIELD EXTENSIONS

5. We have followed above the Liouville proof pattern. Ostrowski
uses a method which contains a genuinely novel idea and constitutes
a noteworthy addition to Liouville's technique. We shall give an
account of it.

We consider $\Im$ as above. Let there be given any finite set of
functions $u_1, \ldots, u_p$ which are meromorphic in $\mho$ and algebraic over

3. Let $\mathfrak{J}'$ be the totality of rational combinations of the u and functions in $\mathfrak{J}$. Then $\mathfrak{J}'$ is a differential field, because the derivative of each u_i is rational in that u_i and functions in $\mathfrak{J}$. We call $\mathfrak{J}'$ an *algebraic extension* of $\mathfrak{J}$. By I, 12, every function in $\mathfrak{J}'$ is algebraic over $\mathfrak{J}$.

Now let u be analytic in $\mathfrak{A}$ and algebraic over $\mathfrak{J}$. Suppose that e^u is not algebraic over $\mathfrak{J}$. Let $\mathfrak{J}'$ be the differential field consisting of all rational combinations of e^u and u' with coefficients in $\mathfrak{J}$.* We call $\mathfrak{J}'$, or any algebraic extension of $\mathfrak{J}'$, an *exponential extension* of $\mathfrak{J}$. Similarly, let u, algebraic over $\mathfrak{J}$, be analytic in $\mathfrak{A}$ and suppose besides that log u is uniform in $\mathfrak{A}$ but not algebraic over $\mathfrak{J}$. Let $\mathfrak{J}'$ be the differential field consisting of all rational combinations of log u and u'/u with coefficients in $\mathfrak{J}$. We call $\mathfrak{J}'$, or any algebraic extension of $\mathfrak{J}'$, a *logarithmic extension* of $\mathfrak{J}$.

Given two distinct differential fields $\mathfrak{J}$ and $\mathfrak{J}_1$ if there exists a finite sequence of differential fields,

$$\mathfrak{J}, \mathfrak{J}', \ldots, \mathfrak{J}^{(n)} = \mathfrak{J}_1,$$

each after the first an exponential or logarithmic extension of its predecessor, we call $\mathfrak{J}_1$ an *extension of $\mathfrak{J}$ of positive rank*. If $\mathfrak{J}_1$ is such an extension of $\mathfrak{J}$ and if n, as above, is the smallest number of extensions which will permit the passage from $\mathfrak{J}$ to $\mathfrak{J}_1$, we call n the *rank of $\mathfrak{J}_1$ with respect to $\mathfrak{J}$*.

Now let y(x) be algebraic over $\mathfrak{J}$ and let the integral u of y be elementary, but not algebraic, over $\mathfrak{J}$. From our study of the structure of u, it is seen that we can find an area $\mathfrak{A}_1$ and an extension $\mathfrak{J}_1$ of $\mathfrak{J}$, relative to $\mathfrak{A}_1$ which contains u. After forming an algebraic extension of $\mathfrak{J}$, one brings in, in succession, monomials analytic in $\mathfrak{A}_1$ and functions algebraic in such monomials. Having such an area $\mathfrak{A}_1$ in which extensions containing u exist, we use an extension $\mathfrak{J}_1$ whose rank, r, is as small as possible. For all we can say offhand, r may depend on the area $\mathfrak{A}_1$ which is chosen; r will not increase if $\mathfrak{A}_1$ is shrunk. We suppose $\mathfrak{A}_1$ to be such that r is as small as possible. Let the successive extensions which produce $\mathfrak{J}_1$ from $\mathfrak{J}$ be $\mathfrak{J}', \mathfrak{J}'', \ldots,$ $\mathfrak{J}^{(r)} = \mathfrak{J}_1$. We shall show that the extensions are all *logarithmic* extensions and that

$$u = v_0 + c_1 \log v_1 + \ldots + c_r \log v_r$$

with each v_i algebraic over $\mathfrak{J}$.

* Note that as u' is algebraic over $\mathfrak{J}$, the higher derivatives of u are rational in u' and functions in $\mathfrak{J}$.

First, let r = 1. Denoting by θ the exponential or logarithm used in building $\mathfrak{J}'$ out of $\mathfrak{J}$, we have

$$u = v(\theta, x)$$

with v algebraic in θ and functions in $\mathfrak{J}$. By the Liouville procedure, we find that

$$u = v_0 + c_1 \log v_1$$

with v_0 and v_1 algebraic over $\mathfrak{J}$. Suppose now that our theorem is proved for r = 1, ..., s. Let r = s + 1. We simply replace $\mathfrak{J}$ by $\mathfrak{J}'$. $\mathfrak{J}^{(r)}$ is of rank s over $\mathfrak{J}'$. Then

$$u = v_0 + c_1 \log v_1 + ... + c_s \log v_s$$

with each v algebraic over $\mathfrak{J}'$ and hence algebraic in the θ used above and functions of $\mathfrak{J}$. The procedure of Liouville then gives our theorem.

Ostrowski's method of field extensions lends an intriguing Galoisian aspect to the theory of Liouville. As there are no groups in the Liouville theory, the resemblance is not too deep. Ostrowski's method consists, broadly speaking, in expressing u with a minimum total number of monomials of all orders.

GENERALIZATION OF ABEL'S THEOREM

6. We shall now show that the v in (4) may be taken rational in y and functions in $\mathfrak{J}$. Our method will be algebraic, rather than analytic as in II, 8.

First we secure a function t(x), linear in the v with constant coefficients such that the v are rational in t and functions in $\mathfrak{J}$.

Rather than the r + 1 functions v, we consider two functions φ and ψ, analytic at a point $\underline{a}$ in $\mathfrak{A}$ and algebraic over $\mathfrak{J}$. One will see that our conclusion holds for any number of functions. Let φ and ψ satisfy equations with coefficients in $\mathfrak{J}$,

$$\alpha_0 \varphi^p + ... + \alpha_p = 0; \quad \beta_0 \psi^q + ... + \beta_q = 0,$$

the equations being irreducible in $\mathfrak{J}$. These equations have solutions

$$\varphi_1, \ldots, \varphi_p; \quad \psi_1, \ldots, \psi_q$$

analytic in some area close to $\underline{a}$. The φ_i are distinct; so also are the ψ_i. We understand that $\varphi_1 = \varphi$, $\psi_1 = \psi$.

We wish to choose constants h and k in such a way that the pq functions $h \varphi_i + k \psi_j$ are distinct. The difference of two such functions is of the form $h \zeta + k \xi$ with ζ and ξ not both zero. The product of all differences of pairs is a polynomial in h and k. If h and k are taken as constants for which the product is not zero identically in x, the pq functions $h \varphi_i + k \psi_j$ will be distinct.

Now let

$$F(z) = \prod [z - (h \varphi_i + k \psi_j)], \quad i = 1, \ldots, p; \quad j = 1, \ldots, q.$$

The coefficients in F are symmetric in the φ and ψ and are thus in $\mathfrak{I}$. For each i and j, we have

$$F(z) = [z - (h \varphi_i + k \psi_j)] \, G_{ij}(z)$$

with G_{ij} a polynomial in z. Let

$$H(z) = \Sigma \, G_{ij}(z) \, \varphi_i, \quad i = 1, \ldots, p; \quad j = 1, \ldots, q.$$

The coefficients of $H(z)$, by their symmetry, belong to $\mathfrak{I}$. Now, if $F'(z)$ is the derivative of F with respect to z, we have

$$G_{11}(h \varphi_1 + k \psi_1) = F'(h \varphi_1 + k \psi_1),$$

an equation in which the members are not zero, since $F(z)$ has p q distinct factors. For i and j not both unity,

$$G_{ij}(h \varphi_1 + k \psi_1) = 0.$$

Thus

$$H(h \varphi_1 + k \psi_1) = F'(h \varphi_1 + k \psi_1) \, \varphi_1.$$

Then, if $t = h \varphi + k \psi$, φ is rational in t. So also is ψ.

7. We suppose that we have a t as indicated for the v. By I, 12, t is algebraic over $\mathfrak{I}$. Let $\mathfrak{O}$ be the totality of rational combinations of y and functions in $\mathfrak{I}$. $\mathfrak{O}$ is a field in the sense of algebra. As $\mathfrak{O}$ contains $\mathfrak{I}$, t satisfies equations with coefficients in $\mathfrak{O}$. Let t satisfy

$$(6) \qquad \qquad \alpha_0 \, t^p + \ldots + \alpha_p = 0$$

with coefficients in $\mathfrak{O}$ and irreducible in $\mathfrak{O}$. Equation (6) will have p distinct solutions $t_1, \ldots, t_p$, analytic in some area close to the point $\underline{a}$ at which we are given (4). We let $t_1 = t$. From (2) we have

$$(7) \qquad \qquad y = v_0' + c_1 \frac{v_1'}{v_1} + \ldots + c_r \frac{v_r'}{v_r}.$$

81

Now let $v_i = R_i(t, x)$, each R rational in t with coefficients in $\mathfrak{I}$. Then

$$v_i' = \frac{\partial R_i(t, x)}{\partial t} t' + \frac{\partial R_i(t, x)}{\partial x}.$$

By (6) t' is rational in t and functions in $\mathfrak{I}$. On this basis, the second member of (7) may be regarded as a rational combination of t and functions in $\mathfrak{I}$. As (6) is irreducible in $\mathfrak{I}$, we may replace t in (7) by $t_2, \ldots, t_p$ in succession and (7) will continue to hold. This is because an equation with coefficients in $\mathfrak{I}$ which admits one solution of (6) admits all p solutions. We have thus for $j = 1, \ldots, p$,

$$(8) \quad y = \frac{d}{dx} [R_0(t_j, x) + c_1 \log R_1(t_j, x) + \ldots + c_p \log R_p(t_j, x)].$$

We add the p equations (8), observing that $\Sigma R_0(t_j, x)$ is rational in y and functions in $\mathfrak{I}$, as is also the product, for any i, of the $R_i(t_j, x)$, $j = 1, \ldots, p$. We secure a relation (4) with v_i which are rational in y and functions of $\mathfrak{I}$.

A comparison of the above proof with that for the algebraic case in II, 8, brings out an important principle. Where one uses, in an analytical treatment of a problem, the process of analytic continuation, one employs, in an algebraic treatment, the theorem which states that, if $f = 0$ is an irreducible equation and if $g = 0$ admits one solution of $f = 0$, then $g = 0$ admits every solution of $f = 0$. The algebraic theorem just stated is a counterpart of the analytical principle of the permanence of functional equations.

INTEGRALS OF ELEMENTARY FUNCTIONS

8. Let y in (4) be an elementary function as in Chapter I. Let $\mathfrak{I}$ be the differential field generated by x and y, that is, the totality of rational combinations of x, y, and derivatives of y, with constant coefficients. If the integral u of y is elementary, u will be elementary over $\mathfrak{I}$. It will be expressed as in (4) with each v rational in x, y and a certain number of derivatives of y. Now y and its derivatives are algebraic in the monomials in y. Thus, *if the integral of an elementary function y is elementary, the integral can be expressed as in (4), with each v algebraic in the monomials in y.*

APPLICATIONS

9. We now consider special types of functions. Liouville investigated the integral

$$(9) \qquad\qquad u = \int e^{g(x)} y(x) \, dx$$

where g and y are algebraic, with g nonconstant. One works in an area in which definite branches of g and y are taken. Liouville proved that u, when elementary, has the form

$$(10) \qquad\qquad e^{g(x)} w(x) + c$$

with w rational in x, g, and y.

To secure this result, we let $\mathfrak{I}$ be the totality of rational combinations of x, $e^{g(x)}$, g, and y. Then $\mathfrak{I}$ is a differential field, since g', for instance, is rational in x and g. Let u be elementary. Let $\theta = e^{g(x)}$. Then

$$(11) \qquad\qquad u = v_0(\theta, x) + \Sigma c_i \log v_i(\theta, x)$$

with each v rational in θ, x, g, y. Differentiating, we have

$$(12) \qquad \theta y = \frac{\partial v_0}{\partial \theta} \theta g' + \frac{\partial v_0}{\partial x} + \Sigma \frac{1}{v_i} \left[\frac{\partial v_i}{\partial \theta} \theta g' + \frac{\partial v_i}{\partial x} \right]$$

As θ, by I, 5, is transcendental, (12) is an identity in θ. We replace θ by $\mu\theta$ and integrate. Then

$$\mu u = v_0(\mu\theta, x) + \Sigma \log v_i(\mu\theta, x) + \beta(\mu).$$

We differentiate with respect to μ and put $\mu = 1$. We find for u an expression rational in θ, x, g, y. Let such an expression be given by

$$u = v(\theta, x).$$

We find as above

$$v(\mu\theta, x) = \mu v(\theta, x) + \beta(\mu).$$

Differentiation with respect to μ gives, for $\mu = 1$,

$$\theta \frac{\partial v(\theta, x)}{\partial \theta} = v(\theta, x) + c.$$

We replace θ by a variable z and integrate the resulting equation. We find that

$$v(z, x) = -c + z \, \gamma(x) = -c + \frac{z}{z_0} \, [v(z_0, x) + c].$$

Thus

$$u = \theta \, [\frac{v(z_0, x) + c}{z_0}] - c$$

so that u is expressed as in (10).

10. It is noteworthy that the manner of pairing the branches of g and y may affect the character of the integral. Thus if

$$g = (1 + x^{\frac{1}{2}})^2, \quad y = 1 + x^{-\frac{1}{2}}$$

where the same branch of $x^{\frac{1}{2}}$ is used in g and y, the integral is simply e^g. If we use one branch in g and the other, its negative, in y, then u is not elementary. If it were, we would find by subtraction that the integral of $x^{-\frac{1}{2}} e^g$ is elementary. Putting $z = 1 + x^{\frac{1}{2}}$, we would find the integral of e^{z^2} to be elementary. This, it will be seen below, is not so.

11. To develop a practical method for determining whether (9) is or is not elementary, it suffices to consider the case of $g(x) = x$. For u to be elementary, the differential equation

(13) $$w' + w = y$$

must have a solution rational in x and y. The treatment of this question parallels the work of II, 10. We shall examine the case in which y is rational. In that case, if (13) has a rational solution w, the poles of w in the finite plane must be poles of y. If y has a pole of order p at a point $\underline{a}$, w must have at $\underline{a}$ a pole of order p − 1. Thus if y = P/Q with P and Q polynomials, and if w is a rational solution of (13), Qw will be a polynomial. We put w = z/Q and (13) goes over into

(14) $$Q z' + (Q - Q') z = PQ.$$

We have to see whether (14) has a solution z which is a polynomial. The degree of Q − Q' is that of Q. It follows that z must have the same degree as P. Thus, if $\underline{P}$ is of degree m, we substitute a general polynomial of degree m for z in (14) and secure a system of linear equations whose compatibility or incompatibility decides the character of our integral.

12. Let, for example,

(15) $$u = \int e^{-x^2} \, dx.$$

If u were elementary, it would be given by (10) where w is rational and satisfies

(16) $w' - 2 x w = 1$.

As the second member of (16) has no pole, no solution of (16) can have a pole in the finite part of the plane. Furthermore, w cannot be a polynomial. If it were, x w would be of higher degree than w' and 1 and (16) could not hold. Thus the integral (15) is not elementary.

Consider now the integral

(17) $u = \int \frac{e^x}{x} dx$.

For it to be elementary, there would have to be a rational w with

$$w' - \frac{w}{x^2} = \frac{1}{x}.$$

Let such a rational solution exist. Let its expansion at the origin be

$$w = a_0 x^p + a_1 x^{p+1} + \ldots.$$

We find that p = 1. Thus w is a polynomial of positive degree. For the neighborhood of ∞, the expansion of w' begins with a higher power than does that of the other terms. We see that (17) is not elementary.

We consider finally the integral

$$u = \int \frac{dx}{\log x}$$

which occurs in the theory of the distribution of prime numbers. Putting $x = e^t$ we find the integral to go over into (17). It is thus not elementary.

13. Liouville extended, as follows, the result presented in §9. Let

(18) $w = e^{g_1} y_1 + \ldots + e^{g_p} y_p$,

with algebraic g and y, have an elementary integral. Suppose further that no two g_i differ by a constant. Then each $e^{g_i} y_i$ has an elementary integral. It is understood that one of the g may be a constant.

We assume, as we may, that p > 1. Our first step will be to secure functions $h_1, \ldots, h_q$ with $1 \leq q \leq p$ which satisfy the following conditions:

(a) There exists no relation $m_1 h_1 + \ldots + m_q h_q + c = 0$, with c a constant and the m integers not all zero.

(b) Each g differs by a constant from a linear combination, with integral coefficients, of the h.

(c) Each h is some g divided by an integer.

Let $p = 2$. If there exists no relation $m_1 g_1 + m_2 g_2 + c = 0$ with m_1 and m_2 integers, we take $q = 2$, $h_1 = g_1$, $h_2 = g_2$. Let such a relation exist and, to fix our ideas, suppose that $m_2 \neq 0$. We notice that g_1 is not a constant. If it were, g_2 would be a constant and so also would be $g_2 - g_1$. We take $q = 1$ and $h_1 = g_1/m_2$.

Suppose then that we have treated all values of p with $p \leq r$. We examine the case of $p = r + 1$. For $g_1, \ldots, g_r$, we have a set of functions satisfying (a), (b), (c). We denote them by $h_1', \ldots, h_{q'}'$. If g_{r+1} does not differ by a constant from a linear combination of the other g with rational coefficients, we take $q = q' + 1$, $h_1 = h_1'$, $i = 1, \ldots, q'$, and $h_q = g_{r+1}$. Now, suppose that

$$g_{r+1} = \frac{m_1}{n} g_1 + \ldots + \frac{m_r}{n} g_r + c.$$

We take $q = q'$, $h_1 = h_1'/n$, $i = 1, \ldots, q'$.

The property (c) will not be used in what follows; it was important only for guaranteeing (a) at each step of the induction.

Now let $h_1, \ldots, h_q$ be any finite set of algebraic functions satisfying (a) above and not necessarily derived, as in what precedes, from a system $g_1, \ldots, g_p$. Let $\theta_i = e^{h_i}$, $i = 1, \ldots, q$. We shall prove that there exists no algebraic relation

(19) $F(x, \theta_1, \ldots, \theta_q) = 0$

which is not an identity in the x and the θ. This is true for $q = 1$, since θ_1 is not algebraic. We examine the case of $q = r + 1$, supposing the lower cases to have been treated. Suppose that (19) with $q = r + 1$ holds but is not an identity. Then θ_q must figure effectively in (19), else we would have an algebraic relation among $x, \theta_1, \ldots, \theta_r$. Thus

(20) $\theta_{r+1} = G(x, \theta_1, \ldots, \theta_r)$

with G algebraic.* We take logarithmic derivatives in (20). Then

(21) $h_{r+1}' = \frac{1}{G} [G_x + \frac{\partial G}{\partial \theta_1} \theta_1 h_1' + \ldots + \frac{\partial G}{\partial \theta_r} \theta_r h_r'].$

* The details are as in I, 14.

In (21), we have an identity in x, θ_1, ..., θ_r. By the μ-process, we find

$$\theta_{r+1} = e^{\beta(\mu)} G(x, \mu \theta_1, \theta_2, ..., \theta_r)$$

with $\beta(\mu)$ analytic for $\mu = 1$. Then, for a variable z and a constant c,

$$z \frac{\partial G(x, z, \theta_2, ..., \theta_r)}{\partial z} + c G(x, z, \theta_2, ..., \theta_r) = 0.$$

Then

$$(22) \quad G(x, z, \theta_2 ..., \theta_r) = z^{-c} \gamma(x) = z^{-c} z_0^c G(x, z_0, \theta_2, ..., \theta_r).$$

As G is algebraic in z, c is rational. By (20) and (22)

$$(23) \quad \theta_{r+1} \theta_1^c = e^{h_{r+1} + c h_1} = G(x, z_0, \theta_2, ..., \theta_r).$$

Now (23) is a nonidentical algebraic relation among the exponentials of the r functions $h_{r+1} + c h_1$, h_2, ..., h_r, which latter satisfy (a) above. Our statement is proved.

We return now to g_1, ..., g_r as given above and to h_1, ..., h_q as found for them, satisfying (a) and (b). We subtract constants from the g in such a way that each of them becomes a linear combination of the h with integral coefficients. This is balanced by multiplying the y in (18) by constants. Supposing the integral u of w in (18) to be elementary, we have

$$u = v_0 + \Sigma c_i \log v_i$$

with each v algebraic in x, θ_1, ..., θ_q where $\theta_i = e^{h_i}$.

Now w in (18) is a sum of terms each of which is a product of an algebraic function and integral powers of the θ. The customary procedure of Liouville shows that if we replace any θ in w by a constant times itself, we get a function which is integrable in finite terms. Two terms of w involve the θ with distinct sets of exponents. Let us separate w into sets of terms, all terms of any one set involving θ_1 to the same power and distinct sets showing θ_1 in different powers. Let w, thus written, be

$$(24) \quad w = \theta_1^{m_1} A_1 + ... + \theta_1^{m_s} A_s.$$

For every μ, the integral of

$$\mu^{m_1} \theta_1^{m_1} A_1 + ... + \mu^{m_s} \theta_1^{m_s} A_s$$

is elementary. As the m are distinct, we can find numbers μ_1, ..., μ_s

for whioh the determinant whose j^{th} row is $\mu_j^{m_1}, \ldots, \mu_j^{m_s}$ does not
vanish; this follows from the linear independence, as functions of
μ, of the μ^{m_1}. Thus each term in (24) has an elementary integral.
We arrange such a term according to powers of θ_2 and continue. We
find each term in (18) to be integrable in finite terms.

14. How far does Liouville's theory go? To what extent can one
determine whether a given function can be integrated in finite
terms? Mordukhai-Boltovskoi has been the chief investigator of
this question. In his book [13] many types of functions are examined and the general forms of their integrals, when elementary, are
obtained. Numerous special examples are treated in detail. What is
of greatest theoretical interest in Mordukhai-Boltovskoi's work is
his study of elementary functions in which the algebraic operations
employed are all rational. One uses thus e^x and $\log x$, but no irrational algebraic operations. A method is developed for testing for
the integrability of such functions. Details are presented for
functions of orders 1 and 2.

Mordukhai-Boltovskoi uses three types of monomials. They are exponentials, irrational powers, and logarithms. Thus x^a with a irrational is of the first order in his theory, whereas, as will be
seen in IV, 2, it is of the second order in the classification of
Liouville. Liouville's procedure has the merit of employing a minimum number of monomials. For instance, as can be shown by the methods, of the following chapter, $\log x$ cannot be obtained by performing algebraic operations and taking exponentials. However, the
irrational powers have a definite formal utility.

Chapter IV

FURTHER QUESTIONS ON THE ELEMENTARY FUNCTIONS

EXISTENCE OF FUNCTIONS OF ALL ORDERS

1. We shall prove, by means of an example, the existence of functions of order n for every positive integer n. Let $e_1(x)$ denote e^x. We let $e_2(x)$ represent the exponential of $e_1(x)$ and, in general, $e_n(x)$ the exponential of $e_{n-1}(x)$. Liouville [6] showed that the order of $e_n(x)$ is n. We shall establish this result.

If $y(x)$ is of order n, $e^{y(x)}$ will be of order n-1 if, and only if, y is an n-logarithm. One might inquire as to the circumstances under which e^y is of order n. We shall prove, in this connection, the following lemma.

LEMMA: *If $e^{y(x)}$, with $y(x)$ of order n, is of order n, then*

(1) $$y(x) = u(x) + \log v(x)$$

with $\log v(x)$ an n-logarithm and with $u(x)$ of order n-1.

We express $y(x)$ and its exponential, which latter we call $w(x)$, in such a way that the number of n-monomials which appear in at least one of them is as small as possible. Let θ be one of these monomials. We write

$$y(x) = f(\theta, x), \quad w(x) = g(\theta, x),$$

so that

(2) $$f(\theta, x) = \log g(\theta, x).$$

Let $\theta = e^{p(x)}$ with $p(x)$ of order n-1. We find first that

$$f(\mu\theta, x) = \log g(\mu\theta, x) + \beta(\mu),$$

then that, for a variable z,

(3) $$f(z, x) = \log g(z, x) + c \log z + \gamma(x).$$

We are going to prove that $f(z, x)$ is independent of z, that c is a rational number distinct from zero, and that

$$g(z, x) = z^{-c} w_1(x)$$

with w_1 algebraic in the monomials other than θ which figure in w. We shall know thus, in particular, that θ does not figure in y.

89

Suppose that $f(z, x)$ is not independent of z. Then there is a value x_0 of x, close to the point $x = a$ near which we are working, such that $f(z, x_0)$ is a nonconstant function of z. By (3),

(4) $$f(z, x_0) = \log g(z, x_0) + c \log z + \gamma(x_0).$$

We let $h(z) = f(z, x_0) - \gamma(x_0)$ and $k(z) = g(z, x_0)$, so that

(5) $$h(z) = \log k(z) + c \log z.$$

The algebraic functions $h(z)$ and $k(z)$ are given to us through branches analytic at $z = \theta(a)$. By III, 6, we can find a $t(z)$, linear in h and k, such that h and k are rational in t and z and thus rational on the Riemann surface of t. We have thus, everywhere on the surface of t,

(6) $$h'(z) = \frac{k'(z)}{k(z)} + \frac{c}{z}.$$

As $h(z)$ is not constant, there is a place on the surface of t at which it has a pole. Let such a place lie over $z = b$. First let b be finite. Then $h'(z)$ has an expansion, at the pole, in which the lowest exponent of $t - b$ is less than -1. As was seen in II, 11, we cannot get such an exponent from the second member of (6). Let $b = \infty$. Then $h'(z)$ has at ∞ an expansion in which the highest exponent exceeds -1. Again we find a contradiction. Thus, for every x_0 close to a, $f(z, x_0)$ is a constant. We take any such x_0 and consider (5) where $h(z)$ is a constant. Then (6) becomes

(7) $$\frac{k'(z)}{k(z)} + \frac{c}{z} = 0.$$

If c, which is independent of x_0, were zero, $g(z, x_0)$ would be free of z for every x_0 and θ would be absent from w as well as from y. Then $c \neq 0$. This means, by II, 11, that k'/k has a residue for $z = 0$ which is a multiple of c. As a residue of k'/k must be an integer, c is a rational number. By (7), $k(z) = d\, z^{-c}$ with d constant. Then $z^{-c} g(z, x)$ is independent of z. It is a function $w_1(x)$, algebraic in the monomials other than θ which appear in w.

We have $w(x) = \theta^{-c} w_1(x)$. We can now prove that there is no other exponential among the n-monomials in y and w. Suppose that there is one, $\zeta = e^{q(x)}$. Let $y_1(x) = y(x) + c\, p(x)$, where $\theta = e^p$. We have

$$e^{y_1(x)} = w_1(x).$$

Now y_1, like y, is free of θ. Thus the n-monomials which appear in

y_1 and w_1 are those other than θ which appear in y and w. It follows as above that ζ does not appear in y_1 and that $w_1 = \zeta^{-d} w_2$ with d rational and w_2 free of θ and ζ. As

$$w = e^{-cp-dq} w_2,$$

y and w are expressible with fewer n-monomials than were used above.

We have thus $w = e^{-cp(x)} w_1$ with w_1 free of n-exponentials. We shall show now that w contains no n-logarithm. Let θ be such a monomial. We use (2) and find first

$$f(\theta+\mu, x) = \log g(\theta+\mu, x) + \beta(\mu),$$

then

$$f(z, x) = \log g(z, x) + c_1 z + \gamma(x).$$

If $g(z, x)$ were not free of z, we would find, fixing x at a suitable x_0, a nonconstant algebraic function whose logarithm is algebraic. Altogether,

$$w = e^{-cp(x)} w_1$$

with w_1 of order less than n. Then

$$y(x) = -c p(x) + \log w_1(x).$$

Necessarily $\log w_1$ is of order n. The lemma is proved.

We can now show that $e_n(x)$ is of order n. We know that e_1 is of order 1. Suppose that we have accounted for the cases from 1 to n-1 inclusive. We make an induction. Since e_{n-1} is of order n-1, $e_n(x)$ is of one of the orders n-2, n-1, n. Let it have order n-2 or n-1. By our lemma,

(8) $e_{n-1} = u + \log v$

with v of order n-2 and u of order less than n-1. We note that if e_n is of order n-2, we merely take $v = e_n$ and $u = 0$. Differentiating, we find

$$e_{n-1} e_{n-2} \cdots e_1$$

to be of order less than n-1. As the reciprocal of $e_{n-2} \cdots e_1$ is at most of order n-2, e_{n-1} is of order less than n-1. Thus e_n is of order n.

In a similar way, one can prove that the $\underline{n}^{th}$ iterate of $\log x$ is of order n.

IRRATIONAL POWERS

2. We consider the function x^a with $\underline{a}$ irrational. We shall prove, with Liouville, that it is of order 2. Its expression $e^{a \log x}$ shows that it has an infinite number of branches; consequently it is not algebraic. If it were of order 1, we would have, by §1,

$$a \log x = u + \log v$$

with u and v algebraic. Then

$$\frac{a}{x} = u' + \frac{v'}{v}.$$

This is impossible because a/x has irrational residues, u' only zero residues, and v'/v integral residues.

KEPLER'S EQUATION

3. Liouville considered the equation of Kepler

$$(9) \qquad\qquad y - a \sin y = x$$

with $\underline{a}$ a constant distinct from zero, an equation which occurs in the elements of astronomical theory. He proved that the function $y(x)$ defined by the equation is not elementary.

We write

$$(10) \qquad\qquad y = \arcsin \frac{y-x}{a}$$

and recall that

$$\arcsin v = - i \log \left(i v + \sqrt{1 - v^2} \right).$$

Then (10) becomes, for $z = i y$,

$$(11) \qquad\qquad z = \log w(x, z)$$

where w, algebraic in x and z, involves z effectively. Equation (11) is certainly not satisfied by a nonconstant algebraic $z(x)$. Let (11) have an elementary solution $z(x)$ of order $n > 0$. As $w(x, z)$ will be of order n, the lemma of §1 shows that $z = u + \theta$ with θ an n-logarithm and u of order n-1. Then w, as well as z, can be expressed with the single n-monomial θ and, when so expressed, will involve θ effectively. But the work of §1 shows that θ will not be present in w. Thus $z(x)$ is not elementary.

In a similar way, one can treat

$$u(x, y) = \log v(x, y)$$

with u and v algebraic, at least one of them involving y effectively. If one of u and v is free of y, there will be an elementary
solution. If u and v both involve y, we can show, as above, that
there is no elementary solution of positive order. The existence of
algebraic solutions depends on that of constants c for which the
equations $u = \log c$, $v = c$ have a solution $y(x)$ in common.

INVERSES OF ELEMENTARY FUNCTIONS

4. In connection with Kepler's equation, we met a function of y,
namely $y - a \sin y$ whose inverse is not elementary. This raises the
question as to when an elementary function has an elementary inverse. The author treated this question [19]. It was shown that
when $y(x)$, of order $n > 0$, has an elementary inverse, there exist
n monomials of the first order, $\theta_1, \ldots, \theta_n$, such that y is an algebraic function of φ_n where

$$\varphi_1 = \theta_1(x), \quad \varphi_2 = \theta_2[\varphi_1(x)], \quad \ldots, \quad \varphi_n = \theta_n[\varphi_{n-1}(x)].$$

That a function of this type has an elementary inverse is obvious.

This result is obtained after there is proved a theorem on functions of functions. The theorem states that if $\varphi(x)$ is of order m
and if a $\psi(x)$ of order n exists such that the order of $\psi[\varphi(x)]$ is
at most $m + n - 2$, then $\varphi(x)$ is an algebraic function of an n-
monomial.

In the proof of this theorem of composition, there is a lemma
which throws some light on the structure of elementary functions.
By a *logarithmic sum of order* n is meant a function of order n of
the form

$$c_1 \log v_1(x) + \ldots + c_r \log v_r(x)$$

with constant c and with v of orders not exceeding n-1. Of course,
at least one v is of order n-1. The lemma states that if, in the
expression of a function y of order n, the number of n-exponentials
plus the number of logarithmic sums of order n is a minimum, each
n-exponential and each logarithmic sum of order n is algebraic in y,
a certain number of derivatives of y, and the monomials of orders
less than n which appear in the expression for y.

One may inquire as to what happens when y is expressed with a minimum total number of exponentials and logarithmic sums of all orders
from 1 to n. Will each exponential and each sum be algebraic in y,

derivatives of y, and x? An example like $e^{a \log x}$ with $\underline{a}$ irrational shows that the answer is negative. Perhaps to secure an extension of the lemma, one should employ a new type of monomial like, for instance, Mordukhai-Boltovskoi's irrational powers of III, 14.

E. R. Lorch has studied elementary transformations, in two variables, which have elementary inverses [10].

SCOPE OF METHOD

5. The method of Liouville can be applied to categories of functions built with transcendental operations other than the taking of exponentials and logarithms. The operation of integration will be considered in Chapter VI. At this point we should like to consider the possibility of using functions other than e^x and $\log x$ in forming monomials of the first order. We have already mentioned x^a with $\underline{a}$ irrational. This does not give a larger class of functions; rather, it facilitates the study of some problems on elementary functions.

We notice that $\log x$ is an integral of $1/x$. Now suppose that $f(x)$ is a nonelementary function whose derivative is algebraic, for instance, an elliptic integral of the first kind. To present a simple scene, suppose that we build a class of functions using one transcendental operation, that of taking the function f of an expression. Thus $f(v)$ with v of order $n-1$ will be an n-monomial if it is not of order lower than n. Let

$$(12) \qquad F(\theta, \ldots, x),$$

of order n, be expressed with the usual economy, and let its derivative be of order less than n. Then

$$F(\theta+\mu, \ldots, x) = F(\theta, \ldots, x) + \beta(\mu)$$

and it is seen that θ figures linearly in F.

Again, let g be a function which is not elementary but whose inverse, like that of e^x, is the integral of an algebraic function. Then

$$g' = \varphi(g)$$

with φ algebraic. Suppose that we build monomials using the operation g, with perhaps others, and that (12) has a derivative of order less than n. If $\theta = g(v)$, we have

$$(13) \qquad \frac{d}{dx} F(\theta, \ldots, x) = \frac{\partial F}{\partial \theta} \varphi(\theta) v' + \ldots .$$

We replace θ in the second member of (13) by $g(v + \mu)$ and find that

$$F(\theta, \ldots, x) = F(g(v + \mu), \ldots, x) + \beta(\mu).$$

We differentiate with respect to μ and put $\mu = 0$. Then

$$\frac{\partial F}{\partial \theta} \varphi(\theta) = c$$

from which we infer the superfluity of θ in F.

HYPERTRANSCENDENTAL FUNCTIONS

6. It is seen without difficulty that every elementary function satisfies an algebraic differential equation, that is, an equation

(14) $$P(x, y, y', \ldots, y^{(n)}) = 0$$

where P is a polynomial with constant coefficients. Thus, a hyper-transcendental function, one which does not satisfy an equation (14), is not elementary. For instance, the gamma function, proved hyper-transcendental by Hoelder, is not elementary.

PROBLEMS OF THE THEORY OF FUNCTIONS

7. It is impossible to work long with the elementary functions without finding functiontheoretic problems. For instance, one might undertake to determine all elementary functions which are uniform. This may be a complicated question. The author [22] has examined certain expressions of the first order. An exponential sum is a function

$$c_1 e^{a_1 x} + \ldots + c_r e^{a_r x}$$

with constant c and a. Let $w(x)$ be defined by an equation

$$\alpha_0 w^p + \ldots + \alpha_p = 0$$

with each α an exponential sum. It is shown that if w is uniform, or more generally, if it is uniform in a sector of opening greater than π, then w is the quotient of two exponential sums. If the quotient of two exponential sums is an integral function, it is an exponential sum. A factorization theory for exponential sums is obtained. These questions are related to the elegant theory of the zeros of exponential sums developed by Tamarkin, C. E. Wilder, Polya, and Schwengeler.*

* See R. L. Langer, *Bulletin of the American Mathematical Society*, XXXVII (1931), 213.

ELEMENTARY NUMBERS

8. We should like to present a class of problems in which numbers are involved rather than functions.

The classification of numbers as algebraic or transcendental is well known. A number is algebraic if it satisfies an algebraio equation with integral coefficients, not all zero; otherwise the number is transcendental.

Let us define *elementary number*. An algebraic number will be called a *number of order zero*. The exponential of any algebraic number distinct from zero, or a logarithm distinct from zero of an algebraic number, will be called a *monomial of order unity*. A *number of order unity* is one which is not algebraic and satisfies an algebraic equation whose coefficients are polynomials, with integral coefficients, in monomials of order one. Continuing, we secure the *elementary numbers*.

There arises immediately, of course, the problem of the existence of numbers of all orders. Priority should perhaps be given to problems on the character of the roots of simple transcendental equations. One might ask, for instance, whether the equation

$$e^z = z$$

has an elementary root. These are, of course, problems of greater difficulty than those which we have been studying.

Chapter V

SERIES OF FRACTIONAL POWERS

1. In the next chapter, there will be presented a variation of Liouville's technique for the treatment of problems of integration in finite terms. The new procedure will require us to develop a function $f(x, y)$, algebraic in y, in descending powers of y for the neighborhood of infinity. A brief derivation of developments of this type, based on the Weierstrass preparation theorem, has been given by Ostrowski.* We use here rather a version of the Newton polygon process, gaining perhaps, in the self-contained nature of our treatment, what is lost as regards brevity.

2. We work first in the neighborhood of $y = 0$. We deal with an equation

(1) $\qquad A_n(x, y) z^n + A_{n-1}(x, y) z^{n-1} + \ldots + A_0(x, y) = 0$

with $n \geq 1$, where each A is analytic in x and y for every x in a given area $\mathfrak{U}$ and for every y with $|y| \leq \eta$ where $\eta > 0$. We understand that A_n does not vanish identically in x and y.

We assume that (1) is irreducible in the (algebraic) field of the A.

For our purposes, it is desirable to replace (1) by an equation in which the coefficient of z^n is unity. Accordingly, we expand each A in a series of powers of y and divide by A_n. Equation (1) takes the form

(2) $\qquad F(x, y, z) = z^n + B_{n-1}(x, y) z^{n-1} + \ldots + B_0(x, y) = 0$,

where B_i, if not identically zero, has an expansion

(3) $\qquad B_i = a_{0i}(x) y^{\sigma_i} + a_{1i}(x) y^{\sigma_i + 1} + \ldots$

with σ_i an integer, positive, negative, or zero, and with $a_{0i}(x)$ not zero. All series (3) converge for x in some area $\mathfrak{B}$ contained in $\mathfrak{U}$ and for every $y \neq 0$ with $|y| < \eta_1$, where $\eta_1 \leq \eta$. Equation (2) is irreducible in the field of the B.

We are going to show that $F(x, y, z)$ in (2) has a representation

$$\overset{n}{\underset{i=1}{\Pi}} \; [z - P_i(x, y)]$$

* *Mathematische Zeitschrift*, XXXVII (1933), 101.

in which each P is a series of the type

(4) $$c_1(x) \, y^{p_1} + c_2(x) \, y^{p_2} + \ldots$$

where the p are rational numbers, with a common denominator, which increase with their subscripts. The p and the $c(x)$ will depend upon the subscript i of P_i. The n series (4) will converge for every x in some area and for $|y|$ small and distinct from 0.

3. We shall prove first the existence of a single series (4). After that it will be easy to get n such series.

If $B_0 = 0$, the quantity 0 is a series (4) which answers our requirements.* In what follows, we assume that B_0 is not identically zero. We consider the ratio

(5) $$\frac{\sigma_0 - \sigma_i}{i}$$

which exists for every $i \geq 1$ for which $B_i \neq 0$, in particular for $i = n$.** Let p_1 be the greatest of the ratios (5).

Let g_1 be the greatest value of i for which (5) equals p_1. For $i = 0, 1, \ldots, g_1$, we define a function $k_i'(x)$ as follows.

We let $k_0'(x) = a_{00}(x)$. For $i > 0$, if $B_i \neq 0$ and if (5) equals p_1, we let $k_i'(x) = a_{0i}(x)$. If, for an $i > 0$, either $B_i = 0$ or else (5) is less than p_1, we let $k_i'(x) = 0$.

We consider the equation for an unknown function $c(x)$,

(6) $$k_0'(x) + k_1'(x) \, c + \ldots + k_{g_1}'(x) \, c^{g_1} = 0.$$

An area contained in $\mathcal{B}$ can be found in which (6) has g_1 analytic solutions, not necessarily distinct. Let $c_1(x)$ be any one of these solutions. For later purposes we shall select in a special way an area in which $c_1(x)$ will be studied. Let $c_1(x)$ be a solution of (6) of multiplicity s_1. The $\underline{s}_1$ th derivative with respect to c of the first member of (6) does not vanish identically in x for $c = c_1(x)$. We shall work with $c_1(x)$ in an area $\mathcal{B}_1$ throughout which the above $\underline{s}_1$ th derivative is distinct from zero.

4. It may be that $c_1(x) \, y^{p_1}$ causes $F(x, y, z)$ in (2) to vanish identically in x and y when substituted for z. If so, $c_1 \, y^{p_1}$ is a series (4) such as we are seeking. In what follows, we assume that the vanishing does not occur.

We put $z = c_1 \, y^{p_1} + z_1$ in (2). Then

---

* This can happen only if n = 1.
** We understand that $B_n = 1$.

(7) $$F(x, y, z) = F_1(x, y, z_1)$$

where F_1 is a polynomial of degree n in z_1. We may write

(8) $$F_1 = B_n' z_1^n + B_{n-1}' z_1^{n-1} + \ldots + B_0' = 0.$$

Each B' is a series of ascending rational powers of y, the exponents of y having as their denominator the denominator of p_1. The coefficients in the B' are analytic in B_1. B_0' is not identically zero, since $c_1 y^{p_1}$ does not annul F. As to B_n', it is merely unity.

For every i for which $B_i' \neq 0$, we let σ_i' denote the least exponent of y in B_i'. There is a σ_0' and a σ_n', the latter being 0. We denote by p_2 the greatest of the quantities

(9) $$\frac{\sigma_0' - \sigma_i'}{i}$$

5. We are going to prove that $p_2 > p_1$. We shall make, in F of (2), the substitution $z = c y^{p_1}$ where c is an indeterminate. F becomes a collection of terms of the type $c^i a(x) y^b$, where b is rational. We shall show that the least b is σ_0 and we shall find the terms with $b = \sigma_0$. From $B_i z^i$ with $B_i \neq 0$, the lowest term secured is

$$u_{0i}(x) c^i y^{\sigma_1 + i p_1}$$

If $i = 0$, or if $i > 0$ and (5) equals p_1, $\sigma_1 + i p_1$ equals σ_0. If (5) is less than p_1, $\sigma_1 + i p_1$ exceeds σ_0. Thus the least b is σ_0, and the sum of the terms with $b = \sigma_0$ may be written

$$y^{\sigma_0} [k_0'(x) + k_1'(x) c + \ldots + k_{\beta_1}'(x) c^{\beta_1}].$$

The bracket in this expression will be designated by $\varphi_1(x, c)$.

If now τ is the least value of b which exceeds σ_0, we may write

(10) $$F(x, y, c y^{p_1}) = \varphi_1(x, c) y^{\sigma_0} + \psi_1(x, y, c) y^\tau$$

where ψ_1 is a polynomial in c whose coefficients are series of terms of the form $a(x) y^d$ with d rational and nonnegative.

In (10), we put

$$c = c_1(x) + z_1 y^{-p_1}.$$

The first member of (10) becomes $F(x, y, c_1 y^{p_1} + z_1)$, which equals $F_1(x, y, z_1)$. Hence

(11) $$F_1(x, y, z_1) = \varphi_1(x, c_1 + z_1 y^{-p_1}) y^{\sigma_0} + \psi_1(x, y, c_1 + z_1 y^{-p_1}) y^\tau.$$

Expanding in powers of z_1, we find

$$(12) \qquad B_i' = y^{\sigma_0 - ip_1} \frac{\varphi_i^{(i)}(x, c_1)}{i!} + y^{\tau - ip_1} \frac{\psi_1^{(i)}(x, y, c_1)}{i!}$$

where the superscript (i) denotes i differentiations with respect to c.

Let us consider a B_i' which is not zero. As $\tau > \sigma_0$, we find that if $\varphi_i^{(i)}(x, c_1)$ does not vanish identically in x, we have

$$(13) \qquad \sigma_i' = \sigma_0 - i\,p_1$$

but that, if the vanishing does occur,

$$(14) \qquad \sigma_i' > \sigma_0 - i\,p_1.$$

In particular, (14) holds for i = 0, so that $\sigma_0' > \sigma_0$.

On the other hand, s_1 being, as at the end of §3, the multiplicity of the solution $c_1(x)$ of (6), we find (13) to hold for i = s_1.

We have

$$(15) \qquad \frac{\sigma_0' - \sigma_i'}{i} = \frac{\sigma_0' - \sigma_0}{i} + \frac{\sigma_0 - \sigma_i'}{i}.$$

For i = s_1, the second term in the second member of (15) is p_1. The first term in the second member is positive, since $\sigma_0' > \sigma_0$. Thus p_2, the greatest value of (9), exceeds p_1.

6. Let g_2 be the greatest value of i for which (9) equals p_2. We denote by $k_0''(x)$ the coefficient of $y^{\sigma_0'}$ in B_0'. For i > 0, if $B_i' \neq 0$ and if (9) equals p_1, we let $k_1''(x)$ denote the coefficient of $y^{\sigma_i'}$ in B_i'. In the other cases with i > 0, we let $k_1''(x)$ denote 0. We consider the equation

$$(16) \qquad k_0''(x) + k_1''(x)\, c + \ldots + k_{g_2}''(x)\, c^{g_2} = 0.$$

The coefficients in (16) are analytic if $\mathcal{B}_1$ of §3. In some area in $\mathcal{B}_1$, (16) has g_2 analytic solutions. Let $c_2(x)$ be one of these. We assign an area $\mathcal{B}_2$ to c_2 in such a way that if c_2 is of multiplicity s_2, the s_2th derivative with respect to c of the first member of (16) is distinct from zero throughout $\mathcal{B}_2$ for c = $c_2(x)$. We understand that $\mathcal{B}_2$ lies with its boundary inside of $\mathcal{B}_1$.

It may be that $c_2 y^{p_2}$ annuls F_1 when substituted for z_1. In that case

$$c_1 y^{p_1} + c_2 y^{p_2}$$

is a series (4) which annuls F. If $c_2 \, y^{p_2}$ does not annul F_1, we put $z_1 = c_2 \, y^{p_2} + z_2$, we write

$$F_1(x, y, z_1) = F_2(x, y, z_2),$$

and we give F_2 the treatment accorded to F and F_1.

7. It may be that at some stage in our process we are led to a finite series.

$$c_1(x) \, y^{p_1} + \ldots + c_r(x) \, y^{p_r}$$

which annuls F. In what follows, we assume that this does not happen, so that we are led to consider an infinite sequence of terms

(17) $$c_1 \, y^{p_1}, \ldots, c_r \, y^{p_r}, \ldots$$

where the p increase with their subscripts and the c are all analytic at some point $\underline{a}$ contained in $\mathcal{B}_1$, $\mathcal{B}_2$, ...

8. We shall be able to build a series (4) out of (17) through the proof of various facts. We shall show that the p have a common denominator. The $c(x)$, analytic at $\underline{a}$, will be seen to be continuable over an area $\mathfrak{C}$ containing $\underline{a}$. The series

(18) $$c_1(x) \, y^{p_1} + \ldots + c_r(x) \, y^{p_r} + \ldots$$

will be found to converge for x in $\mathfrak{S}$ and for $|y|$ small and not zero. If $\underline{s}$ is the common denominator of the p_i, the replacement of $y^{1/s}$ by v in (18) will produce a function of x and v, analytic for x in $\mathfrak{C}$ and for $|v|$ small and not zero.

9. Towards proving the above statements, we shall show that g_2, the degree of (16), does not exceed the multiplicity s_1 of c_1. This will show that $g_2 \leq g_1$.

We inspect (15). The term $(\sigma_0 - \sigma_i')/i$ is a maximum for $i = s_1$. The term $(\sigma_0' - \sigma_0)/i$ is less for $i > s_1$ than for $i \leq s_1$. Hence the greatest value of i for which the first member of (15) is a maximum cannot exceed s_1. Then $g_2 \leq s_1 \leq g_1$.

10. We can now prove that the p have a common denominator. From § 9 we see that for j large, say for $j > e$, where e is some integer, the g_j have a common value, say q. Let j have any fixed value greater than e. The equation

(19) $$k_0^{(j)} + \ldots + k_q^{(j)} \, c^q = 0$$

which determines c_j, must have q equal solutions. The first member of (19) is therefore of the form

(20)
$$k_q^{(j)} \, (c - c_j)^q.$$

This means, because $c_j \neq 0$, that $k_1^{(j)}$ in (19) is not zero. It follows that the ratio

$$\frac{\sigma_0^{(j-1)} - \sigma_i^{(j-1)}}{i}$$

attains its maximum value p_j for $i = 1$. This means that we can use, for the denominator of p_j, the common denominator of the exponents of y in $F_{j-1}(x, y, z_{j-1})$. The same denominator can be used for the p_i with $i > j$.

It follows that the p_i increase towards $+\infty$ with i.

11. We shall show that the series (18) is a *formal* solution of $F(x, y, z) = 0$. This will help us to establish analyticity properties of (18).

By §5, the quantity $\sigma_0^{(j)}$ increases with j. As the $\sigma_0^{(j)}$ all have a common denominator, namely, that of the p, it must be that $\sigma_0^{(j)}$ increases towards $+\infty$ with j. Now the substitution

(21)
$$z = c_1 \, y^{p_1} + \ldots + c_j \, y^{p_j} + z_j$$

converts $F(x, y, z)$ into

(22)
$$F_j(x, y, z_j) = z_j^{(n)} + \ldots + B_0^{(j)}.$$

In (22), $B_0^{(j)}$ is the result of substituting $c_1 \, y^{p_1} + \ldots + c_j \, y^{p_j}$ for z in $F(x, y, z)$. The lowest exponent of y in $B_0^{(j)}$ is $\sigma_0^{(j)}$, which is large if j is large. It follows that (18) satisfies (2) formally.

12. Moving towards the completion of our investigation of (18), we shall prove that q in (19) is unity. Let $q > 1$. For j large we have, by §10,

$$p_{j+1} = \sigma_0^{(j)} - \sigma_1^{(j)} = \frac{\sigma_0^{(j)} - \sigma_q^{(j)}}{q}.$$

Hence

(23)
$$\sigma_1^{(j)} = \sigma_q^{(j)} + (q - 1) \, p_{j+1}.$$

It is easy to see from (2), (21), and (22) that the set of all numbers $\sigma_i^{(j)}$, for all i and j, is bounded from below. Indeed, if σ is the least of the σ_i associated with (2), no $\sigma_i^{(j)}$ is less than

$\sigma - n |p_1|$. From (23) it follows, because $q > 1$, that $\sigma_1^{(j)}$ tends towards $+ \infty$ with j. Now, by (22),

$$(24) \qquad \frac{\partial F_j(x, y, z_j)}{\partial z_j} = n z_j^{(n-1)} + \ldots + 2 B_2^{(j)} z_j + B_1^{(j)}.$$

The least exponent of y in $B_1^{(j)}$ is $\sigma_1^{(j)}$. Also $B_1^{(j)}$ is the result of replacing z in $\partial F(x, y, z)/\partial z$ by $c_1 y^{p_1} + \ldots + c_j y^{p_j}$. It follows that (18) is a formal solution of $\partial F/\partial z = 0$. This contradicts the fact that (2) is irreducible. Thus $q = 1$.

13. We have, analogously to (11),

$$(25) \qquad F_j(x, y, z_j) = \varphi_j(x, c_j + z_j y^{-p_j}) y^{\sigma_0^{(j-1)}}$$

$$\psi_j(x, y, c_j + z_j y^{-p_j}) y^{\tau^{(j-1)}}$$

with $\tau^{(j-1)} > \sigma_0^{(j-1)}$. For $j > e$, we have, by (20),

$$(26) \qquad \varphi_j(x, c) = k_1^{(j)}(x) [c - c_j(x)].$$

We put, in (25), $z_j = y^{p_j} u$. Then

$$c_j + z_j y^{-p_j} = c_j + u$$

and the equation $F_j(x, y, z_j) = 0$ gives, for u, the equation

$$(27) \qquad k_1^{(j)}(x) u = - \psi_j(x, y, c_j + u) y^{\tau^{(j-1)} - \sigma_0^{(j-1)}}.$$

Let s be the common denominator of the p in (17). We put $y = v^s$ in (27), and (27) becomes an equation for u in terms of x and v,

$$(28) \qquad k_1^{(j)} u = D_0(x, v) + D_1(x, v) u + \ldots + D_n(x, v) u^n.$$

The D are series of positive integral powers of v with coefficients analytic in $\mathcal{B}_j$ which represent functions analytic for x in $\mathcal{B}_j$ and for $|v|$ small. From (26) we see that $\mathcal{B}_j$ is taken so that $k_1^{(j)}(a) \neq 0$, where $\underline{a}$ is as in §7.

We are now able to solve for u in (28) by means of the implicit function theorem. As $D_i(a, 0) = 0$ for every i and as $k_1^{(j)}(a) \neq 0$, (28) defines u as a function of x and v, analytic for x in some area $\mathfrak{S}$ containing $\underline{a}$, and for $|v|$ small.

Now $F_j(x, y, z_j)$ is annulled formally when z_j is replaced by the 'series

(29) $c_{j+1} y^{p_j+1} + \dots$.

This means that (28) is satisfied by the series obtained by dividing the series in (29) by y^{p_j} and then replacing $y^{1/s}$ by v. But
(28) is satisfied by just one series of positive powers of v with
coefficients analytic at a, namely, the expansion of the solution
of (28) found above. Thus the coefficients in (18) are continuable
over $\mathfrak{C}$, and (18) converges for x in $\mathfrak{C}$ and for $|y|$ small but not
zero. From this follow the properties of (18) stated in §8.

14. We have obtained one of the series (4); let it be called P_1.
By division, we find

$$F(x, y, z) = (z - P_1) G(x, y, z)$$

where G is degree n-1 in z, with coefficients of the type of the B
in (2), except that y may enter in fractional powers. If n > 1, we
can treat G as F was treated, to obtain a second series (4). The
fractional powers of y constitute no difficulty. We cannot use irreducibility for G as we did for F in §12. However, if q > 1 for
G, G will have a multiple formal solution, so that the same will
be true for F.

The proof of the existence of the n series (4) is thus completed.

15. Let us suppose now that the A_i in (1) are polynomials in y.
We are interested in the neighborhood of y $= \infty$. Dividing by A_n, we
obtain an equation (2) with each B_i a series of *descending* integral
powers of y. Using a transformation y $= 1/t$, we show the existence
of n series (4) where now the p_i *decrease* as their subscripts increase.

The $a_{ji}(x)$ in (3) are rational combinations of the coefficients
of the A. The function $c_1(x)$ in (4), determined by (6), is algebraic in a certain number of the a_{ji}. Again, c_2, determined by (16),
is algebraic in c_1 and some of the a_{ji}. In this way we find that
every $c(x)$ *in* (4) *is algebraic in the coefficients of the* A(x, y).

Chapter VI

INTEGRATION OF DIFFERENTIAL EQUATIONS BY QUADRATURES

INTEGRABILITY BY QUADRATURES

1. In the formal theory of differential equations, one meets frequently equations whose solutions contain integrals which are not elementary functions. For instance, the equation

$$x \, y' + x \, y = 1$$

has the solution

$$y = e^{-x} \int \frac{e^x \, dx}{x} + c \, e^{-x},$$

involving an integral which is not elementary. It is customary to regard an equation as solved when a solution of the above general type can be found for it. In addition to algebraic functions, expo-- nentials, and logarithms, one admits the operation of integration. When for a differential equation the unknown function can be ex- pressed, naturally with arbitrary constants, by means of the opera- tions just indicated, the equation is said to be *integrable by quadratures*. Also, giving a new meaning to the word *terms*, one de- scribes the equation as integrable in finite terms. A *quadrature*, of course, is simply an integration.

Since $\log f(x)$, for any $f(x)$, is an integral of f'/f, we may dis- pense with the operation of taking a logarithm and limit ourselves to algebraic operations, exponentiations, and integrations.

We are dealing now with the representation of the unknown func- tion in explicit form. In the chapters which follow, we shall con- sider problems of implicit representation.

It is our intention to present the results of an investigation of Liouville on the cases of integrability of a type of Riccati equation and of Bessel's equation. For this we first construct, with the operations indicated above, a category of functions.

FUNCTIONS OF LIOUVILLE

2. The variable x will be called a complete monomial of order 0 and every algebraic function of x will be called a function of order 0.

An m.a.f. is called a complete monomial of order 1 if it is not algebraic and if it is either the exponential of an algebraic function or the *integral* of an algebraic function. As has already been observed, a logarithm of an algebraic function is also an integral of an algebraic function. A branch of a complete monomial is called a monomial.

One continues as in Chapter I. The functions of any order n are fluent functions. Their structure is described as in I, 11, integrals replacing logarithms.

The functions to which orders have just been assigned will be called *functions of Liouville.*

EQUATIONS OF RICCATI AND BESSEL

3. The differential equation

$$y' = P(x) + Q(x) \ y + R(x) \ y^2$$

is known as the equation of Riccati. Daniel Bernouilli studied the special Riccati equation

(1) $$y' + y^2 = x^n$$

with n a constant. He found the equation to be integrable by quadratures when $n = -2$ or when $n = -4p/(1 + 2p)$ with p an integer, positive, negative, or zero. Liouville [7], [9], and [23], showed that these are the only cases of integrability by quadratures. In perfectly definite language, these are the only cases in which (1) has even a single solution which is a function of Liouville.

Equation (1) is intimately related to the equation of Bessel

(2) $$x^2 \ y'' + x \ y' + (x^2 - \nu^2) \ y = 0,$$

with ν constant, which, as will be seen, admits a Liouville function distinct from zero as a solution when and only when 2ν is an odd integer.

A THEOREM OF LIOUVILLE

4. The proofs of the results stated in §3 are based on the following theorem.

THEOREM: *If the equation*

(3) $$y' + y^2 = P(x)$$

with P *algebraic, has a special solution which is a Liouville function, it has a special solution which is algebraic.*

In the proof, there will appear the procedure to which reference was made in Chapter V.

One should have well in mind what it means for a function $y(x)$ to satisfy (3). What is meant is that some pair of elements of y and of P, with the same center, satisfy (3).

When $P = 0$, (3) has the solution $1/x$. In what follows, we assume that P is not identically zero.

Let us suppose that (3) is satisfied by Liouville functions but no algebraic function. Let E denote the totality of Liouville functions satisfying (3). Those functions in E which are of a least order m form a class E_1. Of course $m > 0$. We use for each function in E_1 an expression involving as few m-monomials as possible. Let $y(x)$ be a function in E_1 involving no more m-monomials than appear in any other function in E_1.

Now let θ be one of the m-monomials in y. We write $y = F(\theta, x)$. By (3),

$$(4) \qquad F_\theta(\theta, x)\, \theta' + F_x(\theta, x) + [F(\theta, x)]^2 = P(x).$$

If θ is an exponential, e^v, (4) becomes

$$(5) \qquad F_\theta\, \theta\, v' + F_x + F^2 = P.$$

If θ is the integral of a function v,

$$(6) \qquad F_\theta\, v + F_x + F^2 = P.$$

A relation (5) or (6) must be an identity in θ and x. What we are going to do is to expand the first members of (5) and (6) in descending powers of θ, using the results given in V, 15. The question of the subsistence of the relations (5) and (6) for values of the independent variable θ in the neighborhood of ∞ will be examined later.

The function $F(\theta, x)$, with θ an independent variable, is defined by an equation like (1) of Chapter V. The place of θ is taken in that equation by y. The coefficients in the A are polynomials in the monomials other than θ which enter into F. We have thus

$$(7) \qquad F(\theta, x) = c_1(x)\, \theta^{P_1} + c_2(x)\, \theta^{P_2} + \ldots$$

where the p decrease as their subscripts increase. Each c, by V, 15, is algebraic in the monomials other than θ which appear in F. The first member of (5) becomes

$$(8) \qquad [(c_1' + p_1\, c_1\, v')\, \theta^{P_1} + \ldots] + [c_1^2\, \theta^{2P_1} + \ldots]$$

while that of (6) becomes

(9) $$[c_i^1 \theta^{P_1} + \ldots] + [c_i^2 \theta^{2P_1} + \ldots].$$

As P is not zero, the series obtained by simplifying that one of (8) and (9) which is pertinent must start with a term of exponent 0. It follows that, whether θ is an exponential or an integral, $p_1 \geq 0$. If $p_1 > 0$, there will be, both in (8) and in (9), a term in θ^{2P_1} which cannot cancel. Hence $p_1 = 0$, and we obtain from either (5) or (6),

$$c_i^1 + c_i^2 = P.$$

Thus c_1 is a solution of (3). Now c_1 is either of order less than m or else involves fewer m-monomials than $F(\theta, x)$ does. We have here a contradiction which proves the theorem of Liouville.

5. We have now to justify the use of the series (7). Let $\mathcal{B}$ be an area in the plane of x for which $F(\theta, x)$ has a set of expansions (7), the number of expansions being the degree of the equation, analogous to (1) of V, 1, which defines $F(\theta, x)$. Let $K(\theta, x)$ be, for that equation, the product of the discriminant by the coefficient of the highest power of $F(\theta, x)$. We work at a point $\underline{a}$, lying in $\mathcal{B}$, for which

$$K(\theta(a), a) \neq 0.$$

Now K is a polynomial in θ, and thus the function $K(\theta, a)$ of θ can vanish for at most a finite number of values of θ. Suppose now that we have a curve C, in the space of x and θ, defined by equations

$$x = a, \quad \theta = \varphi(t), \quad 0 \leq t \leq 1,$$

with $\varphi(0) = \theta(a)$, the function $\varphi(t)$ having a large modulus for $t = 1$ and assuming nowhere on $(0, 1)$ any of the values of θ which annul $K(\theta, a)$. Then $F(\theta, x)$ can be continued along C from $t = 0$ to $t = 1$. As $\varphi(1)$ has a large modulus, some expansion (7) represents $F(\theta, x)$ in the neighborhood of the terminal point of C, while the first members of (5) and (6) are represented by (8) and (9). The validity of our method is thus established.

The foregoing method was presented by the author [20] in 1926. In some cases, as in that above, it works with surprising speed. A trace of the idea involved appears to exist in Mordukhai-Boltovskoi's book on integration ([13], p. 197), where, in connection with certain rational combinations of a monomial θ, expansions in descending integral powers of θ are used.

Liouville ([4], p. 65, and [8], p. 442) refers to expansions of algebraic functions in ascending or descending powers. He states that he originally used such series in certain proofs but later eliminated them from his work. The proofs to which he refers are proofs, not of general theorems, but rather of results on special functions, obtained as applications of general theorems. He appears to have in mind expansions of algebraic functions of one variable and, perhaps, proofs like that of II, 13. The diffidence towards infinite series is understandable; it was a diffident decade. All that was to disappear soon under the leadership of Cauchy and Weierstrass. It is hard to guess at the types of expansions of algebraic functions which he used, since the general theory of such expansions dates only from Puiseux's investigation of 1854.

APPLICATION TO EQUATIONS OF RICCATI AND BESSEL

6. We take up now Riccati's equation (1) and Bessel's equation (2). We put $y = x^{-1/2} u$, $x = i z$, and (2) goes over into

(10)
$$\frac{d^2u}{dz^2} = (1 + \frac{p(p + 1)}{z^2}) u$$

where $p = v - 1/2$. Putting $v = u'/u$, we have

(11)
$$\frac{dv}{dz} + v^2 = 1 + \frac{p(p + 1)}{z^2}.$$

We now consider (1). For $n = -2$, the substitution $y = v/x$ renders (1) separable, and the solutions of (1) are seen to be elementary functions. In what follows, we suppose that $n \neq -2$. We put $y = w'/w$. Then (1) gives

(12)
$$w'' = x^n w.$$

Let $n = 2q - 2$. Then $q \neq 0$. We put $z = x^q/q$ and (12) becomes

(13)
$$\frac{d^2w}{dz^2} + \frac{q-1}{q} \frac{1}{z} \frac{dw}{dz} - w = 0.$$

We let $p = (1 - q)/(2q)$ and put $w = z^p u$, whereupon (13) goes over into (10).

7. The question of the satisfaction of (1) by a function of Liouville or of (2) by such a function other than 0 reduces thus to the question of the existence of a Liouville function satisfying (11). By §4, we must seek those values of p for which (11) has an algebraic solution. We shall show that an algebraic solution exists when and only when p is an integer. This will validate the statements of §2.

We show first that every algebraic solution of (11) is rational. An algebraic function of z which is not rational must have at least two critical points. That is, for at least two values of z, the Riemann surface of the function must have branch points. At a branch point, a function has an expansion in which fractional powers are effectively present.

We show first that no algebraic solution v of (11) can have a branch point at ∞. Let v have a branch point at ∞, and an expansion

$$(14) \qquad\qquad v = a_1 z^{p_1} + a_2 z^{p_2} + \dots$$

where the p are decreasing fractions. We suppose that no $\underline{a}$ is zero. It is seen from (11) that $p_1 = 0$. Let p_1 be the highest fractional exponent. Then the highest fractional power in v^2 is found in $2 a_1 a_1 z^{p_1}$. But this cannot be cancelled by any term in v', in which the highest fractional power comes from $p_1 a_1 z^{p_1-1}$. Thus p_1 is not a fraction, and there is no branch point at ∞.

We show now that there is no branch point for any finite value $c \neq 0$ of z. Suppose that v has, at c, an expansion

$$v = a_1 (z - c)^{p_1} + \dots .$$

We shall show first that p_1 is not a fraction, then that no p_i is a fraction.

Suppose that p_1 is fractional. The first term in v' is $p_1 a_1 (z - c)^{p_1-1}$ and has a fractional exponent. The right member of (11) is rational, and hence its development in powers of z − c for any c has only integral exponents. Thus there must be a term in v^2 which balances the first term of v'. Hence for some i and j,

$$p_1 - 1 = p_i + p_j.$$

But $p_i + p_j \geq 2 p_1$. Thus $2 p_1 \leq p_1 - 1$ or $p_1 \leq -1$. Now p_1 cannot be less than −1, else the first term of v^2, which is $a_1^2 (z - c)^{2p_1}$, could not be balanced by any term in v' or in the second member of (11). Thus p_1 is integral.

Suppose now that some p_i with $i > 1$ is fractional and indeed the least fractional exponent. The least fractional exponents in v' and v^2 will be found in

$$p_i a_i (z - c)^{p_i-1}, \qquad 2 a_1 a_i (z - c)^{p_1+p_i}$$

We must therefore have $p_1 = -1$ and $p_i = -2 a_1$. The terms of lowest degree in v' and v^2 are respectively,

$$- a_1 (z - c)^{-2}, \quad a_1^2 (z - c)^{-2}$$

The expansion of the second member of (11) about $z = c$ has no negative powers, since $c \neq 0$. Then

$$- a_1 + a_1^2 = 0$$

so that $a_1 = 1$. Hence $p_1 = -2$, and p_1 is not a fraction.

Thus an irrational algebraic solution of (11) could have a critical point only at $z = 0$. Then every algebraic solution of (11) is rational.

8. We are going to show that, if (11) has a rational solution, p is an integer. Let (11) have a solution

$$v = \frac{P(z)}{Q(z)}$$

with P and Q polynomials with no zero in common. From (11), we see that v has no pole at ∞ and that the poles of v in the finite plane are all simple poles. There is certainly a pole at $z = 0$, and if there is a pole at $c \neq 0$, the term in $1/(z - c)$ has unity for coefficient. Let $v(\infty) = h$ and let the zeros of Q be $c_1, \ldots, c_r$. From (11) we see that $h \neq 0$. Then

$$v = h + \frac{k}{z} + \frac{1}{z - c_1} + \ldots + \frac{1}{z - c_r} .$$

We have for k the equation

(15) $$k^2 - k = p(p + 1),$$

so that $k = p + 1$ or $k = -p$. The development of v about ∞ is

$$v = h + \frac{k + r}{z} + \ldots .$$

Now $k + r$ must be zero or the term $2 h (k + r) z^{-1}$ would be present in v^2 and would not be cancelled by v' or the second member of (11). Thus k is an integer. This means, by (15), that p is an integer.

9. We show finally that when p is an integer, (11) has a rational solution. There is no generality lost in assuming that p is positive. When p is either 0 or -1, (11) is satisfied by $v = \pm 1$. If $p < -1$, we let $p' = -p -1$. Then

$$p(p + 1) = p'(p' + 1).$$

We put $v = w'/w$ in (11) so that

$$w'' = (1 + \frac{p(p \pm 1)}{z^2}) w.$$

Now let $w = e^z z^{-p} u$. Then

(16) $$u'' + 2 (1 - \frac{p}{z}) u' - 2 \frac{pu}{z} = 0.$$

Let us determine the power series

$$u = a_0 + a_1 z + \ldots + a_m z^m + \ldots$$

so as to satisfy (16). We find equations

$$- 2 p a_1 - 2 p a_0 = 0$$

$$a_2 (2 - 4 p) + a_1 (2 - 2 p) = 0$$

. . . .

$$a_m [m(m - 1) - 2 m p] + a_{m-1} (2 m - 2 - 2 p) = 0.$$

We are supposing that p is a positive integer. Our equations give a_m in terms of a_{m-1} unless $m - 1 = 2 p$. But when m is $p + 1$, which is less than $2 p + 1$, we find that $a_m = 0$. Thus, if we take $a_m = 0$ for $m > p + 1$, we get a polynomial u which satisfies (16). The logarithmic derivative of the corresponding w is rational and satisfies (11).

This concludes our treatment of the equations of Riccati and Bessel. The existence of a single solution of (1) of the Liouville type is easily seen to imply that all solutions are of that type. The same holds for (2) if we disregard the solution $y = 0$.

FURTHER STUDIES

10. Mordukhai-Boltovskoi [12] investigated the integrability by quadratures of linear differential equations of any order.

In the following chapter we shall go with Mordukhai-Boltovskoi into the question of integrating algebraic differential equations of the first order in terms of elementary functions.

There is another type of problem on integrability by quadratures. The linear equation of the first order

(17) $$y' + P(x) y = Q(x),$$

where P and Q are *any* functions, can be integrated by two quadratures. This raises the problem of finding classes of equations, *involving arbitrary functions*, which can be solved by finite algorithms, with integration among the permitted operations. Such a question was examined by Maximovich [11] in 1885. As the author has not been able to secure Maximovich's paper or any account of it except those given in an abstract in the *Jahrbuch* and in one in the Paris *Comptes Rendus*, he is unable to make a definite statement in regard to it. It appears that, with certain assumptions, Maximovich shows that the linear equations are, essentially, the only general class of equations of the first order which can be integrated in explicit form by quadratures.

Chapter VII

IMPLICIT AND EXPLICIT ELEMENTARY SOLUTIONS OF DIFFERENTIAL EQUATIONS OF THE FIRST ORDER

IMPLICIT REPRESENTATIONS

1. In the preceding chapters we have discussed the possibility of representing functions in *explicit* form by means of certain operations. In solving differential equations, one is perfectly happy to end up with a relation among the unknown function, the variable, and arbitrary constants. For instance, the equation

$$\frac{dy}{dx} (1 - e^y) = 1$$

has for solution

$$y - e^y = x + c.$$

For no value of c is y an elementary function of x. There arise thus questions on the possibility of solving differential equations in finite *implicit* terms. We shall consider some problems of this type.

ELEMENTARY FUNCTIONS OF TWO VARIABLES

2. For our purposes, we must construct the elementary functions of two variables, x and y. We consider an m.a.f. $u(x, y)$. An element of u is a power series $P(x - x_0, y - y_0)$. By the radius of convergence r of P we shall mean the least upper bound, finite or infinite, of those positive numbers ρ which are such that P converges for $|x - x_0| < \rho$, $|y - y_0| < \rho$. An immediate continuation of P is an element secured by developing P in powers of $x - x_1$, $y - y_1$, where $|x_1 - x_0| < r$, $|y_1 - y_0| < r$.

We shall call u *fluent* if for each element $P(x-x_0, y-y_0)$ of u, for each curve

$$(1) \qquad x = \varphi(\lambda), \quad y = \psi(\lambda) \quad (0 \le \lambda \le 1)$$

where $\varphi(0) = x_0$, $\psi(0) = y_0$, and for each $\varepsilon > 0$, there exists a curve

$$(2) \qquad x = \varphi_1(\lambda), \quad y = \psi_1(\lambda) \quad (0 \le \lambda \le 1)$$

114

with $\varphi_1(0) = x_0$, $\psi_1(0) = y_0$, such that

(3) $\qquad |\varphi_1(\lambda) - \varphi_{/}(\lambda)| < \epsilon, \qquad |\psi_1(\lambda) - \psi(\lambda)| < \epsilon$

for $0 \leq \lambda \leq 1$ and such that P can be continued along (2).

In proving the fluency of functions, we shall use the following fact. *Given a* u(x, y), *not identically zero, with an element* P(x - x_0, y - y_0) *continuable along a curve (1), and given any* $\epsilon > 0$, *there is a curve (2) satisfying (3) such that P is continuable along (2) and that* u *is nowhere zero on (2), except perhaps at* (x_0, y_0). To prove this we suppose, as we may, that (1) consists of short straight segments with extremities at (x_0, y_0), (x_1, y_1), ..., (x_n, y_n), the point (x_n, y_n) being the terminal point of (1). We may suppose, furthermore that u is not zero at any (x_1, y_1) with $i > 0$. If each (x_1, y_1) with $i > 0$ is close enough to (x_{1-1}, y_{1-1}), u will be continuable along (1) by a chain of elements $P(x - x_1, y - y_1)$, each after the first an immediate continuation of its predecessor. The first segment is given by

(4) $\qquad x = x_0 + t(x_1 - x_0), \qquad y = y_0 + t(y_1 - y_0), \qquad 0 \leq t \leq 1.$

If x and y are replaced in $P(x - x_0, y - y_0)$ by their expressions in (4), P becomes a function Q(t) analytic for $|t| \leq 1$ and not zero for $t = 1$. Let $t = 0$ be joined to $t = 1$ by a curve $\tau = \mathcal{E}(t), 0 \leq t \leq 1$ with $|\tau - t|$ very small along the curve, Q being distinct from 0 along the curve, except perhaps at the first point. Then $P(x - x_0, y - y_0)$ is analytic along

(5) $\quad x = x_0 + \mathcal{E}(t) (x_1 - x_0), \qquad y = y_0 + \mathcal{E}(t) (y_1 - y_0) \qquad 0 \leq t \leq 1,$

and will not be zero on (5) except perhaps at the first point. One makes similar replacements for the other segments.

3. The variables x and y will be called complete monomials of order zero. An algebraic function u(x, y) will be called a function of order zero. An algebraic u(x, y) is fluent, for, given any curve (1), we can replace it by a curve (2) close to it along which, except perhaps at the first point, the discriminant of the equation defining u and the coefficient of the highest power of u are distinct from zero. If u is algebraic and nonconstant, e^u and log u are called complete monomials of order 0. Of course e^u is fluent; so also, by §2, is log u. A branch of a complete monomial, analytic in a region in the space of x and y, is called a monomial. Continuing, we obtain the *elementary functions of x and y*. The structure of such a function is described as in I, 11. One uses, in the

description, a branch of a function u of order n, analytic in a region $\mathfrak{U}$. One employs, in the place of $\underline{a}$ of I, 11, a point (a, b) in $\mathfrak{U}$. In (I), one uses r_1 algebraic functions of x and y, analytic for $|x - a| < \rho_1$, $|y - b| < \rho_1$. In $(N + 1)$, one uses an algebraic function of $x, y, x_1^!, \ldots, x_{r_n}^{(n)}$. If a function has the structure just described at (a, b), it is said to be of *regular structure* at (a, b).

IMPLICIT SOLUTIONS OF ALGEBRAIC DIFFERENTIAL EQUATIONS

4. Mordukhai-Boltovskoi [14] investigated differential equations of the form

(6) $$y' = f(x, y)$$

with f algebraic in both variables, inquiring as to whether (1) has a solution $g(x, y) = c$ with c an arbitrary constant and g elementary. He proved the following theorem.

THEOREM: *Let* $y' = f(x, y)$ *with* f *algebraic in* x *and* y *have a solution* $g(x, y) = c$ *with* g *elementary. Then it has a solution*

(7) $$\varphi_0(x, y) + a_1 \log \varphi_1(x, y) + \ldots + a_r \log \varphi_r(x, y) = c$$

with each a_i *a constant and each* φ_i *algebraic.*

The form which the φ can be given is discussed in [14].

To prove this theorem, we start by observing that for $g(x, y) = c$, where g is not a constant, to be a solution of (6), it is necessary and sufficient that

(8) $$\frac{\partial g}{\partial x} + f \frac{\partial g}{\partial y} = 0$$

identically in x and y.

Let us suppose that we have a nonconstant elementary solution g of (8) whose order n is as small as possible and which involves as few as possible, say r, n-monomials. We need only consider values of n which exceed 0. Let θ be one of the n-monomials. We write

$$g(x, y) = h(\theta, x, y).$$

Then

(9) $$h_\theta \theta_x + h_x + f(h_\theta \theta_y + h_y) = 0.$$

5. Now, let θ be a logarithm, log v. Then (8) becomes

(10) $h_\theta v^{-1} v_x + h_x + f(h_\theta v^{-1} v_y + h_y) = 0$

and (10) holds identically in θ, x, y.

Let one of the expansions of $h(\theta, x, y)$ for $\theta = \infty$ be

(11) $\alpha_1 \theta^{p_1} + \alpha_2 \theta^{p_2} + \ldots,$

the α being algebraic in the monomials other than θ in g. We assume that $\alpha_1 \neq 0$. We find from (10) that

(12) $(\dfrac{\partial \alpha_1}{\partial x} + f \dfrac{\partial \alpha_1}{\partial y}) \theta^{p_1} + \ldots = 0$

identically in x, y, θ for $|\theta|$ large. The coefficient of θ^{p_1} in (12) is zero. This means that α_1 is a constant. Otherwise α_1 would be a solution of (8) involving fewer than r-monomials of order n. Let $\alpha_1 = a$. If now $p_1 = 0$, we replace g by $g - a$ and use the expansion $\alpha_2 \theta^{p_2} + \ldots,$ with $p_2 < 0$ and $\alpha_2 \neq 0$. Then α_2 is a constant. We therefore make the legitimate assumption that $p_1 \neq 0$. Let $p_1 \gamma$ represent the coefficient of θ^{p_1-1} in (11). The coefficient of θ^{p_1-1} (12) is

$$(p_1 a v^{-1} v_x + p_1 \gamma_x) + f(p_1 a v^{-1} v_y + p_1 \gamma_y)$$

and that coefficient is zero. Hence $a \theta + \gamma$ is a solution of (8). It is not a constant; otherwise θ would be expressible in the other monomials. Suppose now that γ, whose monomials are among those of g, involves a second logarithm, $\zeta = \log w$, of order n. We develop γ with respect to ζ at $\zeta = \infty$. Then

$$a \theta(x, y) + \gamma = \beta_1 \zeta^{p_1} + \ldots.$$

We see as above that β_1 is a constant b, that we may assume that $p_1 \neq 0$, and that $b \zeta + \delta$, with $p_1 \delta$ the coefficient of θ^{p_1-1}, is a solution of (8). Now if $p_1 - 1$ is not zero, δ will not involve θ, and $b\zeta + \delta$ will be too simple a solution of (8). Thus $p_1 = 1$ and $\delta = a \theta + \rho$ with ρ free of θ and ζ. Then $a \theta + b \zeta + \rho$ is a solution of (8). Continuing, we find that (8) has a solution

(13) $a_1 \theta_1 + \ldots + a_s \theta_s + \tau$

where the θ are n-logarithms and τ involves no n-logarithm.

We shall show that no n-exponential appears in τ. Let there be such an exponential, $\theta = e^v$. Let the expression (13) have an expansion given by (11). We find an identity

(14) $\sum\limits_{i=1}^{\infty} [\frac{\partial \alpha_i}{\partial x} + p_i \alpha_i v_x + f(\frac{\partial \alpha_i}{\partial y} + p_i \alpha_i v_y)] \theta^{p_i} = 0.$

Let γ be the coefficient of θ^0 in (11). We see from (13) that

(15) $\gamma = a_1 \theta_1 + \ldots + a_s \theta_s + \xi$

where ξ is free of $\theta_1, \ldots, \theta_s, \theta$. Then (14) gives

$$\gamma_x + f \gamma_y = 0$$

and we have in γ a solution of (8) with fewer than r monomials of order n. Thus, if g contains an n-logarithm, (8) has a solution of the form (7) with each $\log \varphi_i$ an n-logarithm and with φ_0 of order less than n.

6. Suppose now that g contains no n-logarithm. Let $\theta = e^v$ be an n-exponential in g. We use an expansion (11) for g and find the identity (14). Consider some i for which p_i and α_i are both distinct from zero. We equate to zero the coefficient of θ^{p_i} in (14). We multiply the resulting relation through by θ^{p_i}, where $\theta = e^v$ and integrate. We find that $\alpha_i \theta^{p_i}$ is a solution of (8). It is not a constant; otherwise θ would be algebraic in the other monomials in g. We can see now that there is no other exponential among the n-monomials in g. If there were such a monomial, $\zeta = e^w$, we would obtain from $\alpha_i \theta^{p_i}$ a solution of (8)

(16) $\beta_j(x, y) \zeta^{q_j} \theta^{p_i}$

with β_j free of ζ and θ. We could write (16)

$$\beta_j e^{q_j w + p_i v}$$

and (8) would have a solution involving fewer than r n-monomials. Thu Thus α_i is of order less than n. Also

(17) $\log \alpha_i \theta^{p_i} = p_i v + \log \alpha_i$

is a solution of (8). Then $\log \alpha_i$ must be of order n, and we have in (17) a solution of (8) of the type of the first member of (7).

7. We shall now prove that $n = 1$. The integer r being as in §4, there are expressions of the form

(18) $\varphi_0 + a_1 \log \varphi_1 + \ldots + a_r \log \varphi_r,$

representing functions of order n, which satisfy (8). Each φ_i with $i > 0$ is of order $n - 1$ and its logarithm is of order n; φ_0 is of order less than n. With each expression we associate two numbers, first s, the number of (n-1)-monomials appearing in at least one of φ_0, φ_1, ..., φ_r, then t, the number of such monomials in at least one of φ_1, ..., φ_r. We consider those expressions for which s is a minimum and select from them one, g(x, y), for which t is a minimum.

We might use the power series method, but it would be somewhat cumbersome in the present connection. Instead, letting $\theta = e^v$ be one of the (n-1)-monomials, we write

$$g(x, y) = h(\theta, x, y).$$

It is easy to see that $h(\mu\theta, x, y)$ satisfies (8) for μ constant. Also, because f in (8) is free of μ, $h_\mu(\mu\theta, x, y)$ satisfies (8). Then $\theta h_\theta(\theta, x, y)$ satisfies (8). As it is of order less than n, it is a constant. Then it must be a constant when θ is an independent variable. Thus

$$h(\theta, x, y) = a \log \theta + \gamma(x, y)$$

and, in the usual way, we get an expression for g free of θ. Now, let $\theta = \log v$ be one of the (n-1)-monomials in φ_1, ..., φ_r. Then $h(\theta + \mu, x, y)$ is a solution of (8). Thus $h_\theta(\theta, x, y)$ satisfies (8) and is a constant. Then, for θ independent,

$$h(\theta, x, y) = a \theta + \gamma(x)$$

and we find an expression for g in which the n-logarithms are free of θ. This completes the proof of Mordukhai-Boltovskoi's theorem.

EXPLICIT ELEMENTARY SOLUTIONS

8. Applying what precedes, we shall obtain a theorem of Mordukhai-Boltovskoi on algebraic differential equations with solutions possessing *explicit* elementary expressions.

We consider an equation

(19) $$F(x, y, y') = 0$$

with F a polynomial in x, y, y', irreducible in the field of complex constants. Suppose that (19) admits a solution y(x) which is elementary but not algebraic. Such a y(x) cannot be a singular solution of (19), so that (19) can be converted into the form

(20) $$y' = f(x, y)$$

with f analytic at some point (a, y(a)).

Now let $y = h(\theta, x)$ where θ is one of the highest monomials in y.
We find without difficulty that $h((1 + c) \theta, x)$ or $h(\theta + c, x)$ is
a solution of (20) for $|c|$ small according as θ is an exponential
or a logarithm. In either case, h_c cannot vanish for every x when
$c = 0$. If it did, it would vanish identically in θ and x, and h
would be free of θ. The equations

$$y = h((1 + c) \theta, x), \quad y = h(\theta + c, x)$$

can therefore be solved in the neighborhood of some point $(a, y(a))$.
We obtain a relation

$$g(x, y) = c$$

with g analytic at $(a, y(a))$. Of course, g satisfies (8). We see
that g is algebraic in y and in monomials in x.

9. We consider functions $g(x, y)$ algebraic in y and in monomials
in x, which satisfy (8). We do not ask that g be of regular structure, or even analytic, at some point $(a, y(a))$. Whatever pertains
to the given solution $y(x)$ of (20) will be taken care of by special
considerations.

We assume each g to be expressed in such a way that n, the highest of the orders of the monomials in x which it involves, is as
low as possible. We call n the x-order of g.

We now consider a solution g of (8), as in the foregoing, of as
low an x-order n as is possible. Then n cannot be 0. If it were,
$g(x, y)$ would be algebraic in x and y. Since the singularities of
g satisfy an equation algebraic in x and y, and since $y(x)$ as in
§ 8 is not algebraic, g can be continued to some point $(a, y(a))$.
Then $y(x)$ would reduce g to a constant when substituted for y, and
thus $y(x)$ would be algebraic. Then $n > 0$.

10. We now review the argument of §§ 5 and 6, taking g algebraic
in y and of as low an x-order $n > 0$ as it can be with this condition. We are led to one of the following two cases:

Case A. Equation (8) has a solution

(21) $\varphi_0 + a_1 \log \varphi_1 + \ldots + a_r \log \varphi_r$

with each $\log \varphi_i$ an n-logarithm in x and with φ_0 algebraic in y
and of x-order less than n.

Case B. Equation (8) has a solution

(22) αe^v

with e^v an n-exponential in x and α algebraic in y and of x-order
less than n.

We consider Case A. We review §7 and find that $n = 1$, so that the φ_i are algebraic. As above, we can continue the φ_i to points $(a, y(a))$. What is more, because the given $y(x)$ is not algebraic, $\partial\varphi_0/\partial y$ cannot vanish at every $(a, y(a))$. We equate the expression (21) to a constant c and solve for y in terms of x and c. We find a one-parameter family of solutions of (19),

$$y = G(x, a_1 \log \varphi_1 + \cdots + a_r \log \varphi_r - c)$$

where the φ_i are algebraic functions of x and G is algebraic in its two arguments. For some c, we get the given solution $y(x)$.

We now examine Case B. If $n = 1$, we equate αe^v in (22) to a constant c and solve for y. We find that

$$y = G(c\, e^{-v(x)})$$

where v and G are algebraic. In what follows, we assume that $n > 1$.

In each expression αe^v which satisfies (8), a certain number of $(n-1)$-monomials in x appear in at least one of α and v. We choose an expression in which this number is a minimum, say s. The function

(23) $$v + \log \alpha,$$

which we represent by $g(x, y)$, satisfies (8). Let θ be one of the $(n-1)$-monomials. We write $g(x, y) = h(\theta, x, y)$. Let θ be an exponential, e^w. Then $h(\mu\theta, x, y)$ satisfies (8). So does $\theta\, h_\theta(\theta, x, y)$. The latter solution is algebraic in y and of x-order less than n. Hence it is constant and, indeed, identically in θ. If then

$$v = k(\theta, x), \qquad \alpha = k_1(\theta, x, y),$$

we have

$$g(x, y) = b\, w + k(z_0, x) + \log k_1(z_0, x, y) - b \log z_0$$

with b constant. Then

(24) $$e^g = k_1(z_0, x, y)\, e^{k(z_0,\, x) + bw - b \log z_0}.$$

As $k_1(z_0, x, y)$ is algebraic in y and of x-order less than n, the exponential in (24) must be of x-order n. As there are fewer $(n-1)$-monomials in (24) than appear in α and v, it follows that θ cannot be an exponential.

It is now easy to show that $v + \log \alpha$ can be written in the form

$$\log \psi(x, y) + \varphi_0(x) + c_1 \log \varphi_1(x) + \cdots + c_s \log \varphi_s(x)$$

where each $\log \varphi_i$ is an $(n-1)$-logarithm; φ_0 a function of order less than $n-1$; ψ algebraic in y and of x-order less than $n-1$. Then one proves that $n = 2$, so that ψ and the φ are algebraic.

Summarizing the results of Cases A and B, we have the theorem of Mordukhai-Boltovskoi [15].

THEOREM: *If an algebraic differential equation* $F(x, y, y') = 0$ *has a special solution which is an elementary, but not algebraic, function of x, the equation has either a one-parameter family of solutions of the type*

$$y = G(x, a_1 \log \varphi_1(x) + \ldots + a_r \log \varphi_r(x) - c)$$

with c an arbitrary constant, the a *constants, and G and the* φ *algebraic, or the equation has a one-parameter family of solutions*

$$y = G(x, e^{\varphi_0(x)} + a_1 \log \varphi_1(x) + \ldots + a_r \log \varphi_r(x) - c)$$

*of similar description.**

The question of testing for the elementary integrability of $F = 0$ is considered in [15]. This question connects, on the one hand, with investigations of Painlevé on the singularities of solutions of differential equations and, on the other, with fragmentary researches of Darboux, Poincaré, Painlevé, and D'Autonne on algebraic differential equations with algebraic solutions.

* Note that φ_0 is also algebraic.

Chapter VIII

FURTHER IMPLICIT PROBLEMS

INTEGRALS OF ELEMENTARY FUNCTIONS

1. Let $y(x)$ be an elementary function of x and let w be an integral of y. We propose to examine the circumstances under which there exists a relation

(1) $F(w, x) = 0$

with F elementary in w and x and not identically zero. We shall prove that, when such a relation exists, $w(x)$ is elementary [18]. For instance, if the inverse of an integral of an elementary function is elementary, the integral satisfies an equation (1). Thus, *if the inverse of an integral of an elementary function is elementary, the integral is itself elementary.* For instance, the elliptic function which is the inverse of the integral of the first kind in II, 13, is not elementary. As any two elliptic functions with the same period parallelogram are algebraically related, it follows that no nonconstant elliptic function is elementary.

FORMULATION OF PROBLEM

2. Our first task is to decide on the functiontheoretic assumptions to be made in regard to F in (1). Naturally we would wish to avoid assumptions which might suggest that our investigation lacks finality.

Of course, we shall be working with a branch of $w(x)$, analytic in some area in the plane of x. Pairing values of x and w gives a set of points $\mathfrak{M}$ in the space of w and x. It would be convenient to assume that F is of regular structure, as in VII, 3, at a point $w = b$, $x = a$ of $\mathfrak{M}$ and that (1) is satisfied on $\mathfrak{M}$ for a neighborhood of (b, a). Here a critic might object. For instance, the equation

(2) $(w - x)^{1/2} = 0$

serves well to determine w as equal to x, and still the first member of (2) has a singularity wherever $w = x$. Should we not therefore admit equations (1) in which F has singularities on $\mathfrak{M}$? We are going to show that an equation (1) of greatest conceivable generality can be reduced to one in which F is of regular structure at a point (b, a) on $\mathfrak{M}$.

123

3. When, in what follows, we say that F(w, x) assumes the value zero at a point (w', x'), we shall mean that there exists an element of F with center at (w', x') which assumes there the value zero. There may be other elements of F with centers at (w', x') which are not zero at (w', x'). When we say that F has a singularity at (w', x'), we shall mean that there exists a sequence of elements of F, each an immediate continuation of its predecessor (as VII, 2), whose centers approach (w', x') and whose radii of convergence approach zero.

In specifying the nature of F, one will certainly assume that, (w', x') lying on $\mathfrak{P}$, F either assumes the value zero at (w', x') or has a singularity there. One will feel at ease if one is free to assume that F is zero at some points of $\mathfrak{P}$ and has singularities at the remaining points. One will feel the responsibility of explaining what it means for F to be zero at a singular point. Actually this last matter is unessential; the mere existence of the singularities will permit us to show that w(x) is elementary.

In our discussion, it will be convenient to employ an uncountable set of points of $\mathfrak{P}$ rather than all of $\mathfrak{P}$. We thus assume that there is an elementary function of w and x which is not identically zero and which satisfies at least one of the following two conditions:

(a) *It is zero at an uncountable set of points of* $\mathfrak{P}$.

(b) *It has singularities at an uncountable set of points of* $\mathfrak{P}$.
We shall show that there is an elementary function, not identically zero, of regular structure at a point (b, a) on $\mathfrak{P}$, which vanishes on $\mathfrak{P}$ in a neighborhood of (b, a).

4. From among all functions which satisfy (a) or (b), we select one of least order. We denote the function selected by F(w, x) and its order by n.

Suppose first that F is algebraic. Let F be defined by

(3) $\alpha_0(w, x) u^P + \ldots + \alpha_p(w, x) = 0$.

Wherever F is zero, $\alpha_p = 0$. Where F has a singularity, $\alpha_0 D$ with D the discriminant of (3) is zero. Thus $\alpha_0 \alpha_p D$ vanishes at an uncountable set of points of $\mathfrak{P}$. If we replace w in $\alpha_0 \alpha_p D$ by w(x), we get a function of x which clearly has to vanish identically in x. Thus $\alpha_0 \alpha_p D$ is a function such as we are seeking.

5. Let us suppose now that n > 0 and that F satisfies (b). We shall show that F can be replaced by a function of order n for which (a) holds.

F is given to us with regular structure at some point (b, a), not

necessarily on $\mathfrak{N}$. At (b, a), F has an element P_0. Let (w', x') be
one of the points on $\mathfrak{N}$ at which F has a singularity. Then there is
a curve C, joining (b, a) to (w', x') along which P_0 can be continued to any point preceding (w', x'), but not to (w', x') itself.
We form a chain of elements

$$P_0, P_1, \ldots, P_r, \ldots$$

with P_{i+1} an immediate continuation of P_i, the center (w_i, x_i) of
P_i approaching (w', x') as i increases. Then the radius of convergence of P_i approaches zero as i increases. Taking advantage of
fluency, we join (b, a) to a point $(\overline{w}_1, \overline{x}_1)$ close to (w_1, x_1) by a
curve following C closely, the joining curve being such that each
complete monomial which appears in the structure of F can be con-
tinued along it. We take similarly a curve joining $(\overline{w}_1, \overline{x}_1)$ to a
point $(\overline{w}_2, \overline{x}_2)$ close to (w_2, x_2). We continue in this manner, seeing to it that $(\overline{w}_i, \overline{x}_i)$ approaches (w', x') as i increases. We form
in this way a curve C', joining (b, a) to (w', x'), along which F
and each complete monomial can be continued to any point which pre-
cedes (w', x'). At $(\overline{w}_i, \overline{x}_i)$, the continuation of P_0 along C' will
be an element Q_i. We suppose each $(\overline{w}_i, \overline{x}_i)$ to be so close to
(w_i, x_i) that Q_i is an immediate continuation of P_i. Then the radius of convergence of Q_i tends toward zero as i increases.

Let the equation which determines F at (b, a) be (𝕽) with each α
a polynomial in monomials. Two things may happen. Either one of the
complete monomials in the structure of F has a singularity at
(w', x') revealed by C', or else α_0 D, where D is the discriminant
of (𝕽), assumes the value zero at (w', x'). We are going to show
that for an uncountable set of points (w', x') on $\mathfrak{N}$, α_0 D is zero.
This will bring us back to the situation (a), because we have taken
(𝕽) so that D does not vanish identically.

6. Suppose it is not true that α_0 D is zero at an uncountable set
of points of $\mathfrak{N}$. Then at least one of the complete monomials must
have singularities at an uncountable set of points of $\mathfrak{N}$. Let θ be
such a monomial. Let first θ = e^v with v of order less than n. We
use (b, a) as above and a curve C joining it to a (w', x') on $\mathfrak{N}$ at
which θ has a singularity relative to C. We modify C slightly, as
in § 5, so that v can be continued along it to any point which precedes (w', x'). We see that v has a singularity at (w', x'). This
contradicts the assumption that no function of order less than n
satisfies one of the conditions (a), (b). Suppose now that θ = log v.
Then v can be continued along C (modified if necessary), to any

point preceding (w', x'). If v does not have a singularity at (w', x'), it must be that v is zero there; otherwise θ could not have a singularity at (w', x') relative to C. We have the contradiction that v satisfies one of (a), (b). Thus $\alpha_0 D$ satisfies (a).

7. There is a function of order n which satisfies (a). We let F be such a function. We are going to show that F has a branch which is analytic at a point on $\mathfrak{R}$ and equal to zero on $\mathfrak{R}$ for a neighborhood of that point.

Let us consider the uncountable subset of points of $\mathfrak{R}$ at which F is zero. At each point F has an element, with center at the point, which is zero at its center. There must exist a positive ε such that, for some uncountable subset E of these points, the radii of convergence of the elements of F exceed ε. If no such ε existed, the elements at the centers of which F vanishes could be denumerated.

As is well known and easy to see, there exists some point (d, c), every neighborhood of which contains an uncountable subset of points of E. Obviously we may assume that (d, c) is on $\mathfrak{R}$ and that $w(x)$ is analytic for $|x - c| < \varepsilon$. Consider the neighborhood of (d, c) given by

$$|w - d| < \frac{\varepsilon}{3}, \qquad |x - c| < \frac{\varepsilon}{3}$$

and the uncountable subset E' of E which lies in this neighborhood.

We now form for (d, c) the immediate continuations of the elements of F with centers at the points of E'. The radius of convergence of each of these elements exceeds $2\varepsilon/3$, and each continuation equals zero at the center of the element from which it is obtained.

An infinite number of these immediate continuations must be identical. Otherwise there would be an uncountable set of distinct elements of F with centers at (d, c) and we would have a contradiction of the theorem of Poincaré and Volterra which states that an analytic function can have at most a countable set of elements with the same center.*

It follows that one of the elements with center at (d, c) vanishes for an infinite number of points of E'. Substituting for w

---

* This theorem is an immediate consequence of the fact that if P_0 is an element of F, and P a continuation of P_0, we can form a chain P_0, P_1, ..., $P_n = P$ with P_{i+1} an immediate continuation of P_i, the elements P_1, ..., P_{n+1} having centers at points (w, x) at which, for w and x, the real part and the coefficient of i are rational.

in this element the function $w(x)$, we secure a function of x analytic at c which vanishes for an infinite number of points close to c and is therefore zero. This is the situation described at the beginning of the present section.

8. We shall now show that if F is suitably selected, there is a point on $\mathfrak{N}$, close to (d, c), at which F is of regular structure. We consider the equation (3) which defines F. We wish to show that we can replace F by α_p in (3). This will permit us to assume that F is a polynomial in monomials and will simplify the present discussion. Equation (3) holds at a point (b, a), not necessarily on $\mathfrak{N}$. Let (w', x') be any point on $\mathfrak{N}$ close to (d, c). There is a curve C, joining (b, a) to (w', x'), along which F is continued into the branch analytic at (d, c) secured above. We suppose C to be such that every complete monomial in the structure of F can be continued along it to any point preceding (w', x'). If there is a monomial which cannot be continued to (w', x'), that monomial has a singularity at (w', x'). As we saw above, this can happen only for a countable set of points (w', x'). We may therefore suppose, moving (d, c) if necessary, that each coefficient in (3) can be continued, together with F, from (b, a) to (d, c). Thus the continuation of α_p obtained at (d, c) vanishes on $\mathfrak{N}$ for a neighborhood of (d, c). We replace F by the m.a.f. of which α_p is a branch.

We now need only one detail to secure a point (d, c) on $\mathfrak{N}$ at which F is of regular structure. We use a curve C joining (b, a) to (d, c) along which every monomial can be continued. Every monomial in F has one of the forms e^v or $\log v$. As we saw in §6, we may suppose (d, c) and C to be such that the v for each monomial is continuable along C to (d, c). We consider the equations (3) for the various v. If $\alpha_0 D$ in each such equation is not zero at (d, c), F has regular structure at (d, c). Now, by the minimal nature of n, each $\alpha_0 D$ can vanish only for a countable set of points of $\mathfrak{N}$.

We may thus assume that $F(w, x)$ in (1) has regular structure at (d, c). In what follows, we write (b, a) for (d, c).

9. In discussing (1), we shall have to refer to those *monomials in the structure of F which involve w*. A complete monomial will be said to involve w if its partial derivative with respect to w is not identically zero. The structural scheme for F contains such a scheme for every nonomial used in building F. We must arrange matters in such a way that if a monomial θ does not involve w, its structural scheme uses no monomials involving w. The reason is that we do not wish the differentiation of θ to introduce monomials involving w.

First, it is easy to see by induction that a monomial of any order n free of w is an elementary function of x in the sense of Chapter I; that is, it can be constructed without the use of the variable w. It will, in fact, be a monomial of order n as a function of x alone.

Now let the monomials free of w in the structure of F be $\theta_1, \ldots, \theta_p$. For each θ_i we secure a structural scheme as in I, 11, using a single point a_1 close to $\underline{a}$ as in (b, a). Shifting $\underline{a}$ if necessary, we assume that $a_1 = a$. We adjoin to the structure of F all monomials just secured for $\theta_1, \ldots, \theta_p$.

In the formal work below, most of what it is essential to bear in mind relative to F(w, x) is as follows. We have a finite number of monomials. F is algebraic in a certain subset of them. Some of them are free of w; those have derivatives which are algebraic in monomials free of w.

COMPLETION OF PROOF

10. We suppose given an elementary F(w, x), not identically zero, of regular structure at (b, a) on $\mathfrak{M}$, which vanishes on $\mathfrak{M}$ in the neighborhood of (b, a). As we know, we may take F as a polynomial in monomials; we shall not insist on this.

We consider the class A of all such functions; (b, a) may be different for different functions. We suppose that each function has a structure in which the maximum, r, of the orders of those monomials which involve w is as low as possible. We assume further that s, the number of r-monomials involving w, is as small as possible.

We consider those functions in A for which r has a least value, say r_0. From this subset we choose one, F, for which s has a minimum value s_0. Let θ be one of the r_0-monomials in F which involve w. We write

(4) $F(w, x) = G(\theta, w, x).$

The successive partial derivatives of G with respect to θ cannot all vanish at (b, a). If they did, we could replace θ in (4) by $\theta(b, a)$ and s_0 would not be a minimum. Suppose then that the first j derivatives in θ vanish on $\mathfrak{M}$ for a neighborhood of (b, a), but that the derivative of order j + 1 is distinct from zero at some point close to (b, a). We assume, as we may, that this point is (b, a) itself. We designate the jth derivative by H and write, for the w(x) of §1,

(5) $$H[\theta(w, x), w, x] = 0.$$

For the neighborhood of (b, a), we may solve for θ in (5) and write

(6) $$\theta = f(w, x).$$

If $r_0 = 0$, θ is w and (6) states that $w(x)$ is an elementary function of x. In what follows, we suppose that $r_0 > 0$.

Suppose that $\theta = e^v$ with v of order $r_0 - 1$. For w and x independent variables close to (b, a) we write (6) in the form

(7) $$v(w, x) = \log f(w, x).$$

Now, letting w in (7) be $w(x)$, we find, differentiating with respect of x,

(8) $$v_w(w, x)\, y(x) + v_x(w, x) = \frac{f_w(w, x)\, y(x) + f_x(w, x)}{f(w, x)}.$$

In (8) $y(x)$ is the function of which $w(x)$ is an integral. We suppose, as we may, that y has regular structure at $\underline{a}$.

Now (8) must be an identity in w and x for a neighborhood of (b, a). Otherwise it would give a function in A with $r \leqq r_0$ and with $s < s_0$ if $r = r_0$.

The two functions

(9) $$v(w + \mu, x), \qquad \log f(w + \mu, x)$$

are analytic in w, x, μ for $w = b$, $x = a$, $\mu = 0$.

Let us imagine that in (9) we replace w by $w(x)$. The two functions of (9) become functions of x whose derivatives are given by the two members of (8) with w replaced by $w(x) + \mu$. We therefore have

(10) $$v[w(x) + \mu, x] = \log f[w(x) + \mu, x] + \beta(\mu)$$

with β analytic at $\mu = 0$. Then

(11) $$v_w(w, x) = \frac{f_w(w, x)}{f(w, x)} + c$$

for a neighborhood of (b, a) on $\mathfrak{R}$, and therefore identically in w and x. Integrating, we have in w and x

(12) $$v(w, x) = \log f(w, x) + cw + \gamma(x).$$

To determine $\gamma(x)$ we consider a (b', a') close to (b, a) for which (7) does not hold. Fon $w = b'$, (12) gives

(13) $\gamma(x) = v(b', x) - \log f(b', x) - c b'$.

By (7) and (12) we have for $w = w(x)$,

(14) $c w = - \gamma(x)$.

Now c is not zero. If it were, (13) would imply that

$$v(b', x) - \log f(b', x)$$

vanishes for every x, whereas it is not zero for $x = a'$.

By (14), $w(x)$ is elementary.

If $\theta = \log v$, (6) becomes

$$f(w, x) = \log v(w, x)$$

and we find again that $w(x)$ is elementary.

We might have used, in what precedes, a different classification
of functions of w and x, letting w, and every elementary function
of x alone, be monomials of order zero. This would have spared us
the considerations relative to monomials involving w. Having already classified our functions in Chapter VII, we found it easier
to proceed as above. A classification of the new type will be used
below in connection with Bessel's equation.

In the introduction to my paper, [18], I mentioned the problem of
determining whether the integral of an elementary function is elementary if it is one of a set of functions which satisfy a set of
elementary equations. I stated that the formal elements of the
proof for one function could be carried over to answer this question affirmatively. I wish to withdraw this statement. I do not
have the details now, if indeed I ever did. In particular, the
question as to whether the integral of an elementary function may
be represented parametrically by elementary expressions may be of
interest.

LINEAR DIFFERENTIAL EQUATIONS OF THE SECOND ORDER

11. In Chapter VI, we determined when a solution of Bessel's equation is a function of Liouville. We shall now examine the question
of implicit representations.

We shall construct Liouville functions of w and x, letting w and
every Liouville function of x be complete monomials of order 0. A
function of order 0 is an algebraic combination of complete monomials of order 0. An m.a.f. $u(w, x)$ is a complete monomial of order 1
if it is not a function of order 0 and if either

(a) $u = e^v$ with v of order 0

or

·(b) u_w and u_x are both functions of order 0.

The construction continues in the usual way.

12. The differential equation

$$y'' + 2 P(x) y' + Q(x) y = 0$$

goes over, under the transformation $y = e^{-\int P\,dx}$, into an equation

(15) $$w'' = \varphi(x)\, w.$$

Accordingly, we deal with an equation (15). We prove the theorem.*

THEOREM: *If w, a solution of the equation $w'' = \varphi(x)\, w$, in which $\varphi(x)$ is a Liouville function, satisfies an equation $F(w, x) = 0$, where F is a Liouville function not identically zero, then w is a Liouville function of x.*

We proceed formally; one has seen how rigor is secured.

Differentiating $F = 0$ with respect to x, we find

(16) $$F_w\, w' + F_x = 0.$$

The first member of (16) is a function algebraic in w' and in Liouville monomials in w and x which vanishes when w is the given solution of (15). Of course F is also such a function, but (16) gives a better example.

With every expression for every such function we associate two numbers, first r, the maximum of the orders of the monomials in w and x, and secondly s, the number of r-monomials. We take those expressions for which r is a minimum and select from them an $f(w', w, x)$ for which s is a minimum. We write, for the given solution of (15),

(17) $$f(w', w, x) = 0$$

and denote the two integers for the first member of (17) by r and s.

13. We are going to prove that $r = 0$. Let $r > 0$ and let θ be one of the r-monomials in f. Solving (17) for θ; we find

(18) $$\theta - g(w', w, x) = 0$$

where g is algebraic in w' and monomials and has $s - 1$ monomials of order r.

If $\theta = e^v$ with v of order $r - 1$, we write

---

* Ritt [21]. I cannot agree with a remark of Mordukhai-Boltovskoi [13] to the effect that the question now being considered is equivalent to one previously settled by him. The two questions seem quite distinct and not reducible to each other in any obvious way.

(19) $v - \log g(w', w, x) = 0.$

We let the first member of

(20) $G(w', w, x) = 0$

represent the first member of (18), or that of (19), according as
θ is an integral or an exponential.

Differentiating (20) and using (15), we have

(21) $\varphi(x) \, w \, G_{w'} + G_w \, w' + G_x = 0$

which holds identically in w', w, x.

Let μ be any constant. The derivative with respect to x of

$$G(\mu \, w', \, \mu \, w, \, x)$$

will be the first member of (21) with w' and w replaced by $\mu \, w'$
and $\mu \, w$. Thus

$$G(\mu \, w', \, \mu \, w, \, x) = \beta(\mu).$$

We differentiate with respect to μ and put $\mu = 1$. Then

(22) $w' \, G_{w'} + w \, G_w = c$

identically in w', w, x. A particular solution of (22) is $c \log w$,
and w'/w is a solution of the equation obtained from (22) by putting $c = 0$. Then

(23) $G = c \log w + H(\frac{w'}{w}, x).$

For the given solution of (15), we have

(24) $c \log w + H(\frac{w'}{w}, x) = 0.$

We put $\mu = w'/w$ and differentiate with respect to x. Then

(25) $c \, u + [\varphi(x) - u^2] \, H_u + H_x = 0.$

14. Let θ be an integral. As, in this case, G is algebraic in w',
H must be algebraic in u and in Liouville monomials in x. Then (25)
is an identity in u and x. From (23), we see that $c \neq 0$. We shall
show that no H which is algebraic in u can satisfy (25), thus making untenable. in the case in which θ is an integral, the assumption that $r > 0$. Let one of the developments of H about $u = \infty$ be

(26) $\alpha_1(x) \, u^{P_1} + \alpha_2(x) \, u^{P_2} + \dots$

with no α zero. Let $p_1 \neq 0$. The expansion of $(\varphi - u^2)$ H_u begins with an exponent $p_1 + 1$, whereas in H_x the first exponent does not exceed p_1. Hence the first term of $(\varphi - u^2)$ H_u must balance with $c u$, an impossibility when $p_1 \neq 0$. If $p_1 = 0$, $(\varphi - u^2)$ H_u has $p_2 + 1$ for first exponent, and H_x begins with a zero or negative power. As $c \neq 0$, we have to balance $c u$ in (25). This is impossible because $p_2 + 1 < 1$. Thus no algebraic H satisfies (25).

15. Let θ be an exponential. Then e^G is algebraic in w'. Then, if K represents e^H with H as in (23), K is algebraic in w'/w. We find from (25),

$$(27) \qquad\qquad c u K + (\varphi - u^2) K_u + K_x = 0.$$

We consider an expansion (26) for K, and substitute it into (27). The first equation obtained is $c - p_1 = 0$. By (24) we have, for the given solution of (15),

$$(28) \qquad\qquad w^{p_1} K(\tfrac{w'}{w}, x) = 1.$$

As (28) is algebraic in w' and w, we have a final contradiction of the assumption that $r > 0$.

16. We consider (17) with f algebraic in w' and w. If (17) does not involve w', it determines $w(x)$ as a Liouville function of x. If (17) does involve w', we find

$$(29) \qquad\qquad w' = H(w, x)$$

with H algebraic in w and Liouville monomials in x.

Differentiating (29), we have

$$(30) \qquad\qquad \varphi(x)\, w = H_w\, H + H_x.$$

If (30) is not an identity in w and x, it determines w as a Liouville function. Let (30) be an identity. We consider a development of H about $w = \infty$,

$$H = \alpha_1(x)\, w^{p_1} + \dots.$$

The α are functions of Liouville. When we substitute into (30) we find that $p_1 = 1$ and that

$$\alpha_1' + \alpha_1^2 = \varphi.$$

If we let u be the exponential of the integral of α_1, we find that $u'' = \varphi(x)\, u$. Thus u, a nonzero Liouville function, satisfies (15).

The knowledge of one nonzero solution of (15) permits us to reduce (15) to a linear equation of the first order and to get its complete solution, with two constants, by quadratures. This completes the proof of our theorem.

REFERENCES*

[1] Abel, N. H. *Oeuvres complètes.* See also Serret, *Cours d'algèbre supérieure*, Vol. II. Paris, 1928.

[2] Koenigsberger, L. "Bemerkungen zu Liouville's Classificirung der Transcendenten," *Mathematische Annalen*, XXVIII (1887), 482–92.

[3] Liouville, J. "Sur la détermination des intégrales dont la valeur est algébrique," *Journal de l'école polytechnique*, XIV (1833), Section 23, 124–93.

[4] ————"Mémoire sur les transcendantes elliptiques de première et de seconde espèce, considerées comme fonctions de leur amplitude," *Journal de l'école polytechnique*, XIV (1833), Section 24, 57–83.

[5] ————"Mémoire sur l'intégration d'une classe de fonctions transcendantes," *Journal für die reine und angewandte Mathematik*, XIII (1835), 93–118.

[6] ————"Mémoire sur la classification des transcendantes et sur l'impossibilité d'exprimer les racines de certaines équations en fonction finie explicite des coefficients," *Journal de mathématiques, pures et appliquées*, II (1837), 56–104; III, 523–46.

[7] ————"Mémoire sur l'intégration d'une classe d'équations différentielles du second ordre en quantités finies explicites," *Journal de mathématiques, pures et appliquées*, IV (1839), 423–56.

[8] ————"Mémoire sur les transcendantes elliptiques de première et de seconde espèce, considerées comme fonction de leur module," *Journal de mathématiques, pures et appliquées*, V (1840), 34–36, 441–64.

[9] ————"Remarques nouvelles sur l'équation de Riccati," *Journal de mathématiques, pures et appliquées*, VI (1841), 1–13.

[10] Lorch, E. R. "Elementary Transformations," *Annals of Mathematics*, XXXIII (1932), 214–28.

[11] Maximovich, W. P. "Determination of the General Equation of the First Order Which Can Be Integrated in Finite Terms."

---

* This is not a complete list of publications on integration in finite terms.

135

Kasan, 1885. (In Russian). See also *Comptes Rendus de l'Académie des Sciences de Paris*, CI (1885), 809-11.

[12] Mordukhai-Boltovskoi, D. *On the Integration in Finite Terms of Linear Differential Equations*. Warsaw, 1910. (In Russian)

[13] ―――― *On the Integration of Transcendental Functions*. Warsaw, 1913. (In Russian)

[14] ―――― "Researches on the Integration in Finite Terms of Differential Equations of the First Order," *Communications de la société mathématique de Kharkov*, X (1906-1909), 34-64, 231-69. (In Russian)

[15] ―――― "Sur la résolution des équations différentielles du premier ordre en forme finie," *Rendiconti del circolo matematico di Palermo*, LXI (1937), 49-72.

[16] Ostrowski, A. "Sur les relations algébriques entre les intégrales indéfinies," *Acta Mathematica*, LXXVIII (1946), 315-18.

[17] ―――― "Sur l'intégrabilité élémentaire de quelques classes d'expressions," *Commentari Mathematici Helvetici*, XVIII (1946), 283-308.

[18] Ritt, J. F. "On the Integrals of Elementary Functions," *Transactions of the American Mathematical Society*, XXV (1923), 211-22.

[19] ―――― "Elementary Functions and Their Inverses," *Transactions of the American Mathematical Society*, XXVII (1925), 68-90.

[20] ―――― "Simplification de la méthode de Liouville dans la théorie des fonctions élémentaires," *Comptes Rendus de l'Académie des Sciences de Paris*, CLXXXIII (1926), 331-32.

[21] ―――― "On the Integration in Finite Terms of Linear Differential Equations of the Second Order," *Bulletin of the American Mathematical Society*, XXXIII (1927), 51-57.

[22] ―――― "Algebraic Combinations of Exponentials," *Transactions of the American Mathematical Society*, XXXI (1929), 654-79.

[23] Watson, G. N. *Bessel Functions*. Cambridge, 1944, pp. 111-23.

Comments on J.F. Ritt's Book
Integration in Finite Terms

Askold Khovanskii[1]

Contents

1 Preface

I saw J.F. Ritt's book [Rit48] for the first time in 1969 when I was an undergraduate student. I had just started to work on topological obstructions to the representability of algebraic functions by radicals and on an algebraic version of Hilbert's 13th problem on the representability of algebraic functions of several complex variables by composition of algebraic functions of fewer number of variables. My beloved supervisor Vladimir Igorevich Arnold was very interested in these questions.

Ritt's approach, which uses the theory of complex analytic functions and geometry, was very different from a formal algebraic approach. I was very intrigued and I started to read the book, trying to get a feeling for the subject and avoiding all the details on the first reading.

---

[1] This work was partially supported by the Canadian Grant No. 156833-17.

© Springer Nature Switzerland AG 2022
C. G. Raab und M. F. Singer (Hrsg.), *Integration in Finite Terms: Fundamental Sources*, Texts & Monographs in Symbolic Computation,
https://doi.org/10.1007/978-3-030-98767-1_4

My first impression was that the book was brilliant and the presented theory was ingenious. Simultaneously with the reading I obtained the very first results of topological Galois theory. Since then I have spent a few years developing it. I had hoped to return back to the book later, but I never made it (life is life!).

Even a brief reading turned out to be very useful. It helped me to formalize the definition of the Liouvillian classes of functions and the definition of the functional differential fields and their extensions. Later this experience helped me to find an appropriate definition of the class of Pfaffian functions playing a crucial role in the transcendental generalization of real algebraic geometry developed in the book [Kho91].

That is why I was really happy when Michael Singer invited me to write comments for a reprint of the book. I started to read it again after an almost half-century break. I have to confess that it was hard for me to follow all the details. The reason is that Ritt uses an old mathematical language (the book was written about seventy years ago). Nevertheless I still think that the book is brilliant and Liouville's and Ritt's ideas are ingenious.

In section 2 we present modern definitions of Liouvillian classes of functions and modern proofs of Liouville's First Theorem and of Liouville's Second Theorem. I hope that this modern presentation will help readers to better understand the subject and Ritt's book.

In section 3 we present an outline of topological Galois theory, which provides an alternative approach to the problem of solvability of equations in finite terms. We use the definition of classes of functions by the list of basic functions and the list of admissible operations presented in section 2.2.

A few words about our proof of Liouville's First Theorem. All the main ideas of the proof are presented in Ritt's book. I tried to clarify what is hidden behind the integration used by Liouville. I think that there are two statements which are not mentioned explicitly: 1) a closed 1-form with elementary integral whose possible form was found by Liouville is locally invariant under the Galois group action, assuming that the Galois group is connected; 2) A class of closed 1-forms locally invariant under a connected Lie group action can be described explicitly. In fact all the arguments needed to prove the first statement are presented in the book. Liouville used an explicit integration for the description of closed 1-forms locally invariant under a natural action of the additive and the multiplicative groups of complex numbers.

A few words about our proof of Liouville's Second Theorem. In fact we prove its generalization, applicable to linear homogeneous differential equations of any order. Ritt's book essentially contains all the results needed for our version of the proof. Only some statements are missing there (but all arguments needed for their proofs are presented in some form).

Acknowledgement. I would like to thank Michael Singer, who invited me to write comments for a new edition of J.F. Ritt's classic book, for his constant support. I also am grateful to Fedor Kogan, who edited my English, and to my wife Tanya Belokrinitskaya, who helped me to connect a few papers into one text.

2 Solvability of Equations in Finite Terms

2.1 Introduction

Let K be a subfield of the field of meromorphic functions on a connected domain U of the complex line, closed under differentiation (i.e. if $f \in K$ then $f' \in K$). Such a field K with the operation of differentiation $f \to f'$ provides an example of a *functional differential field*.

Liouville's First Theorem suggests conditions on a function f from a functional differential field K which are necessary and sufficient for the representability of an indefinite integral of f in terms of *generalized elementary functions over K*.

Liouville's Second Theorem suggests conditions on a second-order homogeneous linear differential equation whose coefficients belong to a functional differential field K which are necessary and sufficient for its solvability by *generalized quadratures over K*.

Liouville's theory can be generalized to *an abstract differential field K*, whose elements are not necessarily meromorphic functions (see [Kho14], [dPS03]). Abstract algebraic results are not directly applicable to integrals of elementary functions or to solutions of linear differential equations which could be multivalued, could have singularities, and so on. For their applications some extra arguments are needed. Such arguments are presented in Chapter I of Ritt's book.

The contents of this section are as follows.

In section 2.2 we define functional differential fields, generalized elementary functions and generalized quadratures over such fields. The material of section 2.2.1 was inspired by Chapter I of the book [Rit48].

In section 2.3 we prove Liouville's First Theorem, making use of algebraic group actions. Our proof can be considered as a modernization of Liouville's proof presented in Chapter II and Chapter III of the book [Rit48].

In section 2.4 we prove Liouville's Second Theorem and its generalization for homogeneous linear differential equations of any order. Our proof uses a slightly modernized version of the Liouville–Ritt arguments presented in Chapters V and VI of Ritt's book.

2.2 Generalized Elementary Functions and Generalized Quadratures

2.2.1 Introduction

The results presented in this section are inspired by the material of Chapter I of Ritt's book [Rit48]. We discuss here definitions and general statements related to functional and abstract differential fields and classes of their extensions, including generalized elementary extensions and extensions by generalized quadratures. We mainly follow the presentation in the book [Kho14].

The natural definitions of a generalized elementary function and of a function representable by generalized quadratures over K (see definitions 2.6, 2.7, 2.8, and

2.9 below) are hard to deal with. In particular they make use of a non-algebraic operation of composition of functions. Algebraic definitions (see definitions 2.2 and 2.3 below) use solutions of the simplest differential equations instead of composition of functions. We explain how the natural definitions can be reduced to the algebraic ones.

2.2.2 Differential fields and their extensions

Let us start with some purely algebraic definitions.

2.2.2.1 Abstract differential fields

A field F is said to be a *differential field* if an additive map $a \to a'$ is fixed that satisfies the Leibniz rule $(ab)' = a'b + ab'$. The element a' is called the *derivative* of a. An element $y \in F$ is called *a constant* if $y' = 0$. The set of all constants in F form the *field of constants*. We add to the definition of a differential field an extra condition that *the field of constants is the field of complex numbers* (for our purpose it is enough to consider fields satisfying this condition). An element $y \in F$ is said to be: an *exponential* of a if $y' = a'y$; an *exponential of integral* of a if $y' = ay$; a *logarithm* of a if $y' = a'/a$; an *integral* of a if $y' = a$. In each of these cases, y is defined only up to an additive or a multiplicative complex constant.

Let $K \subset F$ be a differential subfield of F. An element y is said to be an *integral over K* if $y' = a \in K$. An *exponential of integral over K*, a *logarithm over K*, and an *integral over K* are defined similarly.

Suppose that a differential field K and a set M lie in some differential field F. *The adjunction* of the set M to the differential field K is the minimal differential field $K\langle M \rangle$ containing both the field K and the set M. We will refer to the transition from K to $K\langle M \rangle$ as *adjoining* the set M to the field K.

2.2.2.2 Generalized elementary extensions

Let $F \supset K$ be an extension of a differential field K.

Definition 2.1. A differential field F is said to be a *generalized elementary extension*[2] of a differential field K if $K \subset F$ and there exists a chain of differential fields $K = F_0 \subset \cdots \subseteq F_n \supset F$ such that $F_{i+1} = F_i \langle y_i \rangle$ for every $i = 0, \ldots, n-1$, where y_i is an exponential, a logarithm, or an algebraic element over F_i.

An element $a \in F$ is a *generalized elementary element* over K, $K \subset F$, if it is contained in a certain generalized elementary extension of the field K.

The following lemma is obvious.

Lemma 2.1. *An extension $K \subset F$ is a generalized elementary extension if and only if there exists a chain of differential fields $K = F_0 \subseteq \cdots \subseteq F_n \supset F$ such that for every $i = 0, \ldots, n-1$, either F_{i+1} is a finite extension of F_i, or F_{i+1} is a pure transcendental extension of F_i obtained by adjoining finitely many exponentials and logarithms over F_i.*

---

[2] These are also called *Liouvillian extensions*.

2.2.2.3 Extensions by generalized quadratures   Let $F \supset K$ be an extension of a differential field K.

Definition 2.2. A differential field F is said to be an *extension of a differential field K by generalized quadratures* if $K \subset F$ and there exists a chain of differential fields $K = F_0 \subseteq \cdots \subseteq F_n \supset F$ such that $F_{i+1} = F_i < y_i >$ for every $i = 0, \ldots, n-1$, where y_i is an exponential of integral, an integral, or an algebraic element over F_i. An element $a \in F$ is *representable by generalized quadratures* over K, $K \subset F$, if it is contained in a certain generalized extension of the field K by elementary generalized quadratures.

Definition 2.3. An extension F of a differential field K is said to be:

1) a *generalized extension by integral* if there are $y \in F$ and $f \in K$ such that $y' = f$, y is transcendental over K, and F is a finite extension of the field $K\langle y \rangle$;
2) a *generalized extension by an exponential of integral* if there are $y \in F$ and $f \in K$ such that $y' = fy$, y is transcendental over K, and F is a finite extension of the field $K\langle y \rangle$.

The following lemma is obvious.

Lemma 2.2. *An extension $K \subset F$ is an extension by generalized quadratures if there is a chain $K = F_0 \subset \cdots \subseteq F_n$ such that $F \subset F_n$ and for every $i = 0, \ldots, n-1$, F_{i+1} is a finite extension of F_i, or F_{i+1} is a generalized extension by integral of F_i, or F_{i+1} is a generalized extension by exponential integral of F_i.*

2.2.3 Functional differential fields and their extensions

Let K be a subfield of the field F of all meromorphic functions on a connected domain U of the Riemann sphere $\mathbb{C}^1 \cup \infty$ with the fixed coordinate function x on $\mathbb{C}^1$. Suppose that K contains all complex constants and is stable under differentiation (i.e. if $f \in K$, then $f' = df/dx \in K$). Then K provides an example of a *functional differential field*.

Let us now give a general definition.

Definition 2.4. Let U, x be a pair consisting of a connected Riemann surface U and a non-constant meromorphic function x on U. The map $f \to df/\pi^* dx$ defines the derivation in the field F of all meromorphic functions on U (the ratio of two meromorphic 1-forms is a well-defined meromorphic function). A *functional differential field* is any differential subfield of F (containing all complex constants).

The following construction helps to *extend* functional differential fields. Let K be a differential subfield of the field of meromorphic functions on a connected Riemann surface U equipped with a meromorphic function x. Consider any connected Riemann surface V together with a nonconstant analytic map $\pi : V \to U$. Fix the function $\pi^* x$ on V. The differential field F of all meromorphic functions on V with the differentiation $\varphi' = d\varphi/\pi^* dx$ contains the differential subfield $\pi^* K$ consisting of functions of the form $\pi^* f$, where $f \in K$. The differential field $\pi^* K$ is isomorphic to the differential field K, and it lies in the differential field F. For a suitable choice of the surface V, an extension of the field $\pi^* K$, which is isomorphic to K, can be done within the field F.

141

Suppose that we need to extend the field K, say, by an integral y of some function $f \in K$. This can be done in the following way. Consider the covering of the Riemann surface U by the Riemann surface V of an indefinite integral y of the form $f dx$ on the surface U. By the very definition of the Riemann surface V, there exists a natural projection $\pi : V \to U$, and the function y is a single-valued meromorphic function on the surface V. The differential field F of meromorphic functions on V with the differentiation $\varphi' = d\varphi/\pi^* dx$ contains the element y as well as the field $\pi^* K$ isomorphic to K. That is why the extension $\pi^* K \langle y \rangle$ is well defined as a subfield of the differential field F. We mean this particular construction of the extension whenever we talk about extensions of functional differential fields. The same construction allows us to adjoin a logarithm, an exponential, an integral or an exponential of integral of any function f from a functional differential field K to K. Similarly, for any functions $f_1, \ldots, f_n \in K$, one can adjoin a solution y of an algebraic equation $y^n + f_1 y^{n-1} + \cdots + f_n = 0$ or all the solutions $y_1, \ldots, y_n$ of this equation to K (the adjunction of all the solutions $y_1, \ldots, y_n$ can be implemented on the Riemann surface of the vector-function $\mathbf{y} = y_1, \ldots, y_n$). In the same way, for any functions $f_1, \ldots, f_{n+1} \in K$, one can adjoin the n-dimensional $\mathbb{C}$-affine space of all solutions of the linear differential equation $y^{(n)} + f_1 y^{(n-1)} + \cdots + f_n y + f_{n+1} = 0$ to K. (Recall that a germ of any solution of this linear differential equation admits an analytic continuation along a path on the surface U not passing through the poles of the functions $f_1, \ldots, f_{n+1}$.)

Thus, *all the above-mentioned extensions of functional differential fields can be implemented without leaving the class of functional differential fields.* When talking about extensions of functional differential fields, we always mean this particular procedure.

The differential field of all complex constants and the differential field of all rational functions of one variable can be regarded as differential fields of functions defined on the Riemann sphere.

2.2.4 Classes of functions and operations on multivalued functions

An indefinite integral of an elementary function is a function rather than an element of an abstract differential field. In functional spaces, for example, apart from differentiation and algebraic operations, an absolutely non-algebraic operation is defined, namely, composition. Furthermore, functional spaces provide more means of writing "explicit formulas" than abstract differential fields. Besides, we should take into account that functions can be multivalued, can have singularities, and so on.

In functional spaces, it is not hard to formalize the problem of unsolvability of equations in explicit form. One can proceed as follows: fix a class of functions and say that an equation is solvable explicitly if its solution belongs to this class. Different classes of functions correspond to different notions of solvability.

2.2.4.1 Defining classes of functions by lists of data A class of functions can be introduced by specifying a list of *basic functions* and a list of *admissible operations*. Given the two lists, the class of functions is defined as the set of all functions that can be obtained from the basic functions by repeated application of admissible operations. Below, we define the class of *generalized elementary functions* and the class of *generalized elementary functions over a functional differential field K* in exactly this way.

Classes of functions which appear in the problems of integrability in finite terms contain multivalued functions. Thus the basic terminology should be made clear. We work with multivalued functions "globally", which leads to a more general understanding of classes of functions defined by lists of basic functions and of admissible operations. A multivalued function is regarded as a single entity. *Operations on multivalued functions* can be defined. The result of such an operation is a set of multivalued functions; every element of this set is called a function obtained from the given functions by the given operation. A *class of functions* is defined as the set of all (multivalued) functions that can be obtained from the basic functions by repeated application of admissible operations.

2.2.4.2 Operations on multivalued functions

Let us define, for example, the sum of two multivalued functions on a connected Riemann surface U.

Definition 2.5. Take an arbitrary point a in U, any germ f_a of an analytic function f at the point a and any germ g_a of an analytic function g at the same point a. We say that the multivalued function φ on U generated by the germ $\varphi_a = f_a + g_a$ is *representable as the sum of the functions f and g.*

For example, it is easy to see that exactly two functions of one variable are representable in the form $\sqrt{x} + \sqrt{x}$, namely, $f_1 = 2\sqrt{x}$ and $f_2 \equiv 0$. Other operations on multivalued functions are defined in exactly the same way. *For a class of multivalued functions, being stable under addition means that, together with any pair of its functions, this class contains all functions representable as their sum.* The same applies to all other operations on multivalued functions understood in the same sense as above.

In the definition given above, not only the operation of addition plays a key role but also the operation of analytic continuation hidden in the notion of multivalued function. Indeed, consider the following example. Let f_1 be an analytic function defined on an open subset V of the complex line $\mathbb{C}^1$ and admitting no analytic continuation outside of V, and let f_2 be an analytic function on V given by the formula $f_2 = -f_1$. According to our definition, the zero function is representable in the form $f_1 + f_2$ *on the entire complex line.* By the commonly accepted viewpoint, the equality $f_1 + f_2 = 0$ holds inside the region V but not outside.

Working with multivalued functions globally, we do not insist on the existence of *a common region* where all necessary operations would be performed on single-valued branches of multivalued functions. A first operation can be performed in a first region, then a second operation can be performed in a second, different region on analytic continuations of functions obtained in the first step. In essence, this more general understanding of operations is equivalent to including analytic continuation in the list of admissible operations on the analytic germs.

2.2.5 Generalized elementary functions

In this section we define the generalized elementary functions of one complex variable and the generalized elementary functions over a functional differential field. We also discuss a relation of these notions with generalized elementary extensions of differential fields. First we'll present needed lists of basic functions and of admissible operations.

List of basic elementary functions

1. All complex constants and an independent variable x.
2. The exponential, the logarithm, and the power x^α where α is any constant.
3. The trigonometric functions sine, cosine, tangent, cotangent.
4. The inverse trigonometric functions arcsine, arccosine, arctangent, arccotangent.

Lemma 2.3. *Basic elementary functions can be expressed through the exponentials and the logarithms with the help of complex constants, arithmetic operations and compositions.*

Lemma 2.3 can be considered as a simple exercise. Its proof can be found in [Kho14].

List of some classical operations

1. The operation of composition takes functions f, g to the function $f \circ g$.
2. The arithmetic operations take functions f, g to the functions $f + g$, $f - g$, fg, and f/g.
3. The operation of differentiation takes the function f to the function f'.
4. The operation of integration takes the function f to a solution of the equation $y' = f$ (the function y is defined up to an additive constant).
5. The operation of taking an exponential of integral takes the function f to a solution of the equation $y' = fy$ (the function y is defined up to a multiplicative constant).
6. The operation of solving algebraic equations takes functions $f_1, \ldots, f_n$ to the function y such that $y^n + f_1 y^{n-1} + \cdots + f_n = 0$ (the function y is not quite uniquely determined by the functions $f_1, \ldots, f_n$ since an algebraic equation of degree n can have n solutions).

Definition 2.6. The class of *generalized elementary functions of one variable* is defined by the following data:

List of basic functions: basic elementary functions.

List of admissible operations: Compositions, Arithmetic operations, Differentiation, Operation of solving algebraic equations.

Theorem 2.1. *A (possibly multivalued) function of one complex variable belongs to the class of generalized elementary functions if and only if it belongs to some generalized elementary extension of the differential field of all rational functions of one variable.*

Theorem 2.1 follows from Lemma 2.3 (all needed arguments can be found in [Kho14]).

Let K be a functional differential field consisting of meromorphic functions on a connected Riemann surface U equipped with a meromorphic function x.

Definition 2.7. The class of *generalized elementary functions over a functional differential field* K is defined by the following data.

List of basic functions: all functions from the field K.

List of admissible operations: Operation of composition with a generalized elementary function ϕ that takes f to $\phi \circ f$, Arithmetic operations, Differentiation, Operation of solving algebraic equations.

Theorem 2.2. *A (possibly multivalued) function on a Riemann surface U belongs to the class of generalized elementary functions over a functional differential field K if and only if it belongs to some generalized elementary extension of K.*

Theorem 2.2 follows from Lemma 2.3 (all needed arguments can be found in [Kho14]).

2.2.6 Functions representable by generalized quadratures

Here we define functions of one complex variable representable by generalized quadratures and functions representable by generalized quadratures over a functional differential field. We also discuss a relation of these notions with extensions of functional differential fields by generalized quadratures. First we'll present needed lists of basic functions and of admissible operations.

Definition 2.8. The class of functions of one complex variable *representable by generalized quadratures* is defined by the following data:

List of basic functions: basic elementary functions.

List of admissible operations: Compositions, Arithmetic operations, Differentiation, Integration, Operation of taking an exponential of integral, Operation of solving algebraic equations.

Theorem 2.3. *A (possibly multivalued) function of one complex variable belongs to the class of functions representable by generalized quadratures if and only if it belongs to some extension of the differential field of all constant functions of one variable by generalized quadratures.*

Theorem 2.3 follows from Lemma 2.3 (all needed arguments can be found in [Kho14]).

Let K be a functional differential field consisting of meromorphic functions on a connected Riemann surface U equipped with a meromorphic function x.

Definition 2.9. The class of functions representable by *generalized quadratures over a functional differential field K* is defined by the following data:

List of basic functions: all functions from the field K.

List of admissible operations: Operation of composition with a generalized elementary function ϕ that takes f to $\phi \circ f$, Arithmetic operations, Differentiation, Integration, Operation of taking an exponential of integral, Operation of solving algebraic equations.

Theorem 2.4. *A (possibly multivalued) function on a Riemann surface U belongs to the class of generalized quadratures over a functional differential field K if and only if it belongs to some extension of K by generalized quadratures.*

Theorem 2.4 follows from Lemma 2.3 (all needed arguments can be found in [Kho14]).

2.3 Liouville's First Theorem and Actions of Lie Groups

2.3.1 Introduction

In 1833 Joseph Liouville proved the following fundamental result.

LIOUVILLE'S FIRST THEOREM *An integral y of a function f from a functional differential field K is a generalized elementary function over K if and only if y is representable in the form*

$$y(x) = \int_{x_0}^{x} f(t)\, dt = r_0(x) + \sum_{i=1}^{m} \lambda_i \ln r_i(x), \tag{1}$$

where $r_0, \ldots, r_m \in K$ and $\lambda_1, \ldots, \lambda_m$ are complex constants.

For large classes of functions algorithms based on Liouville's Theorem make it possible to either evaluate an integral or to prove that the integral cannot be "evaluated in finite terms". Besides these algorithms Liouville's Theorem sheds light on the arrangement of 1-forms integrable in finite terms among all algebraic 1-forms on an algebraic curve [Kho15].

In this section we prove Liouville's First Theorem. We follow the presentation in the paper [Kho19].

Let $K(y) \supset K$ be an extension obtained by adjoining to a functional differential field K an integral y over K. The differential Galois group G of this extension does not contain enough information to determine if the integral y belongs to a generalized elementary extension of K or not. Indeed, if the integral y does not belong to K then the group G is always the same: it is isomorphic to the additive group of complex numbers. From this fact one can conclude that Galois theory is not sensitive enough to prove Liouville's First Theorem. Nevertheless, Liouville's First Theorem can be proved using differential Galois groups.

The first step towards such a proof was suggested by Abel (see sections 2.3.3.1 and 2.3.3.2). This step is related to algebraic extensions and their finite Galois groups.

A second step (see section 2.3.4) deals with a pure transcendental extension F of a functional differential field K, obtained by adjoining $k + n$ logarithms and exponentials, algebraically independent over K. The differential Galois group of the extension $K \subset F$ is an $(k+n)$-dimensional connected commutative algebraic group G. It has a natural representation as a group of analytic automorphisms of an analytic variety X. Thus G acts not only on the differential field F but also on other objects such as closed 1-forms on X. This action plays a key role in our proof (see section 2.3.4.1).

2.3.2 Outline of an inductive proof

Let us outline an inductive proof of Liouville's Theorem.

Definition 2.10. A function g is a *generalized elementary function of complexity* $\leq k$ if there is a chain $K = F_0 \subset F_1 \subset \ldots \subset F_k$ of functional differential fields such that $g \in F_k$ and for any $0 \leq i < k$ either F_{i+1} is a finite extension of F_i, or F_{i+1} is a pure transcendental extension of F_i obtained by adjoining finitely many exponentials, and logarithms over F_i.

We will prove the following induction hypothesis $I(m)$: *Liouville's Theorem is true for every integral y of complexity* $\leq m$ *over any functional differential field K.* The statement $I(0)$ is obvious: if $y \in K$, then $y = r_0 \in K$. Now let $y' \in K$ and $y \in F_k$. Since $y' \in F_1$, by induction $y = R_0 + \sum_{i=1}^{q} \lambda_i \ln R_i$, where $R_0, R_1, \ldots, R_q \in F_1$. We need to show that y is representable in the form (1) with $r_0, \ldots, r_m \in F_0 = K$. We have the following two cases to consider:

1. F_1 is a finite extension of $F_0 = K$. The statement of the induction hypothesis in this case was proved by Abel and is called Abel's Theorem. We will present its proof in section 2.3.3.
2. F_1 is a pure transcendental extension of $F_0 = K$ obtained by adjoining exponentials and logarithms over K. We will deal with this case in section 2.3.4.

2.3.3 The algebraic case

In section 2.3.3.1 we discuss finite extensions of differential fields. In section 2.3.3.2 we present a proof of Abel's Theorem.

2.3.3.1 An algebraic extension of a functional differential field   Let

$$P(z) = z^n + a_1 z^{n-1} + \cdots + a_n \tag{2}$$

be an irreducible polynomial over K, $P \in K[z]$. Suppose that a functional differential field F contains K and a root z of P.

Lemma 2.4. *The field $K(z)$ is stable under differentiation.*

Proof. Since P is irreducible over K, the polynomial $\frac{\partial P}{\partial z}$ has no common roots with P and is different from zero in the field $K[z]/(P)$. Let M be a polynomial satisfying a congruence $M\frac{\partial P}{\partial z} \equiv -\frac{\partial P}{\partial x}$ (mod P). Differentiating the identity $P(z) = 0$ in the field F, we obtain that $\frac{\partial P}{\partial z}(z)z' + \frac{\partial P}{\partial x}(z) = 0$, which implies that $z' = M(z)$. Thus the derivative of the element z coincides with the value at z of a polynomial M. Lemma 2.4 follows from this fact. □

Let $K \subset F$ and $\hat{K} \subset \hat{F}$ be functional differential fields, and P, $\hat{P}$ irreducible polynomials over K, $\hat{K}$ correspondingly. Suppose that F, $\hat{F}$ contain roots z, $\hat{z}$ of P, $\hat{P}$.

Theorem 2.5. *Assume that there is an isomorphism $\tau : K \to \hat{K}$ of differential fields K, $\hat{K}$ which maps coefficients of the polynomial P to the corresponding coefficients of the polynomial $\hat{P}$. Then τ can be extended in a unique way to the differential isomorphism $\rho : K(z) \to \hat{K}(\hat{z})$.*

Theorem 2.5 could be obtained by the arguments used in the proof of Lemma 2.4.

2.3.3.2 Induction hypothesis for an algebraic extension

Let $z_1, \ldots, z_n$ be the roots of the polynomial P given by (2) and let $F_1 = K\langle z_1 \rangle$. Assume that there is an element $y_1 \in F_1$ such that $y_1' \in K$, $M_i \in K[x]$ and y_1' is representable in the form

$$y_1' = \sum_{i=1}^{q} \lambda_i \frac{(M_i(z_1))'}{M_i(z_1)} + (M_0(z_1))'. \qquad (3)$$

Abel's Theorem *Under the above assumptions the element y_1' is representable in the form (1) with polynomials M_i independent of z_1, i.e. with $M_0, M_1, \ldots, M_q \in K$.*

Proof. Let y_1 be equal to $Q(z_1)$ where $Q \in K[z]$. For any $1 \le j \le n$ let y_j be the element $Q(z_j)$. According to Theorem 2.5 the identity (3) implies the identity

$$y_j' = \sum_{i=1}^{q} \lambda_i \frac{(M_i(z_j))'}{M_j(z_1)} + (M_0(z_j))'. \qquad (4)$$

Since $y_1' \in K$ we obtain n equalities $y_1' = \cdots = y_n'$. To complete the proof it is enough to take the arithmetic mean of n equalities (4). Indeed, the elements $\tilde{M}_i = \prod_{1 \le k \le n} M_i(z_k)$ and $\tilde{M}_0 = \sum_{1 \le k \le n} M_0(z_k)$ are symmetric functions in the roots of the polynomial P, thus $\tilde{M}_0, \ldots \tilde{M}_q \in K$. □

Remark. The proof implicitly uses the Galois group G of the splitting field of the polynomial P over the field K. The group G permutes the roots $y_1, \ldots, y_n$ of P. The element $\tilde{M}_i = \prod_{1 \le k \le n} M_i(z_k)$ and $\tilde{M}_0 = \sum_{1 \le k \le n} M_0(z_k)$ are invariant under the action of G, thus they belong to the field K.

2.3.4 The pure transcendental case

Here we prove the induction hypothesis in the pure transcendental case. First we will state the corresponding Theorem 2.6 and outline its proof.

Let F_1 be a functional differential field obtained by extension of the functional differential field K by adjoining algebraically independent over K functions

$$y_1 = \ln a_1, \ldots, y_k = \ln a_k, z_1 = \exp b_1, \ldots, z_n = \exp b_n \qquad (5)$$

where $a_1, \ldots, a_k, b_1, \ldots, b_k$ are some functions from K. We will assume that F_1 consists of meromorphic functions on a connected Riemann surface U and the differentiation in K_1 using a meromorphic function x on U. Let X be the manifold $U \times G$ where $G = \mathbb{C}^k \times (\mathbb{C}^*)^n$. Consider a map $\gamma : U \to \mathbb{C}^k \times (\mathbb{C}^*)^n$ given by the formula

$$\gamma(p) = y_1(p), \ldots, y_k(p), z_1(p) \ldots, z_n(p) \qquad (6)$$

where the functions y_i, z_j are defined by (5).

Let X be the product $U \times (\mathbb{C})^k \times (\mathbb{C}^*)^n$. Denote by $\Gamma \subset X$ the graph of the map γ. Consider a germ Φ of a complex-valued function at the point $a \in X$.

Definition 2.11. We say that Φ is a *logarithmic type germ* if Φ is representable in the form $\Phi_a = R_0 + \sum_{i=1}^q \lambda_i \ln R_i$, where R_i are germs at the point $a \in X$ of rational functions of $(y_1, \ldots, y_k, z_1, \ldots, z_n)$ with coefficients in K and the λ_j are complex numbers.

Theorem 2.6. *Let Φ be a logarithmic type germ at a point $a = (p_0, \gamma(p_0)) \in \Gamma$. Then the germ of the function $\Phi(p, \gamma(p))$ at the point $p_0 \in U$ is a germ of an integral over K if and only if Φ is representable in the following form*

$$\Phi(p, y, z) = \Phi(p, \gamma(p_0)) + \sum_{i=1}^k c_i(y_i - y_i(p_0)) + \sum_{j=1}^n t_j \ln \frac{z_j}{z_j(p_0)} \qquad (7)$$

where c_i, t_j are complex constants.

Theorem 2.6 proves the induction hypothesis in the pure transcendental case. Indeed $\Phi(p, \gamma(p_0))$ is a germ of a function from the field K and according to (5) the identities $c_i y_i = c_i \ln a_i$ and $t_j \ln z_j = t_j b_j$ hold. We split the claim of Theorem 2.6 into two parts.

First we consider the natural action of the group $G = (\mathbb{C}^k) \times (\mathbb{C}^*)^n$ on $X = U \times G$ and we describe all germs of closed 1-forms locally invariant under this action. Corollary 2.2 claims that each such 1-form is a differential of a function representable in the form (7).

Second we show that if the germ Φ satisfies the conditions of Theorem 2.6 then the germ $d\Phi$ is locally invariant under the action of the group G (see Theorem 2.10).

2.3.4.1 Locally invariant closed 1-forms

Let G be a connected Lie group acting by diffeomorphisms on a manifold X. Let $\pi : G \to \mathrm{Diff}(X)$ be the corresponding homomorphism from G to the group $\mathrm{Diff}(X)$ of diffeomorphisms of X. For a vector ξ from the Lie algebra $\mathcal{G}$ of G the action π associates the vector field V_ξ on X. The germ ω_{x_0} at a point $x_0 \in X$ of a differential form ω on X is *locally invariant under the action π* if for any $\xi \in \mathcal{G}$ the Lie derivative $L_{V_\xi}\omega$ is equal to zero.

Lemma 2.5. *The germ of the differential* $d\varphi_{x_0} = \omega_{x_0}$ *of a smooth function* φ *is locally invariant under the action* π *if and only if for each* $\xi \in G$ *the Lie derivative* $L_{V_\xi}\varphi$ *is a constant* $M(\xi)$ *(which depends on* ξ*).*

Proof. Applying "Cartan's magic formula" $L_{V_\xi}\omega = i_{V_\xi}d\omega + d(i_{V_\xi}\omega)$ we obtain that $L_{V_\xi}\omega = 0$ if and only if $d(L_{V_\xi}\varphi) = 0$, which means that $L_{V_\xi}\varphi$ is constant. $\qquad\square$

The following theorem characterizes locally invariant closed 1-forms more explicitly.

Theorem 2.7. *The germ of the differential* $d\varphi_{x_0} = \omega_{x_0}$ *of a smooth complex-valued function* φ *is locally invariant under the action* π *if and only if there exists a local homomorphism* ρ *of* G *to the additive group* $\mathbb{C}$ *of complex numbers such that for any* $g \in G$ *in a neighborhood of the identity the following relation holds:*

$$\varphi(\pi(g)x_0) = \varphi(x_0) + \rho(g).$$

Proof. For $\xi \in G$ the Lie derivative $L_{V_\xi}\varphi$ is a constant $M(\xi)$ by Lemma 2.3. Let us show that for $\xi \in [G]$, where $[G]$ is the commutator of G, the constant $M(\xi)$ is equal to zero. Indeed, if $\xi = [\tau, \rho]$ then

$$L_{V_\xi}\varphi = L_{V_\rho}L_{V_\tau}\varphi - L_{V_\tau}L_{V_\rho}\varphi = L_{V_\rho}M(\tau) - L_{V_\tau}M(\rho) = 0.$$

Thus the linear function $M : G \to \mathbb{C}$ mapping ξ to $M(\xi)$ provides a homomorphism of G to the Lie algebra of the additive group $\mathbb{C}$ of complex numbers. Let ρ be the local homomorphism of G to $\mathbb{C}$ corresponding to the homomorphism M.

Consider a function ϕ on a neighborhood of the identity in G defined by the following formula: $\phi(g) = \varphi(x_0) + \rho(g)$. By definition, on a neighborhood of the identity the function ϕ has the same differential as the function $\varphi(\pi(g)x)$. The values of these functions at the identity are equal to $\varphi(x_0)$. Thus these functions are equal. $\qquad\square$

Assume that $X = U \times G$, where U is a manifold and an action π is given by the formula $\pi(g)(x, g_1) = (x, gg_1)$. Applying Theorem 2.7 to this action we obtain the following corollary.

Corollary 2.1. *If the germ of the differential* $d\varphi = \omega$ *of a smooth complex-valued function* φ *at a point* $(x_0, g_0) \in U \times G$ *is locally invariant under the action* π *then in a neighborhood of the point* (x_0, g_0) *the following identity holds:*

$$\varphi(x, g) = \varphi(x, g_0) + \rho(gg_0^{-1}), \tag{8}$$

where ρ *is a local homomorphism of* G *to the additive group of complex numbers.*

Proof. This follows from Theorem 2.7 since the element gg_0^{-1} maps the point (x, g_0) to the point (x, g). $\qquad\square$

Let G be the group $\mathbb{C}^k \times (\mathbb{C}^*)^n$, where $\mathbb{C}$ and $\mathbb{C}^*$ are respectively the additive and multiplicative group of complex numbers. We will consider the group $\mathbb{C}^k \times (\mathbb{C}^*)^n$ with coordinate functions $(y,z) = (y_1, \ldots, y_k, z_1, \ldots, z_n)$ assuming that $z_1 \cdots z_n \neq 0$.

Corollary 2.2. *If the assumptions of Corollary 2.1 hold for $G = \mathbb{C}^k \times (\mathbb{C}^*)^n$ in a neighborhood of $(x_0, y_0, z_0) \in U \times (\mathbb{C}^k \times (\mathbb{C}^*)^n)$ then the following identity holds*

$$\varphi(x,y,z) = \varphi(x, y_0, z_0) + \sum_{1 \leq i \leq k} \lambda_i (y_i - (y_0)_i) + \sum_{1 \leq j \leq n} \mu_j \ln \frac{z_j}{(z_0)_j},$$

where $\lambda_1, \ldots, \lambda_k, \mu_1 \ldots, \mu_n$ are complex constants.

Proof. This follows from (8) since any local homomorphism ρ from the group $\mathbb{C}^k \times (\mathbb{C}^*)^n$ to the additive group of complex numbers can be given by the formula

$$\rho(y_1, \ldots, y_k, z_1, \ldots, z_n) = \sum_{1 \leq i \leq k} \lambda_1 y_i + \sum_{1 \leq j \leq n} \mu_j \ln z_j,$$

where λ_i and μ_j are complex constants. □

2.3.4.2 The vector field associated to a logarithmic-exponential extension

We use the notation introduced in section 2.3.4. Let G be the group $\mathbb{C}^k \times (\mathbb{C}^*)^n$ and let X be the product $U \times G$. Consider the map $\gamma : U \to \mathbb{C}^k \times (\mathbb{C}^*)^n$ given by the following formula:

$$y_1 = \ln a_1, \ldots, y_k = \ln a_k, z_1 = \exp b_1, \ldots, z_n = \exp b_n. \tag{9}$$

The map γ satisfies the following differential relation:

$$d\gamma = da_1/a_1, \ldots, da_k/a_k, z_1 db_1, \ldots, z_n db_n.$$

Definition 2.12. Let V be a meromorphic vector field on X defined by the following conditions. If V_a is the value of V at the point

$$a = (p, y_1, \ldots, y_k, z_1, \ldots, z_n) \in X$$

then $\langle dx, V_a \rangle = 1$, $\langle dy_i, V_a \rangle = a_i'(p)/a_i(p)$ for $1 \leq i \leq k$, $\langle dz_j, V_a \rangle = b_j'(p) z_j(p)$ for $1 \leq j \leq n$.

The vector field V is regular on $U^0 \times G$, where U^0 is an open subset in U which does not contain the zeros and poles of the functions $a_1, \ldots, a_k$, the poles of the functions $b_1, \ldots, b_n$, or the zeros and poles of the 1-form dx. By construction the graph $\Gamma = (p, \gamma(p)) \subset X$ of the map γ is an integral curve for the differential equation on X defined by the vector field V.

The following lemmas are obvious.

Lemma 2.6. *The vector field V is invariant under the action π on X. For each element $g \in G$ the curve $g\Gamma \subset X$ of the graph Γ of γ is an integral curve of V.*

Lemma 2.7. *The field $K(y,z)$ of rational functions in $y_1,\ldots,y_k,z_1,\ldots,z_n$ over the field K is invariant under the action π on X. For each vector $\xi \in G$ in the Lie algebra $\mathcal{G}$ of G the Lie derivative $L_{V_\xi} R$ of $R \in K(y,z)$ belongs to $K(y,z)$.*

2.3.4.3 Pure transcendental logarithmic exponential extensions
We will assume below that the components (9) of γ are algebraically independent over K.

Liouville's principle. If a polynomial $P \in K[y_1,\ldots,y_k,z_1,\ldots,z_n]$ vanishes on the graph $\Gamma \subset X$ of the map γ then P is identically equal to zero.

Proof. If P is not identically equal to zero then the components of γ are algebraically dependent over the field K. $\qquad\square$

Theorem 2.8. *The extension $K \subset F_1$ is isomorphic to the extension of K by the field of rational functions $K(y,z)$ in $(y_1,\ldots,y_k,z_1,\ldots,z_n)$ over K considered as the field of functions on X equipped with the differentiation sending $f \in K(y,z)$ to the Lie derivative $L_V f$ with respect to the vector field V introduced in definition 2.12.*

Proof. By assumption, the components (9) of the map γ are algebraically independent over K, thus each function from the extension obtained by adjoining to K these components is representable in a unique way as a rational function from $K(y,z)$. By definition the derivatives of the components (9) coincide with their Lie derivatives with respect to the vector field V. $\qquad\square$

The action π of the group $G = \mathbb{C}^k \times (\mathbb{C}^*)^n$ on X induces the action π^* of G on the space of functions on X containing the field $K(y,z)$. The vector field V is invariant under the action π. Thus π^* acts on $K(y,z) \sim F_1$ by differential automorphisms. It is easy to see that a function $f \in K(y,z)$ is fixed under the action π^* if and only if $f \in K$, i.e. the group G is isomorphic to the *differential Galois group* of the extension $K \subset F_1$. We have proved the following result.

Theorem 2.9. *The differential Galois group of the extension $K \subset F_1$ is isomorphic to the group G. The Galois group is induced on the differential field $K(y,z)$ with the differentiation given by the Lie derivative with respect to the field V by the action of G on the manifold $X = U \times \mathbb{C}^k \times (\mathbb{C}^*)^n$.*

Now we are ready to complete the inductive proof of Liouville's First Theorem.

Theorem 2.10. *Let Φ be a logarithmic type germ at a point $a = (p_0, \gamma(p_0)) \in \Gamma \subset X$. If the germ of the function $\Phi(p, \gamma(p))$ on U at the point $p_0 \in U$ is a germ of an integral f over K then the germ of the differential $d\Phi$ at the point $a \in X$ is locally invariant under the action of π on X.*

Proof. By assumption the restriction of the function $(L_V\Phi - f)$ to Γ is equal to zero. Since the function $(L_V\Phi - f)$ belongs to the field $K(y,z)$ the function $(L_V\Phi - f)$ by Liouville's principle is equal to zero identically on X. In particular, it is equal to zero on the integral curve $g\Gamma$ of the vector field V, where g is an element of the group G. Thus the restriction of the function $L_V\pi(g)^*(\Phi - f)$ to Γ is equal to zero. Since the function f is invariant under the action π^* we obtain that the restriction to Γ of $L_V(\Phi - \pi^*(g)\Phi)$ is equal to zero. Differentiating this identity we obtain that for any $\xi \in G$ the restriction to Γ of $L_V(L_{V_\xi}\Phi)$ is equal to zero. Thus on Γ the function $L_{V_\xi}\Phi$ is constant. Lemma 2.6 implies that the function $L_{V_\xi}\Phi$ belongs to the field $K(x,y)$. Thus the function $L_{V_\xi}\Phi$ is a constant on X by Liouville's principle. Thus the 1-form $d\Phi$ is locally invariant under the action of π by Lemma 2.5. Theorem 2.10 is proved. □

Thus we complete the proof of Theorem 2.6 and the inductive proof of Liouville's First Theorem.

2.4 Liouville's Second Theorem and its Generalizations

2.4.1 Introduction

In 1839 Joseph Liouville proved the following fundamental result.

Liouville's Second Theorem *A second-order homogeneous linear differential equation*

$$y'' + a_1 y' + a_n y = 0 \tag{10}$$

whose coefficients a_1, a_2 belong to a functional differential field K is solvable by generalized quadratures over K if and only if it has a solution of the form $y_1 = \exp z$, where z' is algebraic over K.

Much later this theorem was generalized for homogeneous linear differential equations of any order. Consider an equation

$$y^{(n)} + a_1 y^{(n-1)} + \cdots + a_n y = 0 \tag{11}$$

whose coefficients a_i belong to K.

Theorem 2.11. *If the equation (11) has a solution representable by generalized quadratures over K then it necessarily has a solution of the form $y_1 = \exp z$, where z' is algebraic over K.*

The following lemma is obvious.

Lemma 2.8. *Assume that equation (11) has a solution y_1 representable by generalized quadratures over K. Then equation (11) can be solved by generalized quadratures over K if and only if the linear differential equation of order $(n-1)$ over the differential field $K(y_1)$ obtained from (11)) by the reduction of order using the solution y_1 is solvable by generalized quadratures over $K(y_1)$.*

153

Indeed on one hand each solution of the equation obtained from (11) by the reduction of order using y_1 can be expressed in the form $(y/y_1)'$, where y is a solution of (11). On the other hand any solution y of the equation $(y/y_1)' = u$, where u is represented by generalized quadratures over $K(y_1)$, is representable by generalized quadratures over K, assuming that y_1 is representable by generalized quadratures over K.

Thus Theorem 2.11 provides the following criterion for solvability of equation (11) by generalized quadratures.

Theorem 2.12. *Equation (11) is solvable by generalized quadratures over K if and only if the following conditions hold:*

1) equation (11) has a solution y_1 of the form $y_1 = \exp z$, where $z' = f$ is algebraic over K;

2) the linear differential equation of order $(n-1)$ over $K(y_1)$ obtained from (11) by the reduction of order using the solution y_1 is solvable by generalized quadratures over $K(y_1)$.

For $n = 2$, Theorem 2.12 is equivalent to Liouville's Second Theorem because linear differential equations of first order are automatically solvable by quadratures.

The standard proof (E. Picard and E. Vessiot, 1910) of Theorem 2.11 uses differential Galois theory and is rather involved (see [dPS03]).

In the case when the equation (11) is a Fuchsian differential equation and K is the field of rational function of one complex variable, Theorem 2.12 has a topological explanation (see section 3.3 and [Kho14]), which allows us to prove a much stronger version of this result. But in the general case Theorem 2.12 does not have a similar visual interpretation.

Maxwell Rosenlicht [Ros73] proved the following theorem in 1973.

Theorem 2.13. *Let n be a positive integer, and let Q be a polynomial in several variables with coefficients in a differential field K and of total degree less than n. Then if the equation*

$$u^n = Q(u, u', u'', \dots) \tag{12}$$

has a solution representable by generalized quadratures over K, it has a solution algebraic over K.

The logarithmic derivative $u = y'/y$ of any solution of the equation (11) satisfies the *generalized Riccati equation of order $n-1$ associated with* (11), which is a particular case of the equation (12). Rosenlicht showed that Theorem 2.11 easily follows from Theorem 2.12 applied to the corresponding generalized Riccati equation (see section 2.4.2). In modern differential algebra abstract fields equipped with an operation of differentiation are considered. Rosenlicht's proof of Theorem 2.12 is not elementary: it is applicable to abstract differential fields of characteristic zero and makes use of valuation theory.[3]

The logarithmic derivative $u = y'/y$ of any solution y of the homogeneous linear differential equation of second order (10) satisfies the Riccati equation

$$u' + a_1 u + a_2 + u^2 = 0. \tag{13}$$

To prove Liouville's Second Theorem Liouville and Ritt first proved Theorem 2.12 for the Riccati equation (13). To do that, Ritt (in his simplification of Liouville's proof) considered a special one-parameter family of solutions of (13) and used an expansion of these solutions as functions of the parameter into convergent Puiseux series (see Chapter V in [Rit48]). Ritt used a generalization of the following theorem based on ideas suggested by Newton.

Consider an algebraic function $z(y)$ defined by an equation $P(y, z) = 0$, where P is a polynomial with coefficients in a subfield K of $\mathbb{C}$. Then all branches of the algebraic function $z(y)$ at the point $y = \infty$ can be developed into convergent Puiseux series whose coefficients belong to a finite extension of the field K.

The *generalized Newton's Theorem* claims that a similar result holds if instead of a numerical field of coefficients one takes a field K whose element are meromorphic functions on a connected Riemann surface. In Ritt's book [Rit48] this result is proved in the same way as its classical version using Newton's polygon method.

Ritt's proof is written in old mathematical language and does not fit into our presentation. Theorem 2.15 provides an exact statement of the generalized Newton's Theorem. It is presented without proof: the main arguments proving it are well known and classical. One can also obtain a proof by modifying Ritt's exposition. Theorem 2.15 plays a crucial role in section 2.4.3. For the sake of completeness I will present its modern proof in a separate paper.

In this section we discuss a proof of Theorem 2.12 which does not rely on valuation theory. It generalizes Ritt's arguments (makes use of a Puiseux expansion via the generalized Newton's Theorem) and provides an elementary proof of the classical Theorem 2.11. We follow the presentation in the paper [Kho18].

2.4.2 The generalized Riccati equation

Here we define the generalized Riccati equation and reduce Theorem 2.11 to Theorem 2.12. In this section we also generalize Theorem 2.11 to nonlinear homogeneous equations (this generalization will not be used in the next sections).

Assume that u is the logarithmic derivative of a meromorphic function y (not identically equal to zero), i.e. the relation $y' = uy$ holds.

---

[3] According to Michael Singer the valuation theory used in Rosenlicht's proof is a fancy way of using power series methods (private communication).

Definition 2.13. Let D_n be a polynomial in u and in its derivatives $u, u', \ldots, u^{(n-1)}$ up to order $(n-1)$, defined by induction by the following conditions:

$$D_0 = 1; \quad D_{k+1} = \frac{\mathrm{d}D_k}{\mathrm{d}x} + uD_k.$$

Lemma 2.9. *1) The polynomial D_n has integral coefficients and $\deg D_n = n$. The degree n homogeneous part of D_n is equal to u^n (i.e. $D_n = u^n + \tilde{D}_n$, where $\deg \tilde{D}_n < n$).*
2) If y is a function whose logarithmic derivative is equal to u (i.e. if $y' = uy$) then for any $n \geq 0$ the relation $y^{(n)} = D_n(u)y$ holds.

Both claims of Lemma 2.9 can be easily checked by induction.

Consider a homogeneous linear differential equation (11) whose coefficients a_i belong to a differential field K.

Definition 2.14. The equation

$$D_n + a_1 D_{n-1} + \cdots + a_n D_0 = 0 \tag{14}$$

of order $n-1$ is called the *generalized Riccati equation* for the homogeneous linear differential equation (11).

Lemma 2.10. *A non-identically equal to zero function y satisfies the linear differential equation (11) if and only if its logarithmic derivative $u = y'/y$ satisfies the generalized Riccati equation (14).*

Proof. Let y be a nonzero solution of (11) and let u be its logarithmic derivative. Then dividing (11) by y and using the identity $y^{(k)}/y = D_k(u)$ we obtain that u satisfies (14). If u is a solution of (14) then multiplying (4) by y and using the identity $y^{(k)} = D_k(u)y$ we obtain that y is a nonzero solution of (11). □

Corollary 2.3. *1) Equation (11) has a nonzero solution representable by generalized quadratures over K if and only if equation (14) has a solution representable by generalized quadratures over K.*
2) Equation (11) has a solution y of the form $y = \exp z$ where $z' = f$ is an algebraic function over K if and only if equation (14) has an algebraic solution over K.

Proof. 1) A nonzero function y is representable by generalized quadratures over K if and only if its logarithmic derivative $u = y'/y$ is representable by generalized quadratures over K.

2) A function y is equal to $\exp z$ where $z' = f$ if and only if its logarithmic derivative is equal to f. □

The generalized Riccati equation (14) satisfies the conditions of Theorem 2.12. Thus Theorem 2.11 follows from Theorem 2.12 and from Corollary 2.3.

Let us generalize the results of this section. Consider an order n homogeneous equation

$$P(y, y', \ldots, y^{(n)}) = 0 \tag{15}$$

where P is a degree m homogeneous polynomial in $n + 1$ variables $x_0, x_1, \ldots, x_n$ over a functional differential field K.

Definition 2.15. The equation

$$P(D_0, D_1, \ldots, D_n) = 0 \tag{16}$$

of order $n - 1$ is called the *generalized Riccati equation* for the homogeneous equation (15).

Lemma 2.11. *A non-identically equal to zero function y satisfies the homogeneous equation (15) if and only if its logarithmic derivative $u = y'/y$ satisfies the generalized Riccati equation (16).*

Corollary 2.4. *1) Equation (15) has a nonzero solution representable by generalized quadratures over K if and only if equation (16) has a solution representable by generalized quadratures over K.*
2) Equation (15) has a solution y of the form $y = \exp z$ where $z' = f$ is an algebraic function over K if and only if equation (16) has an algebraic solution over K.

Lemma 2.11 and Corollary 2.4 can be proved in exactly the same way as Lemma 2.10 and Corollary 2.3

Let us define the ξ-*weighted degree* $\deg_\xi x^P$ of the monomial $x^P = x_0^{p_0} \cdots x_n^{p_n}$ by the following formula:

$$\deg_\xi x^P = \sum_{i=0}^{i=n} i p_i.$$

We will say that a polynomial $P(x_0, \ldots, x_n)$ satisfies the ξ-*weighted degree condition* if the sum of coefficients of all monomials in P having the biggest ξ-weighted degree is not equal to zero. A polynomial P having a unique monomial with the biggest ξ-weighted degree automatically satisfies this condition. For example a degree m polynomial P containing a term $a x_n^m$ with $a \neq 0$ automatically satisfies the ξ-weighted degree condition.

Theorem 2.14. *Consider the homogeneous equation (15) with the polynomial P satisfying the ξ-weighted degree condition. If this equation has a solution representable by generalized quadratures over K then it necessarily has a solution of the form $y_1 = \exp z$, where z' is algebraic over K.*

Proof. It is easy to check that if the polynomial P satisfies the ξ-weighted degree condition then the generalized Riccati equation (16) satisfies the conditions of Theorem 2.12. Thus Theorem 2.14 follows from Theorem 2.12 and Corollary 2.4. □

Remark. There exists a complete analog of Galois theory for linear homogeneous differential equations (see [dPS03]). Theorem 2.11 can be proved using this theory. The differential Galois group of a nonlinear homogeneous differential equation (15) could be very small and for such equations a complete analog of Galois theory does not exist. Thus Theorem 2.14 cannot be proved in a similar way.

2.4.3 Finite extensions of fields of rational functions

Here we discuss finite extensions of the field $K(y)$ of rational functions over a subfield K of the field of meromorphic function on a connected Riemann surface U. We also state Theorem 2.15 (the generalized Newton's Theorem) which plays a crucial role in this section.

Let F be an extension of $K(y)$ by a root z of a degree m polynomial $P(z) \in (K[y])[z]$ over the ring $K[y]$, irreducible over the field $K(y)$. Let X be the product $U \times \mathbb{C}^1$, where $\mathbb{C}^1$ is the standard complex line with the coordinate function y. An element of the field $K(y)$ can be considered as a meromorphic function on X. One can associate with the element $z \in F$ a multivalued algebroid function on X defined by the equation $P(z) = 0$. Let $D(y)$ be the discriminant of the polynomial P. Let $\Sigma \subset U \times \mathbb{C}^1 = X$ be the hypersurface defined by the equation $p_m(y) \cdot D(y) = 0$, where $p_m(y)$ is the leading coefficient of the polynomial P.

Lemma 2.12. *1) About a point $x \in X \setminus \Sigma$ the equation $P(z) = 0$ defines m germs z_i of analytic functions whose values at x are simple roots of the polynomial P.*

2) Let x be the point $(a, y) \in U \times \mathbb{C}^1 \setminus \Sigma$. Then the field F is isomorphic to the extension $K_a(z_i)$ of the field K_a of germs at $a \in U$ of functions from the field K (considered as germs at $x = (a, y)$ of functions independent of y) extended by the germ z_i at x satisfying the equation $P(z) = 0$.

Proof. Statement 1) follows from the Implicit Function Theorem. Statement 2) follows from 1). □

Below we state Theorem 2.15, which is a generalization of Newton's Theorem about the expansion of an algebraic functions as a convergent Puiseux series. It is stated without proof (see the comments in section 2.4.1).

We use the notations introduced at the beginning of this section. Let z be an element satisfying a polynomial equation $P(z) = 0$ over the ring $K[y]$, where K is a subfield of the field of meromorphic functions on U. Then there exists *a finite extension K_P of the field K associated with the element z* such that the following theorem holds.

Theorem 2.15. *There is a finite covering $\pi : U_P \rightarrow U \setminus O_P$, where $O_P \subset U$ is a discrete subset, such that the following properties hold:*

1) the extension K_P can be realized by a subfield of the field of meromorphic functions on U_P containing the field $\pi^* K$ isomorphic to K;

2) there is a continuous positive function $r : U \setminus O_P \rightarrow \mathbb{R}$ such that in the open domain $W \subset (U \setminus O_P) \times \mathbb{C}^1$ defined by the inequality $|y| > r(a)$ all m germs of z_i at a point (a, y) can be developed into a convergent Puiseux series

$$z_i = z_{i_k} y^{\frac{k}{p}} + z_{i_{k-1}} y^{\frac{k-1}{p}} + \cdots \tag{17}$$

whose coefficients z_{i_j} are germs of analytic functions at the point $a \in U_P$ having analytical continuation as regular functions on U_P belonging to the field K_P.

For the special case when K is a subfield of the field of complex numbers, it is natural to refer to Theorem 2.15 as Newton's Theorem. One can consider K as a field of constant functions on any connected Riemann surface U, choose O_P to be the empty set, U_P to be equal to U, the projection $\pi : V_P \to U$ to be the identity map, and the function $r : U \to \mathbb{R}$ to be a big enough constant. In this case Theorem 2.15 states that an algebraic function z has a Puiseux expansion at infinity whose coefficients belong to a finite extension K_P of the field K. This statement can be proved by Newton's polygon method.

Let F be an extension of $K(y)$ by a root z of the polynomial P and let K_P be the finite extension of the field K introduced in Theorem 2.15. The extension F_P of the field $K_P(y)$ by z is easy to deal with. Denote the product $U_P \times \mathbb{C}^1$ by X_P.

Lemma 2.13. *Let $x \in X_P$ be the point $(a, y_0) \in U_P \times \mathbb{C}^1$. Then the field F_P is isomorphic to the extension $K_{P,a}(z_i)$ of the field $K_{P,a}$ of germs at $a \in U_P$ of functions from the field K_P (considered as germs at $x = (a, y_0)$ of functions independent of y) extended by the germ at x of the function z_i defined by (17).*

Lemma 2.13 follows from Theorem 2.15.

2.4.4 Extension by one transcendental element

Let U be a connected Riemann surface and let K be a differential field of meromorphic functions on U. Let $\mathbb{C}^1$ be the standard complex line with the coordinate function y. Elements of the field $K(y)$ of rational functions over K can be considered as meromorphic functions on $X = U \times \mathbb{C}^1$.

In the field $K(y)$ there are two natural operations of differentiation. The first operation $R(y) \to \frac{\partial R}{\partial x}(y)$ is defined as follows: the derivative $\frac{\partial}{\partial x}$ of the independent variable y is equal to zero, and the derivative $\frac{\partial}{\partial x}$ of an element $a \in K$ is equal to its derivative a' in the field K. For the second operation $R(y) \to \frac{\partial R}{\partial y}(y)$ the derivative of an element $a \in K$ is equal to zero and the derivative of the independent variable y is equal to one.

Let $K \subset F$ be differential fields and let $\theta \in F$ be a transcendental element over K. Assume that $\theta' \in K\langle \theta \rangle$. Under this assumption the field $K\langle \theta \rangle$ has the following description.

Lemma 2.14. *1) The map $\tau : K\langle \theta \rangle \to K(y)$ such that $\tau(\theta) = y$ and $\tau(a) = a$ for $a \in K$ provides an isomorphism between the field $K\langle \theta \rangle$ considered without the operation of differentiation and the field $K(y)$ of rational functions over K.*
2) If $\tau(\theta') = w \in K(y)$ then for any $R \in K(y)$ and $z \in K\langle \theta \rangle$ such that $\tau(z) = R$ the following identity holds

$$\tau(z') = \frac{\partial R}{\partial x} + \frac{\partial R}{\partial y} w. \tag{18}$$

Proof. The first claim of the lemma is straightforward. The second claim follows from the chain rule. □

Let $\Theta \subset X = U \times \mathbb{C}^1$ be the graph of the function $\theta : U \to \mathbb{C}^1$. The following lemma is straightforward.

Lemma 2.15. *The differential field $K\langle\theta\rangle$ is isomorphic to the field $K(y)|_\Theta$ obtained by restriction to Θ of functions from the field $K(y)$ equipped with the differentiation given by (18). For any point $a \in \Theta$, the differential field $K\langle\theta\rangle$ is isomorphic to the differential field of germs at $a \in \Theta$ of functions from $K(y)|_\Theta$.*

2.4.5 An extension by integral

Here we consider extensions of transcendence degree one of a differential field K containing an integral y over K which does not belong to K, $y \notin K$.

2.4.5.1 A pure transcendental extension by integral Let θ be an integral over K, i.e. $\theta' = f \in K$. Assume that θ is a transcendental element over K.[4]

Lemma 2.16. *1) The field $K\langle\theta\rangle$ is isomorphic to the field $K(y)$ of rational functions over K equipped with the following differentiation*

$$R' = \frac{\partial R}{\partial x} + \frac{\partial R}{\partial y} f. \tag{19}$$

2) For every complex number $\rho \in \mathbb{C}$ the map $\theta \to \theta + \rho$ can be extended to the unique isomorphism $G_\rho : K\langle\theta\rangle \to K\langle\theta\rangle$ which fixes elements of the field K.
3) Each isomorphism of $K\langle\theta\rangle$ over K is an isomorphism G_ρ for some $\rho \in \mathbb{C}$. Thus the Galois group of $K\langle\theta\rangle$ over K is the additive group of complex numbers $\mathbb{C}$.

Proof. Claim 1) follows from Lemma 2.14. For any $\rho \in \mathbb{C}$ the element $\theta_\rho = \theta + \rho$ is a transcendental element over K and θ'_ρ is equal to f. Thus claim 2) holds. Claim 3) follows from 2) because if $y' = f$ then $y = \theta_\rho$ for some $\rho \in \mathbb{C}$. □

2.4.5.2 A generalized extension by integral According to Lemma 2.16 the differential field $K\langle\theta\rangle$ is isomorphic to the field $K(y)$ with the differentiation given by (19). Let F be an extension of $K\langle\theta\rangle$ by an element $z \in F$ which satisfies some equation $\tilde{P}(z) = 0$ where $\tilde{P}$ is an irreducible polynomial over $K\langle\theta\rangle$. The isomorphism between $K\langle\theta\rangle$ and $K(y)$ transforms the polynomial $\tilde{P}$ into some polynomial P over $K(y)$. Below we use the notation from section 2.4.3 and deal with the multivalued algebroid function z on X defined by $P(z) = 0$.

Assume that at a point $x \in X$ there are germs of analytic functions z_i satisfying the equation $P(z_i) = 0$. Let θ_ρ be the function $(\theta + \rho) : U_P \to \mathbb{C}^1$ and let $\Theta_\rho \subset X = U \times \mathbb{C}^1$ be its graph. The point $x = (p, q) \in U \times \mathbb{C}^1$ belongs to the graph $\Theta_{\rho(x)}$ for $\rho(x) = q - \theta(p)$.

Let $K(y)|_{\Theta_{\rho(x)}}$ be the differential field of germs at the point $x \in \Theta_{\rho(x)}$ of restrictions to $\Theta_{\rho(x)}$ of functions from the field $K(y)$ equipped with the differentiation given by (19).

---

[4] It is easy to check that if an integral θ over K does not belong to K, then θ is a transcendental element over K (see [Kho14]). We will not use this fact.

Lemma 2.17. *The differential field F is isomorphic to the finite extension of the differential field $K(y)|_{\Theta_{\rho(x)}}$ obtained by adjoining the germ at $x \in \Theta_{\rho(x)}$ of the restriction to $\Theta_{\rho(x)}$ of an analytic germ z_i satisfying $P(z_i) = 0$.*

Proof. For the trivial extension $F = K\langle\theta\rangle$ Lemma 2.17 follows from Lemmas 2.15 and 2.16. Theorem 2.5 allows one to complete the proof for non-trivial finite extensions F of $K\langle\theta\rangle$. $\qquad\qquad\square$

According to section 2.4.3, with the polynomial P over $K(y)$ one can associate the finite extension K_P of the field K and the Riemann surface U_P such that Theorem 2.15 holds. Since K is a functional differential field the field K_P has the natural structure of a functional differential field. Below we will apply Lemma 2.17, taking instead of K the field K_P and considering the extension $F_P \supset K_P\langle\theta\rangle$ by the same algebraic element $z \in F$. The use of K_P instead of K allows one to apply the expansion (17) for z_i.

Theorem 2.16. *Let $x \in X_P = U_P \times \mathbb{C}^1$ be a point (a, y_0) with $|y| >> 0$. The differential field F_P is isomorphic to the extension of the differential field of germs at the point $a \in U_P$ of functions from the differential field K_P by the following germs: by the germ at a of the integral $\theta_{\rho(x)}$ of the function $f \in K$, and by a germ at a of the composition $z_i(\theta_\rho)$ where z_i is a germ at x of a function given by a Puiseux series (17).*

Proof. Theorem 2.16 follows from Lemma 2.17 and Theorem 2.15. $\qquad\qquad\square$

2.4.5.3 Solutions of equations in a generalized extension by integral

Here we discuss Lemma 2.18, providing an important step for our proof of Theorem 2.12. We will use the notation from sections 2.4.5.1 and 2.4.5.2.

Let $T(u, u', \ldots, u^{(n)})$ be a polynomial in an independent function u and its derivatives with coefficients from the functional differential field K. Consider the equation

$$T(u, u', \ldots, u^{(N)}) = 0. \tag{20}$$

In general the derivative of the highest order $u^{(N)}$ cannot be represented as a function of other derivatives via the relation (20). Thus even the existence of local solutions of (20) is problematic and we have no information about the global behavior of its solutions.

Assume that (20) has a solution z in a generalized extension by integral $F \supset K\langle\theta\rangle$ of K. The solution z has a nice global property: it is a meromorphic function on a Riemann surface V with a projection $\pi : V \to U$, which proves a locally trivial covering above $U \setminus O$, where $O \subset U$ is discrete subset.

Moreover, the existence of a solution z implies the existence of a family $z(\rho)$ of similar solutions depending on a parameter ρ: one obtains such a family of solutions by using an integral $\theta + \rho$ instead of the integral θ (see Lemma 2.17). If the parameter ρ has a large absolute value $|\rho| >> 0$ then for a point $a \in U_P$ the germ $z(\rho)$ can be expanded as a Puiseux series in θ_ρ:

$$z_i(\rho) = z_{i_k}\theta_\rho^{\frac{k}{p}} + z_{i_{k-1}}\theta_\rho^{\frac{k-1}{p}} + \cdots \tag{21}$$

The series is convergent and so it can be differentiated using the relation $\theta'_\rho = f$.

Lemma 2.18. *If $z'_{i_k} \neq 0$ then the leading term of the Puiseux series for $z_i(\rho)'$ is $z'_{i_k}\theta_\rho^{\frac{k}{p}}$. Otherwise the leading term has degree $< \frac{k}{p}$. The leading term of the derivative of any order of z_{i_k} has degree $\leq \frac{k}{p}$.*

Let us plug into the differential polynomial $T(u, u', \cdots, u^{(N)})$ the germ (21) and develop the result into a Puiseux series in θ_ρ. If the germ $z_i(\rho)$ is a solution of the equation (20) then all terms of this Puiseux series are equal to zero. In particular the leading coefficient is zero. This observation is an important step toward proving Theorem 2.12.

2.4.6 An extension by exponential of integral

Here we consider extensions of transcendence degree one of a differential field K containing an exponential integral y over K which is not algebraic over K.

2.4.6.1 A pure transcendental extension by an exponential integral
Let θ be an exponential of integral over K, i.e. $\theta' = f\theta$ where $f \in K$. Assume that θ is a transcendental element over K.[5]

Lemma 2.19. *1) The field $K\langle\theta\rangle$ is isomorphic to the field $K(y)$ of rational functions over K equipped with the following differentiation*

$$R' = \frac{\partial R}{\partial x} + \frac{\partial R}{\partial y}fy. \tag{22}$$

2) For every complex number $\mu \in \mathbb{C}^*$ not equal to zero the map $\theta \to \mu\theta$ can be extended to the unique isomorphism $G_\mu : K\langle\theta\rangle \to K\langle\theta\rangle$ which fixes elements of the field K.

3) Each isomorphism of $K\langle\theta\rangle$ over K is an isomorphism G_μ for some $\mu \in \mathbb{C}^*$. Thus the Galois group of $K\langle\theta\rangle$ over K is the multiplicative group of complex numbers $\mathbb{C}^*$.

Proof. Statement 1) follows from Lemma 2.14. For any $\mu \in \mathbb{C}^*$ the element $\theta_\mu = \mu\theta$ is a transcendental element over K and $\theta'_\mu = f\theta$. Thus 2) holds. Statement 3) follows from 2) because if $y' = fy$ and $y \neq 0$ then $y = \theta_\mu$ for some $\mu \in \mathbb{C}^*$. □

---

[5] It is easy to check that if an exponential of integral θ over K is algebraic over K, then θ is a radical over K, i.e. $\theta^k \in K$ for some positive integral k (see [Kho14]). We will not use this fact.

2.4.6.2 A generalized extension by exponential of integral According to Lemma 2.19 the differential field $K\langle\theta\rangle$ is isomorphic to the field $K(y)$ with the differentiation given by (22). Let F be an extension of $K\langle\theta\rangle$ by an element $z \in F$ which satisfies some equation $\tilde{P}(z) = 0$ where $\tilde{P}$ is an irreducible polynomial over $K\langle\theta\rangle$. The isomorphism between $K\langle\theta\rangle$ and $K(y)$ transforms the polynomial $\tilde{P}$ into some polynomial P over $K(y)$. Below we use the notation from section 2.4.3 and we deal with the multivalued algebroid function z on X defined by $P(z) = 0$.

Assume that at a point $x \in X$ there are germs of an analytic function z_i satisfying the equation $P(z_i) = 0$. Let θ_μ be the function $(\mu\theta) : U_P \to \mathbb{C}^1$ and let $\Theta_\mu \subset X = U \times \mathbb{C}^1$ be its graph. The point $x = (p, q) \in U \times \mathbb{C}^1$ where $q \neq 0$ belongs to the graph $\Theta_{\mu(x)}$ if $\mu(x) = q \cdot \theta(p)^{-1}$.

Let $K(y)|_{\Theta_{\mu(x)}}$ be the differential field of germs at the point $x \in \Theta_{\mu(x)}$ of restrictions on $\Theta_{\mu(x)}$ of functions from the field $K(y)$ equipped with the differentiation given by (22).

Lemma 2.20. *The differential field F is isomorphic to the finite extension of the differential field $K(y)|_{\Theta_{\mu(x)}}$ obtained by adjoining the germ at $x \in \Theta_{\mu(x)}$ of the restriction to $\Theta_{\mu(x)}$ of an analytic germ z_i satisfying $P(z_i) = 0$.*

Proof. For the trivial extension $F = K\langle\theta\rangle$ Lemma 2.20 follows from Lemmas 2.15 and 2.19. Theorem 2.5 allows one to complete the proof for non-trivial finite extensions F of $K\langle\theta\rangle$. □

According to section 2.4.3 with the polynomial P over $K(y)$ one can associate the finite extension K_P of the field K and the Riemann surface U_P such that Theorem 2.15 holds. Since K is a functional differential field the field K_P has the natural structure of a functional differential field. Below we will apply Lemma 2.20, taking instead of K the field K_P and considering the extension $F_P \supset K_P\langle\theta\rangle$ by the same algebraic element $z \in F$. The use of K_P instead of K allows us to apply the expansion (17) for z_i.

Theorem 2.17. *Let $x \in X_P = U_P \times \mathbb{C}^1$ be a point (a, y_0) with $|y| >> 0$. The differential field F_P is isomorphic to the extension of the differential field of germs at the point $a \in U_P$ by an exponential of an integral $\theta_{\mu(x)}$, where $\theta'_{\mu(x)} = f\theta_{\mu(x)}$ for the function $f \in K$, and by a germ at a of the composition $z_i(\theta_\mu)$ where z_i is a germ at x of a function given by a Puiseux series (17).*

Proof. Theorem 2.17 follows from Lemma 2.20 and Theorem 2.15. □

2.4.6.3 Solutions of equations in a generalized extension by exponential of integral Here we discuss Lemma 2.21, providing an important step toward our proof of Theorem 2.12. Assume that (20) has a solution z in a generalized extension by exponential of integral $F \supset K\langle\theta\rangle$ of K. The solution z has a nice global property: it is a meromorphic function on a Riemann surface V with a projection $\pi : V \to U$ which proves a locally trivial covering above $U \setminus O$, where $O \subset U$ is a discrete subset.

Moreover, the existence of a solution z implies the existence of a family $z(\mu)$ of similar solutions depending on a parameter $\mu \in \mathbb{C}^*$: one obtains such a family of solutions by using an exponential of integral $\mu\theta$ instead of the exponential of integral θ (see 2.20). If the parameter μ has a large absolute value $|\mu| >> 0$ then for a point $a \in U_P$ the germ $z(\mu)$ can be expanded as a Puiseux series in θ_μ:

$$z_i(\mu) = z_{i_k}\theta_\mu^{\frac{k}{p}} + z_{i_{k-1}}\theta_\mu^{\frac{k-1}{p}} + \cdots \tag{23}$$

The series is convergent and so it can be differentiated using the relation $\theta_\mu' = f\theta_\mu$.

Lemma 2.21. *If* $z_{i_k}' + \frac{k}{p}z_{i_k} \neq 0$ *then the leading term of the Puiseux series for* $z_i(\mu)'$ *is* $(z_{i_k}' + \frac{k}{p}z_{i_k})\theta_\mu^{\frac{k}{p}}$. *Otherwise the leading term has degree* $< \frac{k}{p}$. *The leading term of the derivative of any order of* z_{i_k} *has degree* $\leq \frac{k}{p}$.

Let us plug into the differential polynomial $T(u, u', \ldots, u^{(N)})$ the germ (23) and develop the result as a Puiseux series in θ_μ. If the germ $z_i(\mu)$ is a solution of the equation (20) then all terms of this Puiseux series are equal to zero. In particular the leading coefficient is zero. This observation is an important step towards the proof of Theorem 2.12.

2.4.7 Proof of Rosenlicht's theorem

Here we complete an elementary proof of Theorem 2.12 discovered by Maxwell Rosenlicht [Ros73]. We will first prove the simpler Theorems 2.18 and 2.19 of a similar nature.

Theorem 2.18. *Assume that the equation (12) over a functional differential field K has a solution $z \in F$ where F is a generalized extension by integral of K. Then (12) has a solution in the algebraic extension K_P of K associated with the element $z \in F$.*

Proof. If the constant term of the differential polynomial $T(u, u', u'', \ldots) = u^n - Q(u, u', u'', \ldots)$ is equal to zero, then (12) has solution $u \equiv 0$ belonging to K. In this case we have nothing to prove.

Below we will assume that the constant term T_0 of T is not equal to zero. Thus the differential polynomial T has two special terms: the term u^n which is the only term of highest degree n and the term T_0 which is the only term of smallest degree zero.

Assume that (12) has a solution z in a generalized extension by integral $F \supset K\langle\theta\rangle$ of K. According to section 2.4.5.3 the existence of such a solution z implies the existence of a family $z(\rho)$ of germs of solutions depending on a parameter ρ such that when the absolute value $|\rho|$ is big enough $z(\rho)$ can be expanded as a Puiseux series (21) in θ_ρ.

We will show that the degree $\frac{k}{p}$ of the leading term in (21) is equal to zero and the leading coefficient $z_{i_0} \in K_P$ satisfies (12). This will prove Theorem 2.18.

According to Lemma 2.18 the leading term of the derivative of any order of z_{i_k} has degree $\leq \frac{k}{p}$. Thus the leading term of the Puiseux series obtained by plugging (21) instead of u into the differential polynomial Q has degree $< n\frac{k}{p}$. The leading term of the Puiseux series obtained by raising (21) to the n-th power is equal to

$n\frac{k}{p}$. If $\frac{k}{p} > 0$ this term cannot be canceled after plugging (21) instead of u into the differential polynomial T. Thus the degree $\frac{k}{p}$ cannot be positive.

Let us plug (21) into the differential polynomial $T(u, u', \dots) - T_0$. We will obtain a Puiseux series of negative degree if $\frac{k}{p} < 0$. Thus the term T_0 in the sum $(T - T_0) + T_0$ cannot be canceled. Thus $\frac{k}{p}$ cannot be negative.

We have proved that $\frac{k}{p} = 0$. If in this case we plug (21) into the differential polynomial $T(u, u', \dots)$ we obtain a Puiseux series having only one term of nonnegative degree which is equal to zero. From Lemma 2.18 it is easy to see that this term is equal to $T(z_{i_0}, z'_{i_0}, \dots)\theta_\rho^0$. Thus $z_{i_0} \in K_P$ is a solution of (12). Theorem 2.18 is proved. □

Theorem 2.19. *Assume that equation (12) over a functional differential field K has a solution $z \in F$ where F is a generalized extension by an exponential of integral of K. Then (12) has a solution in the algebraic extension K_P of K associated with the element $z \in F$.*

Proof. Theorem 2.19 can be proved in exactly the same way as Theorem 2.18. Instead of Lemma 2.18 one has to use Lemma 2.20. In the case when the leading term of the Puiseux expansion of z_μ has degree zero, the leading coefficient of its derivative is equal to z'_{i_0} (see Lemma 2.20). That is why the case $\frac{k}{p} = 0$ in Theorem 2.19 can be treated in exactly the same way as in Theorem 2.18. □

Now we are ready to prove Theorem 2.12.

Proof (of Theorem 35). By assumption, equation (12) has a solution $z \in F$, where F is an extension of K by generalized quadratures. By definition there is a chain $K = F_0 \subset \cdots \subseteq F_m$ such that $F \subset F_m$ and, for every $i = 0, \dots, m-1$, F_{i+1} is a finite extension of F_i, or F_{i+1} is a generalized extension by integral of F_i, or F_{i+1} is a generalized extension by exponential integral of F_i. We prove Theorem 2.12 by induction on the length m of the chain of extension. For $m = 1$, Theorem 2.12 follows from Theorem 2.18, or from Theorem 2.19. Assume that Theorem 2.12 is true for $m = k$. A chain $F_0 \subset F_1 \subset \cdots \subset F_{k+1}$ provides the chain $F_1 \subset \cdots \subset F_{k+1}$ of extensions of length k for the field F_1. Thus (12) has an algebraic solution z over the field F_1. The extension $F_0 \subset \tilde{F}_1$, where $\tilde{F}_1$ is the extension of F_1 by the element z, is either an algebraic extension, or extension by generalized integral or extension by generalized exponential of integral. Thus for the extension $F_0 \subset \tilde{F}_1$, Theorem 2.12 holds. We have completed the inductive proof of Theorem 2.12. □

3 Topological Galois Theory

3.1 Introduction

In this section we present an outline of topological Galois theory based on the book [Kho14]. The theory studies topological obstructions to solvability of equations "in finite terms", i.e. to their solvability by radicals, by elementary functions, by quadratures and by functions belonging to other Liouvillian classes.

As was discovered by Camille Jordan, the Galois group of an algebraic equation over the field of rational functions of several complex variables has a topological meaning: it is isomorphic to the monodromy group of the algebraic function defined by this algebraic equation. Therefore the monodromy group is responsible for the representability of an algebraic function by radicals.

In section 3.2 we present a topological proof of the nonrepresentability of algebraic functions by radicals. This proof is based on my old paper [Kho71]. It contains a germ of topological Galois theory.

Not only algebraic functions have a monodromy group. It is defined for any solution of a linear differential equation whose coefficients are rational functions and for many more functions, for which the Galois group does not make sense.

It is thus natural to try to use the monodromy group for these functions instead of the Galois group to prove that they do not belong to a certain Liouvillian class. This particular approach is implemented in topological Galois theory (see [Kho14]), which has a one-dimensional version and a multidimensional version.

In sections 3.3 and 3.4 we present an outline of the one-dimensional version and an outline of the multidimensional version. These sections contain definitions, statements of results and comments on them. Essentially no proofs are presented.

3.2 On the Representability of Algebraic Functions by Radicals

3.2.1 Introduction

This section is dedicated to a self-contained simple proof of the classical criteria for representability of algebraic functions of several complex variables by radicals. It also contains criteria for representability of algebroidal functions by composition of single-valued analytic functions and radicals, and a result related to Hilbert's 13th problem.

Consider an algebraic equation

$$P_n y^n + P_{n-1} y^{n-1} + \cdots + P_0 = 0 \tag{24}$$

whose coefficients $P_n, \ldots, P_0$ are polynomials of N complex variables $x_1, \ldots, x_N$. Camille Jordan discovered that the Galois group of equation (24) over the field $\mathcal{R}$ of rational functions of $x_1, \ldots, x_N$ has a topological meaning (see theorem 3.2 below): it is isomorphic to the *monodromy group* of equation (24).

According to Galois theory, equation (24) is solvable by radicals over the field R if and only if its Galois group is solvable. If equation (24) is irreducible it defines a multivalued algebraic function $y(x)$. Galois theory and Theorem 3.2 imply the following criteria for representability of an algebraic function by radicals, which consists of two statements:

1) *If the monodromy group of an algebraic function $y(x)$ is solvable, then $y(x)$ is representable by radicals.*

2) *If the monodromy group of an algebraic function $y(x)$ is not solvable, then $y(x)$ is not representable by radicals.*

We reduce the first statement to linear algebra (see Theorem 3.5 below) following the book [Kho14].

We prove the second statement topologically without using Galois theory. Vladimir Igorevich Arnold found the first topological proof of this statement [Ale04]. We use another topological approach (see Theorem 3.6 below) based on the paper [Kho71]. This paper contains the first result of topological Galois theory [Kho14] and it gave a hint for its further development.

3.2.2 Monodromy groups and Galois groups

Consider equation (24). Let $\Sigma \subset \mathbb{C}^N$ be the hypersurface defined by the equation $P_n J = 0$, where P_n is the leading coefficient and J is the discriminant of equation (24). The *monodromy group* of equation (24) is the group of all permutations of its solutions which are induced by motions around the singular set Σ of equation (24). Below we discuss this definition more precisely.

At a point $x_0 \in \mathbb{C}^N \setminus \Sigma$ the set Y_{x_0} of all germs of analytic functions satisfying equation (24) contains exactly n elements, i.e. $Y_{x_0} = \{y_1, \ldots, y_n\}$. Indeed, if $P_n(x_0) \neq 0$ then for $x = x_0$ equation (24) has n roots counted with multiplicities. If in addition $J(x_0) \neq 0$ then all these roots are simple. By the implicit function theorem each simple root can be extended to a germ of a regular function satisfying equation (24).

Consider a closed curve γ in $\mathbb{C}^N \setminus \Sigma$ beginning and ending at the point x_0. Given a germ $y \in Y_{x_0}$ we can continue it along the loop γ to obtain another germ $y_\gamma \in Y_{x_0}$. Thus each such loop γ corresponds to a permutation $S_\gamma : Y_{x_0} \to Y_{x_0}$ of the set Y_{x_0} that maps a germ $y \in Y_{x_0}$ to the germ $y_\gamma \in Y_{x_0}$. It is easy to see that the map $\gamma \to S_\gamma$ defines a homomorphism from the fundamental group $\pi_1(\mathbb{C}^N \setminus \Sigma, x_0)$ of the domain $\mathbb{C}^N \setminus \Sigma$ with base point x_0 to the group $S(Y_{x_0})$ of permutations. The *monodromy group* of equation (24) is the image of the fundamental group in the group $S(Y_{x_0})$ under this homomorphism.

Remark. Instead of the point x_0 one can choose any other point $x_1 \in \mathbb{C}^N \setminus \Sigma$. Such a choice will not change the monodromy group up to an isomorphism. To fix this isomorphism one can choose any curve $\gamma : I \to \mathbb{C}^N \setminus \Sigma$, where I is the segment $0 \leq t \leq 1$ and $\gamma(0) = x_0$, $\gamma(1) = x_1$, and identify each germ y_{x_0} of solution of (24) with its continuation y_{x_1} along γ.

Instead of the hypersurface Σ one can choose any bigger algebraic hypersurface D, $\Sigma \subset D \subset \mathbb{C}^N$. Such a choice will not change the monodromy group: one can slightly move a curve $\gamma \in \pi_1(\mathbb{C}^N \setminus \Sigma, x_0)$ without changing the map S_γ in such a way that γ will not intersect D.

The field of rational functions of $x_1, \ldots, x_N$ is isomorphic to the field $\mathcal{R}$ of germs of rational functions at the point $x_0 \in \mathbb{C}^N \setminus \Sigma$. Consider the field extension $\mathcal{R}\langle y_1, \ldots, y_n \rangle$ of $\mathcal{R}$ by the germs $y_1, \ldots y_n$ at x_0 satisfying equation (24).

Lemma 3.1. *Every permutation S_γ from the monodromy group can be uniquely extended to an automorphism of the field $\mathcal{R}\{y_1, \ldots, y_n\}$ over the field $\mathcal{R}$.*

Proof. Every element $f \in \mathcal{R}\langle y_1, \ldots, y_n \rangle$ is a rational function of $x, y_1, \ldots, y_n$. It can be continued meromorphically along the curve $\gamma \in \pi_1(\mathbb{C}^m \setminus \Sigma, x_0)$ together with $y_1, \ldots, y_n$. This continuation gives the required automorphism, because the continuation preserves the arithmetic operations and every rational function returns back to its original values (since it is a single-valued function). The automorphism is unique because the extension is generated by $y_1, \ldots, y_n$. □

By definition the *Galois group* of equation (24) is the group of all automorphisms of the field $\mathcal{R}\{y_1, \ldots, y_n\}$ over the field $\mathcal{R}$. According to Lemma 3.1 the monodromy group of equation (24) can be considered as a subgroup of its Galois group. Recall that by definition a multivalued function $y(x)$ is *algebraic* if all its meromorphic germs satisfy the same algebraic equation over the field of rational functions.

Theorem 3.1. *A germ $f \in \mathcal{R}\langle y_1, \ldots, y_n \rangle$ is fixed under the monodromy action if and only if $f \in \mathcal{R}$.*

Proof. A germ $f \in \mathcal{R}\langle y_1, \ldots, y_n \rangle$ is fixed under the monodromy action if and only if f is a germ of a single-valued function. The field $\mathcal{R}\langle y_1, \ldots, y_n \rangle$ contains only germs of algebraic functions. Any single-valued algebraic function is a rational function. □

According to Galois theory Theorem 3.1 can be formulated in the following way.

Theorem 3.2. *The monodromy group of equation (24) is isomorphic to the Galois group of equation (24) over the field $\mathcal{R}$.*

Below we will not rely on Galois theory. Instead we will use Theorem 3.2 directly.

Lemma 3.2. *The monodromy group acts on the set Y_{x_0} transitively if and only if equation (24) is irreducible over the field of rational functions.*

Proof. Assume that there is a proper subset $\{y_1, y_2, \ldots y_k\}$ of Y_{x_0} invariant under the monodromy action. Then the basic symmetric functions $r_1 = y_1 + \cdots + y_k$, $r_2 = \sum_{i<j} y_i y_j, \ldots, r_k = y_1 \cdots y_k$ belong to the field $\mathcal{R}$. Thus $y_1, y_2, \ldots y_k$ are solutions of the degree $k < n$ equation $y^k - r_1 y^{k-1} t + \cdots (-1)^k r_k = 0$. So equation (24) is reducible. On the other hand if equation (24) can be represented as a product of two equations over $\mathcal{R}$ then their roots belong to two complementary subsets of Y_{x_0} which are invariant under the monodromy action. □

Corollary 3.1. *An irreducible equation (24) defines a multivalued algebraic function $y(x)$ whose set of germs at $x_0 \in \mathbb{C}^N \setminus \Sigma$ is the set Y_{x_0} and whose monodromy group coincides with the monodromy group of equation (24).*

Theorem 3.2, Corollary 3.1 and Galois theory immediately imply the following result.

Theorem 3.3. *An algebraic function whose monodromy group is solvable can be represented by rational functions using the arithmetic operations and radicals.*

A stronger version of Theorem 3.3 can be proved using linear algebra (see Theorem 3.5 in the next section).

3.2.3 Action of solvable groups and representability by radicals

In this section, we prove that if a finite solvable group G acts on a $\mathbb{C}$-algebra V by automorphisms, then all elements of V can be expressed in terms of elements of the invariant subalgebra V_0 of G by taking radicals and adding.

This construction of a representation by radicals is based on linear algebra. More precisely we use the following well-known statement: any finite abelian group of linear transformations of a finite-dimensional vector space over $\mathbb{C}$ can be diagonalized in a suitable basis.

Lemma 3.3. *Let G be a finite abelian group of order n acting by automorphisms on a $\mathbb{C}$-algebra V. Then every element of the algebra V is representable as a sum of elements $x_i \in V$, such that x_i^n lies in the invariant subalgebra V_0 of G, i.e., in the fixed-point set of the group G.*

Proof. Consider a finite-dimensional vector subspace L of the algebra V spanned by the G-orbit of an element x. The space L splits into a direct sum $L = L_1 + \cdots + L_k$ of eigenspaces for all operators from G. Therefore, the vector x can be represented in the form $x = x_1 + \cdots + x_k$, where $x_1, \ldots, x_k$ are eigenvectors for all the operators from the group. The corresponding eigenvalues are n-th roots of unity. Therefore, the elements $x_1^n, \ldots, x_k^n$ belong to the invariant subalgebra V_0. □

Definition 3.1. We say that an element x of the algebra V is an *n-th root* of an element a if $x^n = a$.

We can now restate Lemma 3.3 as follows: every element x of the algebra V is representable as a sum of n-th roots of some elements of the invariant subalgebra.

Theorem 3.4. *Let G be a finite solvable group of order n acting by automorphisms on a $\mathbb{C}$-algebra V. Then every element x of the algebra V can be obtained from the elements of the invariant subalgebra V_0 by takings n-th roots and summing.*

We first prove the following simple statement about the action of a group on a set. Suppose that a group G acts on a set X, let H be a normal subgroup of G, and denote by X_0 the subset of X consisting of all points fixed under the action of G.

Lemma 3.4. *The subset X_H of the set X consisting of the fixed points under the action of the normal subgroup H is invariant under the action of G. There is a natural action of the quotient group G/H on the set X_H with fixed-point set X_0.*

Proof. Suppose that $g \in G$, $h \in H$. Then the element $g^{-1}hg$ belongs to the normal subgroup H. Let $x \in X_H$. Then $g^{-1}hg(x) = x$, or $h(g(x)) = g(x)$, which means that the element $g(x) \in X$ is fixed under the action of the normal subgroup H. Thus the set X_H is invariant under the action of the group G. Under the action of G on X_H, all elements of H correspond to the identity transformation. Hence the action of G on X_H reduces to an action of the quotient group G/H. □

We now proceed with the proof of Theorem 3.4.

Proof. (of Theorem 3.4) Since the group G is solvable, it has a chain of nested subgroups $G = G_0 \supset \cdots \supset G_m = e$ in which the group G_m consists of the identity element e only, and every group G_i is a normal subgroup of the group G_{i-1}. Moreover, the quotient group G_{i-1}/G_i is abelian. Let $V_0 \subset \cdots \subset V_m = V$ denote the chain of invariant subalgebras of the algebra V with respect to the action of the groups $G_0, \ldots, G_m$. By Lemma 3.4 the abelian group G_{i-1}/G_i acts naturally on the invariant subalgebra V_i, leaving the subalgebra V_{i-1} pointwise fixed. The order m_i of the quotient group G_{i-1}/G_i divides the order of the group G. Therefore, Lemma 3.3 is applicable to this action. We conclude that every element of the algebra V_i can be expressed with the help of summation and n-th root extraction in terms of the elements of the algebra V_{i-1}. Repeating the same argument, we will be able to express every element of the algebra V in terms of the elements of the algebra V_0 using a chain of summations and n-th root extractions. □

Theorem 3.5. *An algebraic function whose monodromy is solvable can be represented by rational functions by root extractions and summations.*

Proof. One can prove Theorem 3.5 by applying Theorem 3.4 to the monodromy action by automorphisms on the extension $\mathcal{R}\langle y_1, \ldots, y_n \rangle$ with the field of invariants $\mathcal{R}$. □

3.2.4 Topological obstructions to representation by radicals

Let us introduce some notation.

By G^m we denote the m-th commutator subgroup of the group G. For any $m \geq 1$ the group G^m is a normal subgroup of G.

By $F(D, x_0)$ we denote the fundamental group of the domain $U = \mathbb{C}^N \setminus D$ with the base point $x_0 \in U$, where D is an algebraic hypersurface in $\mathbb{C}^N$.

Let $H(D, m)$ be the covering space of the domain $\mathbb{C}^N \setminus D$ corresponding to the subgroup $F^m(D, x_0)$ of the fundamental group $F(D, x_0)$.

We will say that an algebraic function is an *R-function* if it becomes a single-valued function on some covering $H(D, m)$.

Lemma 3.5. *If $m_1 \geq m_2$ and $D_1 \supset D_2$ then there is a natural projection $\rho : H(D_1, m_1) \to H(D_2, m_2)$. Thus if a function y becomes a single-valued function on $H(D_2, m_2)$ then it certainly becomes a single-valued function on $H(D_1, m_1)$.*

Proof. Let $p_* : F(D_1, x_0) \to F(D_2, x_0)$ be the homomorphism induced by the embedding $p : \mathbb{C}^N \setminus D_1 \to \mathbb{C}^N \setminus D_2$. Lemma 3.5 follows from the obvious inclusions: $p_*^{-1}[F^{m_2}(D_2, x_0)] \subset F^{m_2}(D_1, x_0)$ and $F^{m_2}(D_1, x_0) \subset F^{m_1}(D_1, x_0)$. □

Lemma 3.6. *If y_1 and y_2 are R-functions then $y_1 + y_2$, $y_1 - y_2$, $y_1 \cdot y_2$ and y_1/y_2 are also R-functions.*

Proof. Assume that the R-functions y_1 and y_2 become single-valued on the coverings $H(D_1, m_1)$ and $H(D_2, m_2)$. By Lemma 3.5 the functions y_1, y_2 become single-valued on the covering $H(D, m)$ where $D = D_1 \bigcup D_2$ and $m = \max(m_1, m_2)$. Thus

170

the functions $y_1 + y_2$, $y_1 - y_2$, $y_1 \cdot y_2$ and y_1/y_2 also become single-valued on the covering $H(D, m)$. This completes the proof since $y_1 + y_2$, $y_1 - y_2$, $y_1 \cdot y_2$ and y_1/y_2 are algebraic functions. $\qquad\square$

Lemma 3.7. *The composition of an R-function with the degree q radical is an R-function.*

Proof. Assume that the function y defined by (24) is an R-function which becomes a single-valued function on the covering $H(D_1, m)$. Let $D_2 \subset \mathbb{C}^N$ be the hypersurface defined by the equation $P_n P_0 = 0$, where P_n and P_0 are the leading coefficient and the constant term of the equation (24). According to Lemma 3.5 the function y becomes a single-valued function on the covering $H(D, m)$ where $D = D_1 \bigcup D_2$. Let $h_0 \in H(D, m)$ be a point whose image under the natural projection $\rho : H(D, m) \to \mathbb{C}^N \setminus D$ is the point x_0. One can identify the fundamental groups $\pi_1(H(D, m), h_0)$ and $F^m(D, x_0)$.

By definition of D_2 the function y is never equal to zero or to infinity on $H(D, m)$. Hence y defines a map $y : H(D, m) \to \mathbb{C} \setminus \{0\}$. Let $y_* : \pi_1(H(D, m), h_0) \to \pi_1(\mathbb{C} \setminus \{0\}, y(h_0))$ be the induced homomorphism of the fundamental groups. The group $\pi_1(H(D, m), h_0)$ is identified with the group $F^m(D, x_0)$ and the group $\pi_1(\mathbb{C} \setminus \{0\}, y(h_0))$ is isomorphic to $\mathbb{Z}$. So $\ker y_* \subset F^{m+1}(D, x_0)$. Thus all loops from the group $y_*(F^{m+1}(D, x_0))$ do not wind around the origin $0 \in \mathbb{C}$. Hence any germ of $y^{1/q}$ does not change its value after continuation along a loop from the group $F^{m+1}(D, x_0)$. So $y^{1/q}$ is a single-valued function on $H(D, m+1)$. This completes the proof since $y^{1/q}$ is an algebraic function. $\qquad\square$

Lemma 3.8. *An algebraic function y is an R-function if and only if its monodromy group is solvable.*

Proof. Assume that y is defined by (24). Let D be the hypersurface $P_n J = 0$ where P_n is the leading coefficient and J is the discriminant of (24). Let M be the monodromy group of y. Consider the natural homomorphism $p : F(D, x_0) \to M$. If M is solvable then for some natural number m the m-th commutator of M is the identity element e. The function y becomes single-valued on the covering $H(D, m)$ since $F^m(D, x_0) \subset p^{-1}(M^m) = p^{-1}(e)$. Conversely, if y is a single-valued function on some covering $H(D, m)$ then $p(F^m(D, x_0)) = e$. But $p(F^m(D, x_0)) = M^m$. Thus the monodromy group M is solvable. $\qquad\square$

Theorem 3.6. *If an algebraic function has an unsolvable monodromy group then it cannot be represented by compositions of rational functions and radicals*

Proof. Lemma 3.7 and Lemma 3.8 show that the class of R-functions is closed under arithmetic operations and compositions with radicals. Lemma 3.8 shows that the monodromy group of any R-function is solvable. $\qquad\square$

3.2.5 Compositions of analytic functions and radicals

In this section we describe a class of multivalued functions on a domain $U \subset \mathbb{C}^N$ representable by composition of single-valued analytic functions and radicals.

A multivalued function y on U is called an *algebroidal function* on U if it satisfies an irreducible equation

$$y^n + f_{n-1}y^{n-1} + \cdots + f_0 = 0 \tag{25}$$

whose coefficients $f_{n-1}, \ldots, f_0$ are analytic functions on U. An algebroidal function can be considered as a continuous multivalued function on U which has finitely many values.

Theorem 3.7 ([Kho70], [Kho71]). *A multivalued function y in the domain U can be represented by composition of radicals and single-valued analytic functions if and only y is an algebroidal function in U with solvable monodromy group.*

To prove the "only if" part one can repeat the proof of Theorem 3.6 replacing coverings over domains $\mathbb{C}^N \setminus D$ by coverings over domains $U \setminus \tilde{D}$, where $\tilde{D}$ is an analytic hypersurface in U.

To prove Theorem 3.7 in the opposite direction one can use Theorem 3.4 in the same way as it was used in the proof of Theorem 3.5.

3.2.6 Local representability

In this section we describe a local version of Theorem 3.7.

Let y be an algebroidal function in U defined by (25). One can localize the equation (25) at any point $p \in U$, i.e. one can replace the coefficients f_i of equation (25) by their germs at p. After such a localization equation (25) can become reducible, i.e. it can become representable as a product of irreducible equations. Thus an algebroidal function y on an arbitrarily small neighborhood of a point p defines several algebroidal functions, which we will call *ramified germs of y at p*. For a ramified germ of y at p the monodromy group is defined (as the monodromy group of an algebroidal function on an arbitrarily small neighborhood of the point p).

A ramified germ of an algebroidal function y of one variable x on a neighborhood of a point $p \in \mathbb{C}^1$ has a simple structure: its monodromy group is a cyclic group $\mathbb{Z}/m\mathbb{Z}$ and it can be represented as a composition of a radical and an analytic single-valued function: $y(x) = f((x - p)^{1/m}))$, where m is the ramification order of y. The next corollary follows from Theorem 3.7.

Corollary 3.2 ([Kho70], [Kho71]).

1) If a multivalued function y on the domain U can be represented by composition of an algebroidal function of one variable and single-valued analytic functions then the monodromy group of any ramified germ of y is solvable.

2) If the monodromy group of a ramification germ of y at p is solvable then in a small neighborhood of p it can be represented by composition of radicals and single-valued analytic functions.

The *local monodromy group* of an algebroidal function y at a point $p \in U$ is the monodromy group of the equation (25) in an arbitrarily small neighborhood of the point p. The ramified germs of y at the point p correspond to the orbits of the local monodromy group actions. This statement can be proved in the same way as Lemma 3.2.

3.2.7 Application to Hilbert's 13th problem

In 1957 A.N. Kolmogorov and V.I. Arnold proved the following totally unexpected theorem, which gave a negative solution to Hilbert's 13th problem.

Theorem 3.8 (Kolmogorov–Arnold). *Any continuous function of n variables can be represented as the composition of functions of a single variable with the help of addition.*

Hilbert's 13th problem has the following algebraic version which still remains open: *Is it possible to represent any algebraic function of n > 1 variables by algebraic functions of a smaller number of variables with the help of composition and arithmetic operations?*

An *entire algebraic function* y on $\mathbb{C}^N$ is an algebraic function defined on $U = \mathbb{C}^N$ by an equation (25) whose coefficient f_i are polynomials. An entire algebraic function can be considered as a continuous algebraic function.

It turns out that in the Kolmogorov–Arnold Theorem one cannot replace continuous functions by entire algebraic functions.

Theorem 3.9 ([Kho70], [Kho71]). *If an entire algebraic function can be represented as a composition of polynomials and entire algebraic functions of one variable, then its local monodromy group at each point is solvable.*

Proof. This follows from Corollary 3.2. □

Corollary 3.3. *A function $y(a, b)$ defined by the equation $y^5 + ay + b = 0$ cannot be expressed in terms of entire algebraic functions of a single variable by means of composition, addition and multiplication.*

Proof. Indeed, it is easy to check that the local monodromy group of y at the origin is the unsolvable permutation group S_5 (see [Kho70], [Kho71]). □

Division is not a continuous operation and it destroys the locality. One cannot add division to the operations used in Theorem 3.9. It is easy to see that the function $y(a, b)$ from Corollary 3.2 can be expressed in terms of entire algebraic functions of a single variable by means of composition and arithmetic operations: $y(a, b) = g(b/\sqrt[4]{a^5})\sqrt[4]{a}$, where $g(u)$ is defined by the equation $g^5 + g + u = 0$.

The following particular case of the algebraic version of Hilbert's 13th problem still remains open.

Problem. *Show that there is an algebraic function of two variables which cannot be expressed in terms of algebraic functions of a single variable by means of composition and arithmetic operations.*

3.3 One-Dimensional Topological Galois Theory

3.3.1 Introduction

In the one-dimensional version we consider functions from Liouvillian classes as multivalued analytic functions of one complex variable. It turns out that there exist topological restrictions on the way the Riemann surface of a function from a certain Liouvillian class can be positioned over the complex plane. If a function does not satisfy these restrictions, then it cannot belong to the corresponding Liouvillian class.

Besides a geometric appeal, this approach has the following advantage. Topological obstructions relate to branching. It turns out that if a function does not belong to a certain Liouvillian class for topological reasons then it automatically does not belong to a much wider class of functions. This wider class can be obtained if we add to the Liouvillian class all single-valued functions having at most a countable set of singularities and allow them to enter all formulas.

Composition of functions is not an algebraic operation. In differential algebra, this operation is replaced with a differential equation describing it. However, for example, the Euler Γ-function does not satisfy any algebraic differential equation. Hence it is pointless to look for an equation satisfied by, say, the function $\Gamma(\exp x)$ and one cannot describe it algebraically (but the function $y = \exp(\Gamma(x))$ satisfies the equation $y' = \Gamma' y$ over a differential field containing Γ and it makes sense in the differential algebra). The only known results on non-representability of functions by quadratures and, say, the Euler Γ-function are obtained by our method.

On the other hand, our method cannot be used to prove that a particular single-valued meromorphic function does not belong to a certain Liouvillian class.

There are the following topological obstructions to representability of functions by generalized quadratures, k-quadratures and quadratures (see section 3.3.6).

Firstly, the functions representable by generalized quadratures and, in particular, the functions representable by k-quadratures and quadratures may have no more than countably many singular points in the complex plane (see section 3.3.4).

Secondly, the monodromy group of a function representable by quadratures is necessarily solvable (see section 3.3.6). There are similar restrictions for a function representable by generalized quadratures and k-quadratures. However, these restrictions are more involved. To state them, the monodromy group should be regarded not as an abstract group but rather as a transitive subgroup in the permutation group. In other terms, these restrictions make use not only of the monodromy group but rather of the *monodromy pair* of the function consisting of the monodromy group and the stabilizer of some germ of the function (see section 3.3.7).

One can prove that the only reasons for unsolvability in finite terms of Fuchsian linear differential equations are topological (see section 3.3.12). In other words, if there are no topological obstructions to solvability of a Fuchsian equation by generalized quadratures (by k-quadratures, by quadratures), then this equation is solvable by generalized quadratures (by k-quadratures or by quadratures respectively). The proof is based on a linear-algebraic part of differential Galois theory (dealing with linear algebraic groups and their differential invariants).

3.3.2 Solvability in finite terms and Liouvillian classes of functions

An equation is solvable "in finite terms" (or is solvable "explicitly") if its solutions belong to a certain class of functions. Different classes of functions correspond to different notions of solvability in finite terms.

A class of functions can be introduced by specifying a list of *basic functions* and a list of *admissible operations* (see section 2.2.4). Given the two lists, the class of functions is defined as the set of all functions that can be obtained from the basic functions by repeated application of admissible operations (see section 2.2.4). Below, we define classes of functions in exactly this way.

Liouvillian classes of functions, which appear in problems of integrability in finite terms, contain multivalued functions. Thus the operations on multivalued functions have to be defined. Such a definition can be found in section 2.2.4 (note that in the multidimensional case we use a slightly different, more restricted definition of the operations on multivalued functions).

We need the list of *basic elementary functions* (see section 2.2.5). In essence, this list contains functions that are studied in high school and which are frequently used in pocket calculators.

We also use the *classical operations* on functions, such as the arithmetic operations, the operation of composition and so on. The list of such operations is presented in section 2.2.5.

We can now return to the definition of Liouvillian classes of single variable functions.

3.3.2.1 Functions representable by radicals
List of basic functions: all complex constants, an independent variable x. List of admissible operations: arithmetic operations and the operation of taking the n-th root $f^{\frac{1}{n}}$, $n = 2, 3, \ldots$, of a given function f.

3.3.2.2 Functions representable by k-radicals
List of basic functions: all complex constants, an independent variable x. List of admissible operations: arithmetic operations and the operation of taking the n-th root $f^{\frac{1}{n}}$, $n = 2, 3, \ldots$, of a given function f, the operation of solving algebraic equations of degree $\leq k$.

3.3.2.3 Elementary functions
List of basic functions: basic elementary functions. List of admissible operations: compositions, arithmetic operations, differentiation.

3.3.2.4 Generalized elementary functions
This class of functions is defined in the same way as the class of elementary functions. We only need to add the operation of solving algebraic equations to the list of admissible operations.

3.3.2.5 Functions representable by quadratures
List of basic functions: basic elementary functions. List of admissible operations: compositions, arithmetic operations, differentiation, integration.

3.3.2.6 Functions representable by k-quadratures This class of functions is defined in the same way as the class of functions representable by quadratures. We only need to add the operation of solving algebraic equations of degree at most k to the list of admissible operations.

3.3.2.7 Functions representable by generalized quadratures This class of functions is defined in the same way as the class of functions representable by quadratures. We only need to add the operation of solving algebraic equations to the list of admissible operations.

3.3.3 Simple formulas with complicated topology

Developing topological Galois theory I adopted the following plan:

I. To find a wide class of multivalued functions such that:

    a) it is closed under all classical operations;
    b) it contains all entire functions and all functions from each Liouvillian class;
    c) for functions from the class the monodromy group is well defined.

II. To use the monodromy group instead of the Galois group inside the class.

Let us discuss some difficulties that one needs to overcome on this way.

Example 3.1. Consider an elementary function f defined by the following formula:

$$f(z) = \ln \sum_{j=1}^{n} \lambda_j \ln(z - a_j)$$

where a_j are different points in the complex line, and $\lambda_j \in \mathbb{C}$ are constants.

Let Λ denote the additive subgroup of complex numbers generated by the constants $\lambda_1, \dots, \lambda_n$. It is clear that if $n > 2$, then for almost every collection of constants $\lambda_1, \dots, \lambda_n$, the group Λ is everywhere dense in the complex line.

Lemma 3.9. *If the group Λ is dense in the complex line, then the elementary function f has a dense set of logarithmic ramification points.*

Proof. Let g be the multivalued function defined by the formula

$$g(z) = \sum_{j=1}^{n} \lambda_j \ln(z - a_j).$$

Take a point $a \neq a_j$, $j = 1, \dots, n$, and let g_a be one of the germs of g at a. A loop around the points $a_1, \dots, a_n$ adds the number $2\pi i \lambda$ to the germ g_a, where λ is an element of the group Λ. Conversely, every germ $g_a + 2\pi i \lambda$, where $\lambda \in \Lambda$, can be obtained from the germ g_a by analytic continuation along some loop. Let U be a small neighborhood of the point a, such that the germ g_a has a single-valued analytic continuation G on U. The image V of the domain U under the map $G : U \to \mathbb{C}$ is open. Therefore, in the domain V, there is a point of the form $2\pi i \lambda$, where $\lambda \in \Lambda$. The function $G - 2\pi i \lambda$ is one of the branches of the function g over the domain U, and the zero set of this branch in the domain U is nonempty. Hence, one of the branches of the function $f = \ln g$ has a logarithmic ramification point in U. $\square$

The set Σ of singular points of the function f is a *countable set* (see section 3.3.4). Under the assumptions of Lemma 3.9 the set Σ is everywhere dense.

It is not hard to verify that the monodromy group (see section 3.3.7) of the function f has the cardinality of the continuum. This is not surprising: the fundamental group $\pi_1(\mathbb{C} \setminus \Sigma)$ obviously has the cardinality of the continuum provided that Σ is a countable dense set.

One can also prove that the image of the fundamental group $\pi_1(\mathbb{C} \setminus \{\Sigma \cup b\})$ of the complement of the set $\Sigma \cup b$, where $b \notin \Sigma$, in the permutation group is a proper subgroup of the monodromy group of f.

The fact that the removal of one extra point can change the monodromy group makes all the proofs more complicated.

Thus even the simplest elementary functions can have dense singular sets and monodromy groups of cardinality of the continuum. In addition the removal of an extra point can change their monodromy groups.

3.3.4 The class of S-functions

In this section, we define a broad class of functions of one complex variable needed in the construction of topological Galois theory.

Definition 3.2. A multivalued analytic function of one complex variable is called an *S-function* if the set of its singular points is at most countable.

Let us make this definition more precise. Two regular germs f_a and g_b defined at points a and b of the Riemann sphere $\mathbb{S}^2$ are called *equivalent* if the germ g_b is obtained from the germ f_a by analytic continuation along some path. Each germ g_b equivalent to the germ f_a is also called a regular germ of the multivalued analytic function f generated by the germ f_a.

A point $b \in \mathbb{S}^2$ is said to be a *singular point* for the germ f_a if there exists a path $\gamma : [0,1] \to \mathbb{S}^2$, $\gamma(0) = a$, $\gamma(24) = b$ such that the germ has no analytic continuation along this path, but for any τ, $0 \leq \tau < 1$, it admits an analytic continuation along the truncated path $\gamma : [0, \tau] \to S^2$.

It is easy to see that equivalent germs have the same set of singular points. A regular germ is called an *S-germ* if the set of its singular points is at most countable. A multivalued analytic function is called an S-function if each of its regular germs is an S-germ.

Theorem 3.10 (on stability of the class of S-functions). *The class S of all S-functions is stable under the following operations:*

1. *differentiation, i.e. if $f \in S$, then $f' \in S$;*
2. *integration, i.e. if $f \in S$ and $g' = f$, then $g \in S$;*
3. *composition, i.e. if g, $f \in S$, then $g \circ f \in S$;*
4. *meromorphic operations, i.e. if $f_i \in S$, $i = 1, \ldots, n$, the function $F(x_1, \ldots, x_n)$ is a meromorphic function of n variables, and $f = F(f_1, \ldots, f_n)$, then $f \in S$;*
5. *solving algebraic equations, i.e. if $f_i \in S$, $i = 1, \ldots, n$, and $f^n + f_1 f^{n-1} + \cdots + f_n = 0$, then $f \in S$;*
6. *solving linear differential equations, i.e. if $f_i \in S$, $i = 1, \ldots, n$, and $f^{(n)} + f_1 f^{(n-1)} + \cdots + f_n f = 0$, then $f \in S$.*

Remark. The arithmetic operations and exponentiation are examples of meromorphic operations, hence *the class of S-functions is stable under the arithmetic operations and exponentiation.*

Corollary 3.4 (see [Kho14]). *If a multivalued function f can be obtained from single-valued S-functions by integration, differentiation, meromorphic operations, compositions, solutions of algebraic equations and linear differential equations, then the function f has at most a countable number of singular points.*

Corollary 3.5. *A function having uncountably many singular points cannot be represented by generalized quadratures. In particular, it cannot be a generalized elementary function and it cannot be represented by k-quadratures or by quadratures.*

Example 3.2. Consider a discrete group Γ of fractional linear transformations of the open unit ball U having a compact fundamental domain. Let f be a nonconstant meromorphic function on U invariant under the action of Γ. Each point on the boundary ∂U belongs to the closure of the set of poles of f, thus the set Σ of singular points of f contains ∂U. So Σ has the cardinality of the continuum and f cannot be expressed by generalized quadratures.

3.3.5 The monodromy group of an S-function

The *monodromy group* of an S-function f is the group of all permutations of the sheets of the Riemann surface of f which are induced by motions around the singular set Σ of the function f. Below we discuss this definition more precisely.

Let F_{x_0} be the set of all germs of the S-function f at a point $x_0 \notin \Sigma$. Consider a closed curve γ in $\mathbb{S}^2 \setminus \Sigma$ beginning and ending at the point x_0. Given a germ $y \in F_{x_0}$ we can continue it along the loop γ to obtain another germ $y_\gamma \in Y_{x_0}$. Thus each such loop γ corresponds to a permutation $S_\gamma : F_{x_0} \to F_{x_0}$ of the set F_{x_0} that maps a germ $y \in F_{x_0}$ to the germ $y_\gamma \in F_{x_0}$.

It is easy to see that the map $\gamma \to S_\gamma$ defines a homomorphism from the fundamental group $\pi_1(\mathbb{S}^2 \setminus \Sigma, x_0)$ of the domain $\mathbb{S}^2 \setminus \Sigma$ with the base point x_0 to the group $S(F_{x_0})$ of permutations. The *monodromy group* of the S-function f is the image of the fundamental group in the group $S(F_{x_0})$ under this homomorphism.

Remark. Instead of the point x_0 one can choose any other point $x_1 \in \mathbb{S}^2 \setminus \Sigma$. Such a choice will not change the monodromy group up to an isomorphism. To fix this isomorphism one can choose any curve $\gamma : I \to \mathbb{C}^N \setminus \Sigma$, where I is the segment $0 \le t \le 1$ and $\gamma(0) = x_0, \gamma(1) = x_1$, and identify each germ f_{x_0} of f with its continuation f_{x_1} along γ.

3.3.6 Strong non-representability by quadratures

One can prove the following theorem.

Theorem 3.11 (see [Kho14]). *The class of all S-functions having a solvable monodromy group is stable under composition, meromorphic operations, integration and differentiation.*

Definition 3.3. A function f is *strongly nonrepresentable by quadratures* if it does not belong to a class of functions defined by the following data. List of basic functions: basic elementary functions and all single-valued S-functions. List of admissible operations: compositions, meromorphic operations, differentiation and integration.

Theorem 3.11 implies the following corollary.

Corollary 3.6 (Result on quadratures). *If the monodromy group of an S-function f is not solvable, then f is strongly non-representable by quadratures.*

Example 3.3. The monodromy group of an algebraic function $y(x)$ defined by an equation $y^5 + y - x = 0$ is the unsolvable group S_5. Thus $y(x)$ provides an example of a function with finite set of singular points, which is strongly non-representable by quadratures.

The following corollary contains a stronger result on non-representability of algebraic functions by quadratures.

Corollary 3.7. *If an algebraic function of one complex variable has an unsolvable monodromy group then it is strongly non-representable by quadratures.*

For algebraic functions of several complex variables there is a result similar to Corollary 3.7.

3.3.7 The monodromy pair

The monodromy group of a function f is not only an abstract group but is also a transitive group of permutations of germs of f at a non-singular point x_0.

Definition 3.4. The *monodromy pair of an S-function f* is a pair of groups consisting of the monodromy group of f at x_0 and the stationary subgroup of a certain germ of f at x_0.

The monodromy pair is well defined, i.e. this pair of groups, up to isomorphism, does not depend on the choice of the non-singular point and on the choice of the germ of f at this point. The intersection of the stationary subgroups of all germs of f at x_0 is the identity element since the monodromy group acts transitively on this set.

Definition 3.5. A pair of groups $[\Gamma, \Gamma_0]$ is an *almost normal pair* if there is a finite subset $A \subset \Gamma$ such that the intersection $\bigcap_{a \in A} a(\Gamma_0)a^{-1}$ is equal to the identity element.

Example 3.4. Let y_0 be a germ at a point a of a solution y of an equation

$$y^{(n)} + f_1 y^{(n-1)} + \cdots + f_n y = 0$$

whose coefficients are one-valued S-functions (for example, meromorphic functions on the complex line). Then the monodromy pair of y_0 is almost normal. To see this let L be the linear space of germs at the point a spanned by all germs of the function y at the point a. Since $\dim L \leq n$, one can choose finitely many germs $y_{0,i}$ that generate L. Let $\Gamma_i \subset \Gamma$ be the stationary subgroup of the germ $y_{0,i}$. The groups Γ_i are conjugate since they are the stationary subgroups of different germs of the same multivalued function. The intersection $\cap \Gamma_i$ fixes the space L so it fixes all germs of the function y. This $\cap \Gamma_i$ is the identity element in Γ.

Definition 3.6. A pair of groups $[\Gamma, \Gamma_0]$ is called an *almost solvable pair of groups* if there exists a sequence of subgroups

$$\Gamma = \Gamma_1 \supseteq \cdots \supseteq \Gamma_m, \quad \Gamma_m \subset \Gamma_0,$$

such that for every i, $1 \leq i \leq m - 1$, the group Γ_{i+1} is a normal divisor of the group Γ_i and the factor group Γ_i / Γ_{i+1} is either a commutative group, or a finite group.

Definition 3.7. A pair of groups $[\Gamma, \Gamma_0]$ is called a *k-solvable pair of groups* if there exists a sequence of subgroups

$$\Gamma = \Gamma_1 \supseteq \cdots \supseteq \Gamma_m, \quad \Gamma_m \subset \Gamma_0,$$

such that for every i, $1 \leq i \leq m - 1$, the group Γ_{i+1} is a normal divisor of the group Γ_i and the factor group Γ_i / Γ_{i+1} is either a commutative group, or a subgroup of the group S_k of permutations of k elements.

We say that the group Γ is *almost solvable* or *k-solvable* if the pair $[\Gamma, e]$, where e is the group containing only the unit element, is almost solvable or k-solvable respectively.

We have the following result relating these concepts:

Proposition 3.1. *An almost normal pair of groups $[\Gamma, \Gamma_0]$ is almost solvable or k-solvable if and only if the group Γ is almost solvable or k-solvable respectively.*

To prove Proposition 3.1, we will need the following

Lemma 3.10. *Let G and $H_1 \supset H_2$ be subgroups of an ambient group. Then*

$$\#((G \cap H_1)/(G \cap H_2)) \leq \#(H_1/H_2).$$

Furthermore, if H_2 is a normal subgroup of H_1, then the map $g \in G \cap H_1 \mapsto g H_2$ is an injective homomorphism of $(G \cap H_1)/(G \cap H_2)$ into H_1/H_2.

180

Proof. If $g, h \in G \cap H_1$ belong to distinct cosets of $G \cap H_2$, then they clearly belong to distinct cosets of H_2. Therefore a set of distinct coset representative in $(G \cap H_1)/(G \cap H_2)$ has the same property in H_1/H_2. The second statement follows easily from the first. $\qquad\square$

Lemma 3.11. *Let* $\Gamma \supset \Gamma_1^1 \supset \ldots \supset \Gamma_m^1 \subset \Gamma_0^1$ *and* $\Gamma \supset \Gamma_1^2 \supset \ldots \supset \Gamma_n^2 \subset \Gamma_0^2$ *be chains such that for each* $j = 1, 2$ *and each* i, *either* $\#(\Gamma_i^j/\Gamma_{i+1}^j) \leq k$ *or* Γ_{i+1}^j *is normal in* Γ_i^j *and* $\Gamma_i^j/\Gamma_{i+1}^j$ *is abelian.*

Then successive pairs of groups in the chain $\Gamma \supset \ldots \supset \Gamma_m^1 \supset (\Gamma_m^1 \cap \Gamma_1^2) \supset \ldots \supset (\Gamma_m^1 \cap \Gamma_n^2)$ *have similar properties. Moreover, the last group* $\Gamma_m^1 \cap \Gamma_n^2$ *in the chain is contained in the group* $\Gamma_0^1 \cap \Gamma_0^2$.

Proof. This follows easily from Lemma 3.10. $\qquad\square$

Proof (of Proposition 3.1). We shall only show that if an almost normal pair of groups $[\Gamma, \Gamma_0]$ is k-solvable then Γ is almost k-solvable; the other implications are verified in a straightforward manner. To do this note that a group G has a normal subgroup H such that G/H is embeddable in the group S_k of permutations of k elements if and only if G has a subgroup K such that $\#(G/K) \leq k$. Now the result follows by applying Lemma 3.11 to several decreasing chains of subgroups. $\qquad\square$

3.3.8 Strong non-representability by k-quadratures

One can prove the following theorem.

Theorem 3.12 (see [Kho14]). *The class of all S-functions having a k-solvable monodromy pair is stable under composition, meromorphic operations, integration, differentiation and solutions of algebraic equations of degree* $\leq k$.

Definition 3.8. A function f is *strongly non-representable by k-quadratures* if it does not belong to a class of functions defined by the following data. List of basic functions: basic elementary functions and all single-valued S-functions. List of admissible operations: compositions, meromorphic operations, differentiation, integration and solutions of algebraic equations of degree $\leq k$.

Theorem 3.12 implies the following corollary.

Corollary 3.8 (Result on k-quadratures). *If the monodromy pair of an S-function* f *is not k-solvable, then* f *is strongly non-representable by k-quadratures.*

Example 3.5. The monodromy group of an algebraic function $y(x)$ defined by an equation $y^n + y - x = 0$ is the permutation group S_n. For $n \geq 5$ the group S_n is not an $(n-1)$-solvable group. Thus $y(x)$ provides an example of a function with finite set of singular points which is strongly non-representable by $(n-1)$-quadratures.

This example can be generalized.

Corollary 3.9 (see [Kho14]). *If an algebraic function of one complex variable has a non-k-solvable monodromy group then it is strongly non-representable by k-quadratures.*

Theorem 3.13 (see [Kho14]). *An algebraic function of one variable whose monodromy group is k-solvable can be represented by k-radicals.*

Results similar to Corollary 3.9 and Theorem 3.13 also hold for algebraic functions of several complex variables.

3.3.9 Strong non-representability by generalized quadratures

One can prove the following theorem.

Theorem 3.14 (see [Kho14]). *The class of all S-functions having an almost solvable monodromy pair is stable under composition, meromorphic operations, integration, differentiation and solutions of algebraic equations.*

Definition 3.9. A function f is *strongly non-representable by generalized quadratures* if it does not belong to a class of functions defined by the following data. List of basic functions: basic elementary functions and all single-valued S-functions. List of admissible operations: compositions, meromorphic operations, differentiation, integration and solutions of algebraic equations.

Theorem 3.14 implies the following corollary.

Corollary 3.10 (Result on generalized quadratures). *If the monodromy pair of an S-function f is not almost solvable, then f is strongly non-representable by generalized quadratures.*

Suppose that the Riemann surface of a function f is a universal covering space over the Riemann sphere with n punctures. If $n \geq 3$ then the function f is strongly non-representable by generalized quadratures. Indeed, the monodromy pair of f consists of the free group with $n-1$ generators, and its unit subgroup. It is easy to see that such a pair of groups is not almost solvable.

Example 3.6. Consider the function $z(x)$ which maps the upper half-plane onto a triangle with vanishing angles, bounded by three circular arcs. The Riemann surface of $z(x)$ is a universal covering space over the sphere with three punctures.[6] Thus $z(x)$ is strongly non-representable by generalized quadratures.

Example 3.7. Let K_1 and K_2 be the following elliptic integrals, considered as functions of the parameter x:

$$K_1(x) = \int_0^1 \frac{dt}{\sqrt{(1-t^2)(1-t^2x^2)}} \text{ and } K_2(x) = \int_0^{\frac{1}{x}} \frac{dt}{\sqrt{(1-t^2)(1-t^2x^2)}}.$$

---

[6] It is easy to see that the function $z(x)$ maps its Riemann surface to the open ball whose boundary contains the vertices of the triangle. These properties of the function $z(x)$ play a crucial role in Picard's beautiful proof of his Little Picard Theorem.

The functions $z(x)$ can be obtained from $K_1(x)$ and from $K_2(x)$ by quadratures. Thus both functions $K_1(x)$ and $K_2(x)$ are strongly non-representable by generalized quadratures.

In the next section we will list all polygons G bounded by circular arcs for which the Riemann map of the upper half-plane onto G is representable by generalized quadratures.

3.3.10 Maps of the upper half-plane onto a curved polygon

Consider a polygon G on the complex plane bounded by circular arcs, and the function f_G establishing the Riemann mapping of the upper half-plane onto the polygon G. The Riemann–Schwarz reflection principle allows one to describe the monodromy group L_G of the function f_G and to show that all singularities of f_G are simple enough. This information together with Theorem 3.14 provide a complete classification of all polygons G for which the function f_G is representable in explicit form (see [Kho14]).

If a polygon $\tilde{G}$ is obtained from a polygon G by a linear transformation $w : \mathbb{C} \to \mathbb{C}$ then $f_{\tilde{G}} = w(f_G)$. Thus it is enough to classify G up to a linear transformation.

3.3.10.1 The first case of integrability: the continuations of all sides of the polygon G intersect at one point.

Mapping this point to infinity by a fractional linear transformation, we obtain a polygon G bounded by straight line segments. All transformations in the group $L(G)$ have the form $z \to az + b$. All germs of the function $f = f_G$ at a non-singular point c are obtained from a fixed germ f_c by the action of the group $L(G)$ consisting of the affine transformations $z \to az + b$. The germ $R_c = (f''/f')_c$ is invariant under the action of the group $L(G)$. Therefore, the germ R_c is a germ of a single-valued function R. The singular points of R can only be poles (see [Kho14]). Hence the function R is rational. The equation $f''/f' = R$ is integrable by quadratures. This integrability case is well known. The function f in this case is called the *Christoffel–Schwarz integral*.

3.3.10.2 The second case of integrability: there is a pair of points such that, for every side of the polygon G, these points are either symmetric with respect to this side or belong to the continuation of the side.

We can map these two points to zero and infinity by a fractional linear transformation. We obtain a polygon G bounded by circular arcs centered at the point 0 and intervals of straight rays emanating from 0 (see [Kho14]). All transformations in the group $L(G)$ have the form $z \to az, z \to b/z$. All germs of the function $f = f_G$ at a non-singular point c are obtained from a fixed germ f_c by the action of the group $L(G)$:

$$f_c \to af_c, f_c \to b/f_c.$$

The germ $R_c = (f'_c/f_c)^2$ is invariant under the action of the group $L(G)$. Therefore, the germ R_c is a germ of a single-valued function R. The singular points of R can only be poles (see [Kho14]). Hence the function R is rational. The equation $R = (f'/f)^2$ is integrable by quadratures.

3.3.10.3 The finite nets of circles. To describe the third case of integrability we first need to define the finite nets of circles on the complex plane. The classification of finite groups generated by reflections in the Euclidean space $\mathbb{R}^3$ is well known. Each such group is the symmetry group of one of the following bodies:

1. a regular n-gonal pyramid;
2. a regular n-gonal dihedron, or the body formed by two equal regular n-gonal pyramids sharing the base;
3. a regular tetrahedron;
4. a regular cube or icosahedron;
5. a regular dodecahedron or icosahedron.

All these groups of isometries, except for the group of the dodecahedron or icosahedron, are solvable.

The intersections of the unit sphere, whose center coincides with the barycenter of the body, with the mirrors, in which the body is symmetric, is a certain net of great circles. The stereographic projection of this net of great circles gives a net of circles on the complex plane defined up to a fractional linear transformation. The nets corresponding to the bodies listed above will be called the finite nets of circles.

3.3.10.4 The third case of integrability: every side of a polygon G belongs to some finite net of circles. In this case the function f_G has finitely many branches. Since all singularities of the function f_G are algebraic (see [Kho14]), the function f_G is an algebraic function. For all finite nets but the net of the dodecahedron or icosahedron, the algebraic function f_G is representable by radicals. For the net of the dodecahedron or icosahedron the function f_G is representable by radicals and solutions of degree five algebraic equations (in other words f_G is representable by k-radicals).

3.3.10.5 Strong non-representability. Our results imply the following:

Theorem 3.15 (see [Kho14]). *If a polygon G bounded by circular arcs does not belong to one of the three cases described above, then the function f_G is strongly non-representable by generalized quadratures.*

3.3.11 Non-solvability of linear differential equations

Consider a homogeneous linear differential equation

$$y^{(n)} + r_1 y^{(n-1)} + \cdots + r_n y = 0 \tag{26}$$

whose coefficients r_i are rational functions of the complex variable x. The set $\Sigma \subset \mathbb{C}$ of poles of the r_i is called *the set of singular points* of the equation (26). At a point

$x_0 \in \mathbb{C} \setminus \Sigma$ the germs of solutions of (26) form a $\mathbb{C}$-linear space V_{x_0} of dimension n. The *monodromy group M of the equation (26)* is the group of all linear transformations of the space V_{x_0} which are induced by motions around the set Σ. Below we discuss this definition more precisely.

Consider a closed curve γ in $\mathbb{C} \setminus \Sigma$ beginning and ending at the point x_0. Given a germ $y \in V_{x_0}$ we can continue it along the loop γ to obtain another germ $y_\gamma \in V_{x_0}$. Thus each such loop γ corresponds to a map $M_\gamma : V_{x_0} \to V_{x_0}$ of the space V_{x_0} to itself that maps a germ $y \in V_{x_0}$ to the germ $y_\gamma \in V_{x_0}$. The map M_γ is linear since analytic continuation respects the arithmetic operations. It is easy to see that the map $\gamma \to M_\gamma$ defines a homomorphism of the fundamental group $\pi_1(\mathbb{C} \setminus \Sigma, x_0)$ of the domain $\mathbb{C} \setminus \Sigma$ with the base point x_0 to the group $GL(n)$ of invertible linear transformations of the space V_{x_0}.

The *monodromy group M* of the equation (26) is the image of the fundamental group in the group $GL(n)$ under this homomorphism.

Remark. Instead of the point x_0 one can choose any other point $x_1 \in \mathbb{C} \setminus \Sigma$. Such a choice will not change the monodromy group up to an isomorphism. To fix this isomorphism one can choose any curve $\gamma : I \to \mathbb{C}^N \setminus \Sigma$, where I is the segment $0 \le t \le 1$ and $\gamma(0) = x_0$, $\gamma(1) = x_1$, and identify each germ y_{x_0} of solution of (26) with its continuation y_{x_1} along γ.

Lemma 3.12. *The stationary subgroup in the monodromy group M of the germ $y \in V_{x_0}$ of almost every solution of the equation (26) is trivial (i.e. contains only the unit element $e \in M$).*

Proof. The monodromy group M contains countably many linear transformations M_i. The space $L_i \subset V_{x_0}$ of fixed points of a non-identity transformation M_i is a proper subspace of V_{x_0}. The union L of all subspaces L_i is a measure zero subset of V_{x_0}. The stationary subgroup in M of $y \in V_{x_0} \setminus L$ is trivial. $\square$

Theorem 3.16 (see [Kho14]). *If the monodromy group of equation (26) is not almost solvable (is not k-solvable, or is not solvable) then almost every solution of (26) is strongly non-representable by generalized quadratures (correspondingly, is strongly non-representable by k-quadratures, or is strongly non-representable by quadratures).*

Consider a homogeneous system of linear differential equations

$$y' = Ay \tag{27}$$

where $y = (y_1, \ldots, y_n)$ is the unknown vector-valued function and $A = \{a_{i,j}(x)\}$ is an $n \times n$ matrix whose entries are rational functions of the complex variable x. One can define the *monodromy group* of equation (27) in exactly the same way as it was defined for equation (26).

We will say that a vector-valued function $y = (y_1, \ldots, y_n)$ belongs to a certain class of functions if all of its components y_i belong to this class. For example, the statement "a vector-valued function $y = (y_1, \ldots, y_n)$ is strongly non-representable by generalized quadratures" means that at least one component y_i of y is strongly non-representable by generalized quadratures.

Theorem 3.17. *If the monodromy group of the system (27) is not almost solvable (is not k-solvable, or is not solvable) then almost every solution of (27) is strongly non-representable by generalized quadratures (correspondingly, is strongly non-representable by k-quadratures, or is strongly non-representable by quadratures).*

3.3.12 Solvability of Fuchsian equations

The differential field of rational functions of x is isomorphic to the differential field $\mathcal{R}$ of germs of rational functions at the point $x_0 \in \mathbb{C} \setminus \Sigma$. Consider the differential field extension $\mathcal{R}\langle y_1, \ldots, y_n \rangle$ of $\mathcal{R}$ where the germs $y_1, \ldots y_n$ form a basis in the space V_{x_0} of solutions of the equation (26) at x_0.

Lemma 3.13. *Every linear map M_γ from the monodromy group of equation (26) can be uniquely extended to a differential automorphism of the differential field $\mathcal{R}\{y_1, \ldots, y_n\}$ over the field $\mathcal{R}$.*

Proof. Every element $f \in \mathcal{R}\langle y_1, \ldots, y_n \rangle$ is a rational function of the independent variable x, the germs of solutions $y_1, \ldots, y_n$ and their derivatives. It can be continued meromorphically along the curve $\gamma \in \pi_1(\mathbb{C} \setminus \Sigma, x_0)$ together with $y_1, \ldots, y_n$. This continuation gives the required differential automorphism, since the continuation preserves the arithmetic operations and differentiation, and every rational function of x returns back to its original values (since it is a single-valued function). The differential automorphism is unique because the extension is generated by $y_1, \ldots, y_n$. $\square$

The *differential Galois group* (see [Kho14], [dPS03]) of equation (26) over $\mathcal{R}$ is the group of all differential automorphisms of the differential field $\mathcal{R}\{y_1, \ldots, y_n\}$ over the differential field $\mathcal{R}$. According to Lemma 32 the monodromy group of equation (26) can be considered as a subgroup of its differential Galois group over $\mathcal{R}$.

The differential field of invariants of the monodromy group action is a subfield of $\mathcal{R}\langle y_1, \ldots, y_n \rangle$ consisting of the single-valued functions. In contrast to the algebraic case, in the case of differential equations the field of invariants under the action of the monodromy group can be bigger than the field of rational functions. The reason is that the solutions of differential equations may grow exponentially when approaching the singular points or infinity.

Example 3.8. All solutions of the simplest differential equation $y' = y$ are single-valued exponential functions $y = C \exp x$, which are not rational.

For the wide class of Fuchsian linear differential equations all the solutions, while approaching the singular points and the point at infinity, grow polynomially.

The following Frobenius theorem is an analog for Fuchsian equations of C. Jordan's theorem (see [Kho14]) for algebraic equations.

Theorem 3.18 (Frobenius). *For Fuchsian differential equations the subfield of the differential field $\mathcal{R}\langle y_1, \ldots, y_n \rangle$ consisting of single-valued functions coincides with the field of rational functions.*

A system of linear differential equations (27) is called a *Fuchsian system* if the matrix A has the following form:

$$A(x) = \sum_{i=1}^{k} \frac{A_i}{x - a_i}, \tag{28}$$

where the A_i's are constant matrices. Linear Fuchsian systems of differential equations are in many ways similar to linear Fuchsian differential equations.

In the construction of explicit solutions of linear differential equations the following theorem is needed.

Theorem 3.19 (Lie–Kolchin). *Any connected solvable algebraic group acting by linear transformations on a finite-dimensional vector space over $\mathbb{C}$ is triangularizable in a suitable basis.*

Using the Frobenius Theorem and the Lie–Kolchin Theorem one can prove that the only reasons for unsolvability of Fuchsian linear differential equations and systems of linear differential equations are topological. In other words, if there are no topological obstructions to solvability then such equations and systems of equations are solvable. Indeed, the following theorems hold:

Theorem 3.20 (see [Kho14]). *If the monodromy group of the linear Fuchsian differential equation (26) is almost solvable (is k-solvable, or is solvable) then every solution is representable by generalized quadratures (correspondingly, is representable by k-quadratures, or is representable by quadratures).*

Theorem 3.21 (see [Kho14]). *If the monodromy group of the linear Fuchsian system differential equations (27) is almost solvable (is k-solvable, or is solvable) then every solution is representable by generalized quadratures (correspondingly, is representable by k-quadratures, or is representable by quadratures).*

3.3.13 Fuchsian systems with small coefficients

In general the monodromy group of a given Fuchsian equation is very hard to compute. It is known only for very special equations, including the famous hypergeometric equations. Thus Theorems 3.20 and 3.21 are not explicit.

If the matrix $A(x)$ in the system (27) is triangular then one can easily solve the system by quadratures. It turns out that if the matrix $A(x)$ has the form (28), where the matrices A_i are sufficiently small, then the system (27) with a non-triangular matrix $A(x)$ is unsolvable by generalized quadratures for a topological reason.

Theorem 3.22 (see [Kho14]). *If the matrices A_i are sufficiently small, $\|A_i\| < \varepsilon(a_1, \ldots, a_k, n)$, then the monodromy group of the system*

$$y' = \left(\sum_{i=1}^{k} \frac{A_i}{x - a_i} \right) y \tag{29}$$

is almost solvable if and only if the matrices A_i are triangularizable in a suitable basis.

Corollary 3.11. *If in the assumptions of Theorem 3.22 the matrices A_i are not triangularizable in a suitable basis then almost every solution of the system (29) is strongly non-representable by generalized quadratures.*

3.3.14 Polynomials invertible by radicals

In 1922 J.F. Ritt published (see [Rit22]) the following beautiful theorem, which fits nicely into topological Galois theory.

Theorem 3.23 (J.F. Ritt). *The inverse function of a polynomial with complex coefficients can be represented by radicals if and only if the polynomial is a composition of linear polynomials, the power polynomials $z \to z^n$, Chebyshev polynomials and polynomials of degree at most 4.*

OUTLINE OF PROOF (FOLLOWING [BK16])

1) *Every polynomial is a composition of primitive polynomials:* Every polynomial is a composition of polynomials that are not themselves compositions of polynomials of degree > 1. Such polynomials are called *primitive*. Recall that a permutation group G acting on a non-empty set X is called *primitive* if G acts transitively on X and G preserves no nontrivial partition of X. *A polynomial is primitive if and only if the monodromy group of the inverse of the polynomial acts primitively on its branches.*

2) *Reduction to the case of primitive polynomials:* A composition of polynomials is invertible by radicals if and only if each polynomial in the composition is invertible by radicals. Indeed, if each of the polynomials in the composition is invertible by radicals, then their composition also is. Conversely, if a polynomial R appears in the presentation of a polynomial P as a composition $P = Q \circ R \circ S$ and P^{-1} is representable by radicals, then $R^{-1} = S \circ P^{-1} \circ Q$ is also representable by radicals. Thus it is enough to classify only the primitive polynomials invertible by radicals.

3) *A result on solvable primitive permutation groups containing a full cycle:* A primitive polynomial is invertible by radicals if and only if the monodromy group of the inverse of the polynomial is solvable. Since it acts primitively on its branches and contains a full cycle (corresponding to a loop around the point at infinity on the Riemann sphere), the following group-theoretical result of Ritt is useful for the classification of polynomials invertible by radicals:

Theorem 3.24 (on primitive solvable groups with a cycle). *Let G be a primitive solvable group of permutations of a finite set X which contains a full cycle. Then either $|X| = 4$, or $|X|$ is a prime number p and X can be identified with the elements of the field F_p so that the action of G gets identified with the action of the subgroup of the affine group $\mathrm{AGL}_1(p) = \{x \to ax + b | a \in (F_p)^*, b \in F_p\}$ that contains all the shifts $x \to x + b$.*

4) *Solvable monodromy groups of the inverse of primitive polynomials:* It can be shown by applying the Riemann–Hurwitz formula that among the groups

188

in Theorem 3.24 on primitive solvable groups with a cycle, only the following groups can be realized as monodromy groups of the inverse of primitive polynomials: 1. $G \subset S_4$, 2. The cyclic group $G = \{x \to x + b\} \subset \mathrm{AGL}_1(p)$. 3. The dihedral group $G = \{x \to \pm x + b\} \subset \mathrm{AGL}_1(p)$.

5) *Description of primitive polynomials invertible by radicals:* It can be easily shown (see for instance [Rit22], [BK16]) that the following result holds:

Theorem 3.25. *If the monodromy group of the inverse of a primitive polynomial is a subgroup of the group $\{x \to \pm x + b\} \subset \mathrm{AGL}_1(p)$, then up to a linear change of variables the polynomial is either a power polynomial or a Chebyshev polynomial.*

Thus the polynomials whose inverse have monodromy groups 1–3 are respectively: 1. Polynomials of degree four. 2. Power polynomials up to a linear change of variables. 3. Chebyshev polynomials up to a linear change of variables.

In each of these cases the fact that the polynomial is invertible by radicals follows from the solvability of the corresponding monodromy group or from explicit formulas for its inverse (see for instance [BK16]).

3.3.15 Polynomials invertible by k-radicals

In this section we discuss the following generalization of Ritt's Theorem.

Theorem 3.26. *(see [BK16]) A polynomial invertible by radicals and solutions of equations of degree at most k is a composition of power polynomials, Chebyshev polynomials, polynomials of degree at most k and, if $k \leq 14$, certain primitive polynomials whose inverses have exceptional monodromy groups. A description of these exceptional polynomials can be given explicitly.*

The proofs rely on the classification of monodromy groups of inverses of primitive polynomials obtained by Müller based on group-theoretical results of Feit and on previous work on primitive polynomials whose inverses have exceptional monodromy groups by many authors. Besides the references to these highly involved and technical results an outline of the proof of Theorem 3.26 is not complicated and it resembles the outline of the proof of Ritt's Theorem.

Let us start with some background on representability by k-radicals.

Definition 3.10. Let k be a natural number. A field extension L/K is k-*radical* if there exists a tower of extensions $K = K_0 \subset K_1 \subset \cdots \subset K_n$ such that $L \subset K_n$ and for each i, K_{i+1} is obtained from K_i by adjoining an element a_i which is either a solution of an algebraic equation of degree at most k over K_i, or satisfies $a_i^m = b$ for some natural number m and $b \in K_i$.

Theorem 3.27 (see [Kho14]). *A Galois extension L/K of fields of characteristic zero is k-radical if and only if its Galois group is k-solvable.*

An algebraic function $z = z(x)$ of one or several complex variables is said to be *representable by k-radicals* if the corresponding extension of the field of rational functions is a k-radical extension.

Theorem 2.15 and C. Jordan's Theorem (see sections 3.2.1 and 3.2.2) imply the following corollary.

Corollary 3.12. *An algebraic function is representable by k-radicals if and only if its monodromy group is k-solvable.*

(Note that Theorem 3.13 above coincides with a part of Theorem 3.27.)

Let us outline briefly the main steps in the proof of Theorem 3.26:

OUTLINE OF PROOF OF THEOREM 3.26:
1) Exactly as in Ritt's theorem one can show that a composition of polynomials is invertible by k-radicals if and only if each polynomial in the composition is invertible by k-radicals. Thus one can reduce Theorem 3.13 to the case of primitive polynomials.

2) Feit and Jones completely classified all primitive permutation groups of n elements containing a full cycle.

3) Using this classification and the Riemann–Hurwitz formula, Müller listed all groups of permutations of n elements which are monodromy groups of inverses of degree n primitive polynomials.

4) For each group from Müller's list of groups of permutations of n elements one can determine the smallest k for which it is k-solvable and choose the *exceptional groups* for which k is smaller than n.

5) For each such exceptional group one can explicitly describe polynomials whose inverse has the exceptional monodromy group.

3.4 Multidimensional Topological Galois Theory

3.4.1 Introduction

In this section we present an outline of the multidimensional version of topological Galois theory. The presentation is based on the book [Kho14]. It contains definitions, statements of results and comments on them. Essentially no proofs are presented.

In topological Galois theory for functions of one variable (see section 3.3 and [Kho14]), it is proved that the way the Riemann surface of a function is positioned over the complex line can obstruct the representability of this function "in finite terms" (i.e. its representability by radicals, by quadratures, by generalized quadratures and so on). This not only explains why many algebraic and differential equations are not solvable in finite terms, but also gives the strongest known results on their unsolvability.

In the multidimensional version of topological Galois theory analogous results are proved. But in the multidimensional case all constructions and proofs are much more complicated and involved than in the one-dimensional case (see [Kho14]).

3.4.2 Classes of functions

An equation is solvable "in finite terms" (or is solvable "explicitly") if its solutions belong to a certain class of functions. Different classes of functions correspond to different notions of solvability in finite terms.

A class of functions can be introduced by specifying a list of *basic functions* and a list of *admissible operations*. Given the two lists, the class of functions is defined as the set of all functions that can be obtained from the basic functions by repeated application of admissible operations. Below, we define Liouvillian classes of functions in exactly this way.

Classes of functions which appear in the problems of integrability in finite terms contain multivalued functions. Thus the basic terminology should be made clear.

We understand operations on multivalued functions of several variables in a slightly more restrictive sense than operations on multivalued functions of a single variable (the one-dimensional case is discussed in section 3.3 and in [Kho14]).

Fix a class of basic functions and some set of admissible operations. Can a given function (which is obtained, say, by solving a certain algebraic or a differential equation) be expressed through the basic functions by means of admissible operations? We are interested in various *single-valued branches* of multivalued functions over various domains. Every function, even if it is multivalued, will be considered as the collection of all its single-valued branches. We will only apply admissible operations (such as arithmetic operations and composition) to single-valued branches of the function over various domains. Since we deal with analytic functions, it suffices only to consider small neighborhoods of points as domains.

We can now rephrase the question in the following way: *can a given function germ at a given point be expressed through the germs of basic functions with the help of admissible operations?* Of course, the answer depends on the choice of a point and on the choice of a single-valued germ at this point belonging to the given multivalued function. It turns out, however, that for the classes of functions interesting to us the desired expression is either impossible for every germ of a given multivalued function at every point or the "same" expression serves all germs of a given multivalued function at almost every point of the space.

For functions of one variable, we use a different, extended definition of operations on multivalued functions, in which the multivalued function is viewed as a single object. This definition is essentially equivalent to including the operation of analytic continuation in the list of admissible operations on analytic germs (all the details can be found in [Kho14]). For functions of many variables, we need to adopt a more restrictive understanding of operations on multivalued functions, which is, however, no less (and perhaps even more) natural.

3.4.3 Specifics of the multidimensional case

I was always under impression that a full-fledged multidimensional version of topological Galois theory was impossible. The reason was that, to construct such a version for the case of many variables, one would need to have information on extendability of function germs not only outside their ramification sets but also along these sets. It seemed that there was nothing to extract such information from.

To illustrate the problem consider the following situation. Let f be a multivalued analytic function on $\mathbb{C}^n$ whose set of singular points is an analytic set $\Sigma_f \subset \mathbb{C}^n$. Let f_a be an analytic germ of f at a point $a \in \mathbb{C}^n$. Let $g : (\mathbb{C}^k, b) \to (\mathbb{C}^n, a)$ be an analytic map. Consider a germ φ_b at the point $b \in \mathbb{C}^k$ of the composition $f_a \circ g_b$. One can ask the following questions:

1) Is it true that φ_b is a germ of a multivalued function φ on $\mathbb{C}^k$ whose set of singular points Σ_φ is contained in a proper analytic subset of $\mathbb{C}^k$?

2) Is it true that the monodromy group M_φ of φ corresponding to motions around the set $\Sigma_\varphi \subset \mathbb{C}^k$ can be estimated in terns of the monodromy group M_f of f corresponding to motions around the set $\Sigma_f \subset \mathbb{C}^n$? For example, if M_f is a solvable group is it true that M_φ is also a solvable group?

If the image $g(\mathbb{C}^k)$ is not contained in the singular set Σ_f then the answers to both questions are positive: the set Σ_φ belongs to the analytic set $g^{-1}(\Sigma_f)$ and the group M_φ is a subgroup of a certain factor group of M_f. These statements are not complicated and can be proved by the same arguments as in the one-dimensional topological Galois theory.

Assume that the multivalued function f has an analytic germ f_a at a point a belonging to the singular set Σ_f (some of the germs of the multivalued function f may appear to be nonsingular at singular points of this function). Assume now that the image $g(\mathbb{C}^k)$ is contained in the singular set Σ_f and $a = g(b)$. It turns out that for the germ $\varphi_b = f_a \circ g_b$ the answers to both questions are also positive. In this situation all the proofs are more involved. They use new arguments from multidimensional complex analysis and from group theory.

It turns out that function germs can sometimes be automatically extended along their ramification sets (see [Kho14]). This new statement from complex analysis suggests a positive answer to the first question.

To describe the connection between the monodromy group of the function f and the monodromy groups of the composition $\varphi = f \circ g$, we introduce and develop the notion of pullback closure for groups (see [Kho14]). The use of this operation, in turn, forces us to reconsider all arguments we used in the one-dimensional version of topological Galois theory. As a result we obtain a positive answer to the second question.

3.4.4 Liouvillian classes of multivariate functions

In this section we define Liouvillian classes of functions for the case of several variables. These classes are defined in the same way as the corresponding classes for functions of one variable (see section 3.2.2 and [Kho14]). The only difference is in the details.

We fix an ascending chain of standard coordinate subspaces of strictly increasing dimension: $0 \subset \mathbb{C}^1 \subset \cdots \subset \mathbb{C}^n \subset \cdots$ with coordinate functions $x_1, \ldots, x_n, \ldots$ (for every $k > 0$, the functions $x_1, \ldots, x_k$ are coordinate functions on $\mathbb{C}^k$). Below, we define Liouvillian classes of functions for each of the standard coordinate subspaces $\mathbb{C}^k$.

To define Liouvillian classes, we will need the list of basic elementary functions and the list of classical operations.

List of basic elementary functions

1. All complex constants and all coordinate functions $x_1, \ldots, x_n$ for every standard coordinate subspace $\mathbb{C}^n$.
2. The exponential, the logarithm and the power x^α, where α is any complex constant.
3. Trigonometric functions: sine, cosine, tangent, cotangent.
4. Inverse trigonometric functions: arcsine, arccosine, arctangent, arccotangent.

Let us now turn to the list of classical operations on functions.

List of classical operations

1. *The operation of composition* that takes a function f of k variables and functions $g_1, \ldots, g_k$ of n variables to the function $f(g_1, \ldots, g_k)$ of n variables.
2. *Arithmetic operations* that take functions f and g to the functions $f + g$, $f - g$, fg and f/g.
3. *Operations of partial differentiation with respect to independent variables*. For functions of n variables, there are n such operations: the i-th operation assigns the function $\frac{\partial f}{\partial x_i}$ to a function f of the variables $x_1, \ldots, x_n$.
4. *The operation of integration* that takes k functions $f_1, \ldots, f_k$ of the variables $x_1, \ldots, x_n$, for which the differential one-form $\alpha = f_1 dx_1 + \cdots + f_k dx_k$ is closed, to the indefinite integral y of the form α (i.e. to any function y such that $dy = \alpha$). The function y is determined by the functions $f_1, \ldots, f_k$ up to an additive constant.
5. *The operation of solving an algebraic equation* that takes functions $f_1, \ldots, f_n$ to the function y such that $y^n + f_1 y^{n-1} + \cdots + f_n = 0$. The function y may not be quite uniquely determined by the functions $f_1, \ldots, f_n$, since an algebraic equation of degree n can have n solutions.

We now resume our definition of Liouvillian classes of functions.

3.4.4.1 Functions of n variables representable by radicals. List of basic functions: All complex constants and all coordinate functions. List of admissible operations: composition, arithmetic operations and the operation of taking the m-th root $f^{\frac{1}{m}}$, $m = 2, 3, \ldots$, of a given function f.

3.4.4.2 Functions of n variables representable by k-radicals. This class of functions is defined in the same way as the class of functions representable by radicals. We only need to add the operation of solving algebraic equations of degree $\leq k$ to the list of admissible operations.

3.4.4.3 Elementary functions of n variables.   List of basic functions: basic elementary functions. List of admissible operations: composition, arithmetic operations, differentiation.

3.4.4.4 Generalized elementary functions of n variables.   This class of functions is defined in the same way as the class of elementary functions. We only need to add the operation of solving algebraic equations to the list of admissible operations.

3.4.4.5 Functions of n variables representable by quadratures.   List of basic functions: basic elementary functions. List of admissible operations: composition, arithmetic operations, differentiation, integration.

3.4.4.6 Functions of n variables representable by k-quadratures.   This class of functions is defined in the same way as the class of functions representable by quadratures. We only need to add the operation of solving algebraic equations of degree at most k to the list of admissible operations.

3.4.4.7 Functions of n variables representable by generalized quadratures.   This class of functions is defined in the same way as the class of functions representable by quadratures. We only need to add the operation of solving algebraic equations to the list of admissible operations.

3.4.5 Strong non-representability in finite terms

Topological obstructions to the representability of functions in finite terms relate to branching. It turns out that if a function does not belong to a certain Liouvillian class for topological reasons then it automatically does not belong to a much wider *extended Liouvillian class of functions.*

Such an extended Liouvillian class is defined as follows: its list of admissible operations is the same as in the original Liouvillian class and its list of basic functions is the list of basic function in the original class extended by all single-valued functions of any number of variables having a proper analytic set of singular points.

Definition 3.11. A germ f is a germ of a function belonging to the *extended class of functions representable by quadratures* if it can be represented by germs of basic elementary functions and by germs of single-valued functions, whose set of singular points is a proper analytic set, by means of composition, integration, arithmetic operations and differentiation.

Definition 3.12. A germ f is *strongly non-representable by quadratures* if it is not a germ of a function from the extended class of functions representable by quadratures.

The definition of *strong non-representability of a germ f* by radicals, by k-radicals, by elementary functions, by generalized elementary functions, by k-quadratures and by generalized quadratures is similar to the above definition.

3.4.6 Holonomic systems of linear differential equations

Consider a system of N linear differential equations $L_j(y) = 0$, $j = 1, \ldots, N$,

$$L_j(y) = \sum a_{i_1,\ldots,i_n} \frac{\partial^{i_1 + \cdots + i_n} y}{\partial x_1^{i_1} \ldots \partial x_n^{i_n}} = 0, \tag{30}$$

of an unknown function y, whose coefficients $a_{i_1,\ldots,i_n}$ are analytic functions in a domain $U \subset \mathbb{C}^n$.

The system (30) is *holonomic* if at every point $a \in U$ the $\mathbb{C}$-linear space V_a of germs y_a satisfying the system (30) has finite dimension, $\dim_{\mathbb{C}} V_a = d(a) < \infty$. Holonomic systems can be considered as a multidimensional generalization of linear differential equations in one unknown function of a single variable. Kolchin obtained a generalization of the Picard–Vessiot theory (Galois theory for linear differential equations) to the case of holonomic systems of differential equations [dPS03].

The holonomic system (30) has the following properties:

1) There exists an analytic *singular hypersurface* $\Sigma \subset U$ such that the dimension $d(a) = \dim_{\mathbb{C}} V_a$ is constant $d(a) \equiv d$ on $U \setminus \Sigma$.
2) Let $\gamma : I \to U \setminus \Sigma$ be a continuous map, where I is the unit interval $0 \le t \le 1$ and $\gamma(0) = a$, $\gamma(1) = b$. Then the space V_a of solutions of (30) at the point a admits an analytic continuation along γ and the space obtained by the continuation at the point b is the space V_b of solutions of (30) at the point b.
3) If all equations of the system (30) admit an analytic continuation to some domain W, then the system obtained by such a continuation is a holonomic system in the domain W.

Let $a \notin \Sigma$ be a point not belonging to the hypersurface Σ. Take an arbitrary path $\gamma(t)$ in the domain U originating and terminating at a and avoiding the hypersurface Σ. Solutions of this system admit analytic continuations along the path γ, which are also solutions of the system. Therefore, every such path γ gives rise to a linear map M_γ of the solution space V_a to itself. The collection of linear transformations M_γ corresponding to all paths γ form a group, which is called the *monodromy group of the holonomic system*.

3.4.7 SC-germs

There is a wide class of S-functions in one variable containing all Liouvillian functions and stable under classical operations, for which the monodromy group is defined. The class of S-functions plays an important role in the one-dimensional version of topological Galois theory (see [Kho14, sec. 2]). Is there a sufficiently wide class of multivariate function germs with similar properties?

For a long time, I thought that the answer to this question was negative. In this section the class of SC-germs is defined. This provides an affirmative answer to this question.

A subset A in a connected k-dimensional analytic manifold Y is called *meager* if there exists a countable set of open domains $U_i \subset M$ and a countable collection of proper analytic subsets $A_i \subset U_i$ in these domains such that $A \subset \bigcup A_i$.

The following definition plays a key role in what follows.

Definition 3.13. A germ f_a of an analytic function at a point $a \in \mathbb{C}^n$ is an SC-*germ* if the following condition is fulfilled. For every connected complex analytic manifold Y, every analytic map $G : Y \to \mathbb{C}^n$ and every preimage b of the point a, $G(b) = a$, there exists a meager set $A \subset Y$ such that for every path $\gamma : [0,1] \to Y$ originating at the point b, $\gamma(0) = b$, and intersecting the set A at most at the initial moment, $\gamma(t) \notin A$ for $t > 0$, the germ f_a admits an analytic continuation along the path $G \circ \gamma : [0,1] \to \mathbb{C}^n$.

The following lemma is obvious.

Lemma 3.14. *The class of SC-germs contains all germs of analytic functions on $\mathbb{C}^N \setminus \Sigma$ where Σ is an analytic subset in $\mathbb{C}^N$, N a natural number. In particular, the class contains all analytic germs of S-functions of one variable and all germs of meromorphic functions of many variables.*

The proof of the following Theorem 3.28 uses the results on extendability of multivalued analytic functions along their singular point sets (see [Kho14]).

Theorem 3.28 (on stability of the class of SC-germs). *The class of SC-germs on $\mathbb{C}^n$ is stable under the operation of taking the composition with SC-germs of m-variable functions, the operation of differentiation and integration. It is stable under solving algebraic equations whose coefficients are SC-germs and under solving holonomic systems of linear differential equations whose coefficients are SC-germs.*

Theorem 3.28 implies the following corollary.

Corollary 3.13. *If a germ f is not an SC-germ then f is strongly non-representable by generalized quadratures. In particular, it cannot be a germ of a function belonging to a certain Liouvillian class.*

3.4.8 The monodromy group of an SC-germ

The *monodromy group and the monodromy pair* of an SC-germ f_a can be defined in the same way as for S-functions of one variable. By definition the set $\Sigma \subset \mathbb{C}^n$ of singular points of f_a is a meager set. Take any point $x_0 \in \mathbb{C}^n \setminus \Sigma$ and consider the action of the fundamental group $\pi_1(\mathbb{C}^n \setminus \Sigma, x_0)$ on the set F_{x_0} of all germs equivalent to the germ f_a. The *monodromy group* of f_a is the image of the fundamental group under this action. The *monodromy pair* of f_a is the pair $[\Gamma, \Gamma_0]$ where Γ is the monodromy group and Γ_0 is the stationary subgroup of a germ $f \in F_{x_0}$. Up to isomorphism the monodromy group and the monodromy pair are independent of the choice of the point x_0 and the germ f.

Remark. If an SC-germ f_a is defined at a singular point $a \in \Sigma$ then the *monodromy group of f_a along Σ* is defined: one can consider continuations of f_a along curves γ belonging to Σ and define a singular set $\Sigma_1 \subset \Sigma$ for f_a along Σ. The monodromy group of f_a along Σ corresponds to the action of the fundamental group of $\pi_1(\Sigma \setminus \Sigma_1, x_1)$ on the set of germs at $x_1 \in \Sigma \setminus \Sigma_1$ obtained by continuation of f_a along Σ. If the point a belongs to Σ_1 then one can also define a monodromy group of f_a along Σ_1 and so on. Thus in the multidimensional case one can associate to an SC-germ a hierarchy of monodromy groups. All these monodromy groups (and corresponding monodromy pairs) appear in multidimensional topological Galois theory. But the monodromy group and the monodromy pair we discussed above are most important for our purposes.

3.4.9 Stability of certain classes of SC-germs

One can prove the following theorems.

Theorem 3.29 (see [Kho14]). *The class of all SC-germs having a solvable monodromy group is stable under composition, arithmetic operations, integration and differentiation. This class contains all germs of basic elementary functions and all germs of single-valued functions whose set of singular points is a proper analytic set.*

Theorem 3.30. *(see [Kho14]) The class of all SC-germs having a k-solvable monodromy pair (see [Kho14, sec. 2]) is stable under composition, arithmetic operations, integration, differentiation and solution of algebraic equations of degree at most k. This class contains all germs of basic elementary functions and all germs of single-valued functions whose set of singular points is a proper analytic set.*

Theorem 3.31 (see [Kho14]). *The class of all SC-germs having an almost solvable monodromy pair (see [Kho14, sec. 2]) is stable under composition, arithmetic operations, integration, differentiation and solution of algebraic equations. This class contains all germs of basic elementary functions and all germs of single-valued functions whose set of singular points is a proper analytic set.*

Theorems 3.29–3.29 imply the following corollaries.

RESULT ON QUADRATURES *If the monodromy group of an SC-germ f is not solvable, then f is strongly non-representable by quadratures.*

RESULT ON k-QUADRATURES *If the monodromy pair of an SC-germ f is not k-solvable, then f is strongly non-representable by k-quadratures.*

RESULT ON GENERALIZED QUADRATURES *If the monodromy pair of an SC-germ f is not almost solvable, then f is strongly non-representable by generalized quadratures.*

3.4.10 Solvability and non-solvability of algebraic equations

Consider an irreducible algebraic equation

$$P_n y^n + P_{n-1} y^{n-1} + \cdots + P_0 = 0 \tag{31}$$

whose coefficients $P_n, \ldots, P_0$ are polynomials of N complex variables $x_1, \ldots, x_N$. Let $\Sigma \subset \mathbb{C}^N$ be the singular set of the equation (31) defined by $P_n J = 0$, where J is the discriminant of the polynomial (31).

Theorem 3.32. *(see [Kho14, sec. 2], [Kho71]) Let y_{x_0} be a germ of analytic function at a point $x_0 \in \mathbb{C}^N \setminus \Sigma$ satisfying the equation (31). If the monodromy group of the equation (31) is solvable (is k-solvable) then the germ y_{x_0} is representable by radicals (is representable by k-radicals).*

According to Camille Jordan's theorem (see [Kho71]) the Galois group of the equation (31) over the field $\mathcal{R}$ of rational functions of $x_1, \ldots, x_N$ is isomorphic to the monodromy group of this equation (31). Thus Theorem 3.32 follows from Galois theory (see [Kho14], [Kho71]).

Theorem 3.33 (see [Kho14]). *Let y_{x_0} be a germ of analytic function at a point $x_0 \in \mathbb{C}^N$ satisfying the equation (31). If the monodromy group of the equation is not solvable (is not k-solvable) then the germ y_{x_0} is strongly non-representable by quadratures (is strongly non-representable by k-quadratures).*

Theorem 3.33 follows from the results on quadratures and on k-quadratures from the previous section.

Consider the universal degree n algebraic function $y(a_n, \ldots, a_0)$ defined by the equation

$$a_n y^n + \cdots + a_0 = 0. \tag{32}$$

It is easy to see that the monodromy group of the equation (32) is isomorphic to the group S_n of all permutations of n elements. For $n \geq 5$ the group S_n is unsolvable and it is not a k-solvable group for $k < n$. Thus Theorem 3.33 implies the following strongest known version of the Abel–Ruffini Theorem.

Theorem 3.34 (a version of the Abel–Ruffini Theorem). *Let y_a be a germ of analytic function at a point a satisfying the universal degree $n \geq 5$ algebraic equation. If $n \geq 5$ then the germ y_a is strongly non-representable by $(n-1)$ quadratures. In particular the germ y_a is strongly non-representable by quadratures.*

3.4.11 Solvability and non-solvability of holonomic systems of linear differential equations

Consider a system of N linear differential equations $L_j(y) = 0$, $j = 1, \ldots, N$,

$$L_j(y) = \sum a_{i_1, \ldots, i_n} \frac{\partial^{i_1 + \cdots + i_n} y}{\partial x_1^{i_1} \ldots \partial x_n^{i_n}} = 0, \tag{33}$$

on an unknown function y, whose coefficients $a_{i_1,...,i_n}$ are rational functions of n complex variables $x_1, \ldots, x_n$. Assume that the system (33) is holonomic on $\mathbb{C}^n \setminus \Sigma_1$, where Σ_1 is the union of poles of the coefficients $a_{i_1,...,i_n}$. Let $\Sigma_2 \subset \mathbb{C}^n \setminus \Sigma_1$ be the singular hypersurface of a holonomic system (33).

Every germ y_a of a solution of the system at a point $a \in \mathbb{C}^n \setminus \Sigma$ where $\Sigma = \Sigma_1 \cup \Sigma_2$ admits an analytic continuation along every path avoiding the hypersurface Σ so the monodromy group of the system ((33) is well defined.

Theorem 3.35 (see [Kho14]). *If the monodromy group of the holonomic system (33) is not solvable (not k-solvable, not almost solvable), then a germ y_a of almost every solution at a point $a \in \mathbb{C}^n \setminus \Sigma$ is strongly non-representable by quadratures (is strongly non-representable by k-quadratures, is strongly non-representable by generalized quadratures).*

Theorem 3.35 follows from the results on quadratures, on k-quadratures and on generalized quadratures from section 3.4.9.

A holonomic system is said to be *regular* if near the singular set Σ and near infinity the solutions of the system grow at most polynomially.

Theorem 3.36 (see [Kho14]). *If the monodromy group of a regular holonomic system is solvable (is k-solvable, is almost solvable), then a germ y_a of almost every solution at a point $a \in \mathbb{C}^n \setminus \Sigma$ is representable by quadratures (is representable by k-quadratures, is representable by generalized quadratures).*

3.4.12 Completely integrable systems of linear differential equations with small coefficients

Consider a completely integrable system of linear differential equations of the following form

$$dy = Ay, \tag{34}$$

where $y = y_1, \ldots, y_N$ is an unknown vector-function, and A is an $(N \times N)$-matrix consisting of differential one-forms with rational coefficients on the space $\mathbb{C}^n$ satisfying the condition of complete integrability $dA + A \wedge A = 0$ and having the following form:

$$A = \sum_{i=1}^{k} A_i \frac{dl_i}{l_i},$$

where the A_i are constant matrices, and the l_i are linear (not necessarily homogeneous) functions on $\mathbb{C}^n$.

If the matrices A_i can be simultaneously reduced to triangular form, then system (34), as any completely integrable triangular system, is solvable by quadratures. Of course, there exist solvable systems that are not triangular. However, if the matrices A_i are sufficiently small, then there are no such systems. Namely, the following theorem holds.

Theorem 3.37 (see [Kho14]). *A system (34) that does not reduce to triangular form and such that the matrices A_i have sufficiently small norms is unsolvable by generalized quadratures in the following strong sense. At every point $a \in \mathbb{C}^n$ where the matrix A is regular, and for almost any germ $y_a = (y_1,\ldots,y_N)_a$ of a vector-function satisfying the system (34), there is a component $(y_i)_a$ which is strongly non-representable by generalized quadratures.*

The multidimensional Theorem 3.37 is similar to the one-dimensional Corollary 3.11. Their proofs (see [Kho14]) are also similar. We only need to replace the reference to the (one-dimensional) Lappo-Danilevsky theory with its multidimensional version from [Lek91].

References

[Ale04] Valeriy B. Alekseev. *Abel's theorem in problems and solutions.* Kluwer Academic Publishers, Dordrecht, 2004. Based on the lectures of Professor V. I. Arnold, with a preface and an appendix by Arnold and an appendix by A. Khovanskii.

[BK16] Yuri Burda and Askold G. Khovanskiĭ. Polynomials invertible in k-radicals. *Arnold Math. J.*, 2(1):121–138, 2016.

[Kho70] Askold G. Khovanskiĭ. The representability of algebroidal functions by superpositions of analytic functions and of algebroidal functions of one variable. *Funkcional. Anal. i Priložen.*, 4(2):74–79, 1970.

[Kho71] Askold G. Khovanskiĭ. Superpositions of holomorphic functions with radicals. *Uspekhi Mat. Nauk*, 26(3(159)):213–214, 1971.

[Kho91] Askold G. Khovanskiĭ. *Fewnomials*, volume 88 of *Translations of Mathematical Monographs*. American Mathematical Society, Providence, RI, 1991. Translated from the Russian by Smilka Zdravkovska.

[Kho14] Askold G. Khovanskiĭ. *Topological Galois theory.* Springer Monographs in Mathematics. Springer, Heidelberg, 2014. Solvability and unsolvability of equations in finite terms, Appendices C and D by Khovanskii and Yuri Burda [Yura Burda on title page verso], Translated from the Russian by V. Timorin and V. Kirichenko (Chapters 1–7) and Lucy Kadets (Appendices A and B).

[Kho15] Askold G. Khovanskiĭ. On algebraic functions integrable in finite terms. *Funct. Anal. Appl.*, 49(1):50–56, 2015. Translation of Funktsional. Anal. i Prilozhen. 49 (2015), no. 1, 62–70.

[Kho18] Askold G. Khovanskiĭ. Solvability of equations by quadratures and Newton's theorem. *Arnold Math. J.*, 4(2):193–211, 2018.

[Kho19] Askold G. Khovanskiĭ. Integrability in finite terms and actions of Lie groups. *Moscow Mathematical Journal*, 19(2):329–341, 2019.

[Lek91] Vladimir P. Leksin. The Riemann–Hilbert problem for analytic families of representations. *Mat. Zametki*, 50(2):89–97, 1991.

[dPS03] Marius van der Put and Michael F. Singer. *Galois theory of linear differential equations*, volume 328 of *Grundlehren der Mathematischen Wissenschaften*

[Fundamental Principles of Mathematical Sciences]. Springer-Verlag, Berlin, 2003.

[Rit22] Joseph F. Ritt. On algebraic functions which can be expressed in terms of radicals. *Trans. Amer. Math. Soc.*, 24(1):21–30, 1922.

[Rit48] Joseph F. Ritt. *Integration in Finite Terms. Liouville's Theory of Elementary Methods*. Columbia University Press, New York, N.Y., 1948.

[Ros73] Maxwell Rosenlicht. An analogue of l'Hospital's rule. *Proc. Amer. Math. Soc.*, 37:369–373, 1973.

On the Integration of Elementary Functions which are Built Up Using Algebraic Operations[*]

Robert H. Risch[1]

Abstract This paper gives an algorithm to decide if a function built up from the rational functions using logarithms, exponentials and arbitrary algebraic operations can be integrated in terms of functions built up in a similar manner.

In this paper we show that the general problem of integration in finite terms reduces to a problem in the theory of algebraic functions, namely determining whether a given divisor of an algebraic function field has a power that is a principal divisor. It may be posed as follows: to test whether a point, in the Jacobian of the curve associated to the function field, is of finite order. This latter problem has been shown to be decidable [Ris70].

A previous paper, [Ris69b], showed that one can test, in a finite number of steps, the elementary integrability of an element in a field of the form $K(z, \theta_1, \ldots, \theta_n)$, where K consists of constants and is finitely generated over the rationals, $z' = 1$, and where each θ_i is a monomial over $K(z, \theta_1, \ldots, \theta_{i-1})$. In the concluding remarks to that paper we were pessimistic about the possibility of extending the algorithm to the cases where some of the θ_i are algebraic over $K(z, \theta_1, \ldots, \theta_{i-1})$. Since writing that, we have observed that by studying the Puiseux expansions of elements of $\mathcal{D}(\theta, w)$, θ a monomial, w algebraic over $\mathcal{D}(\theta)$, we can obtain results similar to those in [Ris69b], where we used partial fraction expansions of elements of $\mathcal{D}(\theta)$.

This paper is a revised version of [Ris68], including parts of [Ris69a]. This latter paper reduced the problem of finding elementary antiderivatives of elementary functions to a problem in arithmetic algebraic geometry: the problem of determining if a point on the Jacobian of a curve is of finite order. Risch presented a solution

* The editors would like to thank Ralf Hemmecke and Antonio Jimenez-Pastor for their help in preparing this manuscript for publication.

[1] Robert H. Risch
IBM Thomas J. Watson Research Center, Yorktown Heights, New York, N.Y. 10598

© Springer Nature Switzerland AG 2022
C. G. Raab und M. F. Singer (Hrsg.), *Integration in Finite Terms: Fundamental Sources*, Texts & Monographs in Symbolic Computation,
https://doi.org/10.1007/978-3-030-98767-1_5

of this latter problem in the context of deciding the elementary integrability of an algebraic function in [Ris70].[2] The present paper includes the complete proof.

We assume that the reader is familiar with the notion of elementary extension and related concepts from [Ris69b]. The reader should also be familiar with methods of calculating Puiseux expansions, the concept of a divisor, and the properties of a basis, normal at infinity, for the multiples of a divisor. These may be found in [Bli66], Chapter II, sections 14 and 15 and Chapter III through theorem 21.2, and also in [Coa70], [Dav81] and [Tra84].

In Chevalley's book [Che51] a theory of algebraic functions over general ground fields is presented. Using the concepts and results there we could give an algorithm which would ordinarily involve much less computation than given here: more general p-adic expansions (in powers of polynomials in θ, which are irreducible over the ground field) would replace the Puiseux expansions used here. However, it is not clear from reading Chevalley how to compute the entities involved. Since use of his concepts would necessitate lengthy discussion, we follow Bliss's (or Trager's) procedures.

The author wishes to thank Michael Singer for assembling this paper from some unpublished reports, making some corrections and clarifications, and arranging for its publication.

We can certainly compute effectively with the elements of an algebraically closed field $\mathcal{D}$ of finite transcendence degree over the rationals, i.e., assign distinct symbols to the elements of the field, construct bases for finitely generated subfields, tell when two elements are equal, factor polynomials over it, etc. We will not give a formal justification of the above statement but refer the reader to [vdW53, Ch. 42, p. 134] and [Ris69b, Sections 2 and 3].

We will be able to compute with elements of an algebraic function field $\mathcal{D}(\theta, w)$ if we are given the irreducible polynomial that w satisfies. Thus we can certainly speak of being effectively given a divisor $Q = p_1^{\beta_1} \ldots p_k^{\beta_k}$, for we can compute the first terms of the various Puiseux expansions of w above $a \in \mathcal{D}$ or infinity, until they begin to differ. These initial terms will suffice to differentiate the places of $\mathcal{D}(\theta, w)$ lying above the same place of $\mathcal{D}(\theta)$. Thus the following problem makes sense:

Problem 1 (of the Points of Finite Order). *Let $\mathcal{D}(\theta, w)$ be an algebraic function field of one variable, where $\mathcal{D}$ is algebraically closed and of finite transcendence degree over the rationals. Given an arbitrary divisor $Q = p_1^{\beta_1} \ldots p_k^{\beta_k}$ where $\sum \beta_i = 0$, decide in a finite number of steps whether there is an integer $\gamma \neq 0$ and an element $f \in \mathcal{D}(\theta, w)$ such that Q^γ is the divisor of f, and if so, find such a γ and f.*

A solution of this problem was first outlined in [Ris70] and discussed in [Dav81] and [Tra84]. It is easy to show that for the smallest such positive γ and for any function f such that Q^γ is the divisor of f, we have the following: for any f_1, γ_1 satisfying the condition Q^{γ_1} is the divisor of f_1, there is an integer δ and $c \in \mathcal{D}$ such that $\gamma_1 = \delta \gamma$ and $f^\delta = c f_1$.

---

[2] This and other problems concerning computations in algebraic function fields have been further discussed in [Dav81] and [Tra84].

Using the fact that the problem of the points of finite order is solvable, we are able to show the following two facts (which are generalizations of the two theorems stated at the beginning of Section 3 of [Ris69b]):

(1) For any regular efd e, Σ_e is recursive.
(2) The regular efd's form a recursive subset of all efd's.

Another version of this result is: let $\mathcal{F}$ be an elementary extension of $K(z)$, where K is a finitely generated extension of the rationals and $z' = 1$ and assume that the constants of $\mathcal{F}$ are K. Given $f \in \mathcal{F}$ we can effectively decide if e^f and $\log f$ are monomials over $\mathcal{F}$ and decide if $\int f$ is elementary over $\mathcal{F}$.

These results follow from the theorem of this paper in the same way that the corresponding results in [Ris69b] followed from the main theorem in that paper. In our present theorem, we shall consider fields $\mathcal{F} = \mathcal{D}_n$ defined as follows: $\mathcal{D}_0$ is the algebraic closure of $K(z)$ where K is an algebraically closed field of constants which is of finite transcendence degree over the rationals and $z' = 1$; for $i > 0$, $\mathcal{D}_i$ is the algebraic closure of $\mathcal{D}_{i-1}(\theta_i)$, where θ_i is a monomial over $\mathcal{D}_{i-1}$. For such a field we have the following lemma, which is easily proved by the techniques of computing in algebraic extension fields.

Lemma 1. *Let S_1 be a finite set of simultaneous linear algebraic equations with coefficients in $\mathcal{F}$. Then one can construct a set S_2 of linear algebraic equations with coefficients in K such that $(k_1, \ldots, k_r) \in K^r$ satisfies S_1 if and only if it satisfies S_2.*

The next lemma will be useful to us in studying Puiseux expansions in $\theta = \log f$ at infinity and in $\theta = \exp f$ at 0 or infinity. We consider differential fields $\mathcal{D}(\theta)$ where θ is transcendental over $\mathcal{D}$ and denote the derivation by $'$.

We can define a derivation $\frac{\partial}{\partial z}$ on $\mathcal{D}(\theta)$ by $\frac{\partial \alpha}{\partial z} = \alpha'$ for all $\alpha \in \mathcal{D}$ and $\frac{\partial \theta}{\partial z} = 0$. We define the derivation $\frac{\partial}{\partial \theta}$ on $\mathcal{D}(\theta)$ by $\frac{\partial \alpha}{\partial \theta} = 0$ for all $\alpha \in \mathcal{D}$ and $\frac{\partial \theta}{\partial \theta} = 1$. Both $\frac{\partial}{\partial z}$ and $\frac{\partial}{\partial \theta}$ extend to the field of power series $\mathcal{D}((\theta))$, and $' = \frac{\partial}{\partial z} + \theta' \frac{\partial}{\partial \theta}$.

Lemma 2. *Let $v = 1 + x_1 \theta + x_2 \theta^2 + \ldots$ be in $\mathcal{D}[[\theta]]$. Then*

$$\frac{\frac{\partial v}{\partial \theta}}{v} = \sum_{k=1}^{\infty} (-1)^{k-1} I_k \theta^{k-1}$$

where

$$I_k = \begin{vmatrix} x_1 & 1 & 0 & 0 & \ldots & 0 \\ 2x_2 & x_1 & 1 & 0 & \ldots & 0 \\ 3x_3 & x_2 & x_1 & 1 & \ddots & 0 \\ \vdots & \vdots & \vdots & \ddots & \ddots & 0 \\ (k-1)x_{k-1} & x_{k-2} & \ldots & \ldots & x_1 & 1 \\ kx_k & x_{k-1} & \ldots & \ldots & x_2 & x_1 \end{vmatrix}$$

and

205

$$\frac{\frac{\partial v}{\partial z}}{v} = \sum_{k=1}^{\infty} (-1)^{k-1} J_k \theta^k$$

where

$$J_k = \begin{vmatrix} x_1' & 1 & 0 & 0 & \dots & 0 \\ x_2' & x_1 & 1 & 0 & \dots & 0 \\ x_3' & x_2 & x_1 & 1 & \ddots & 0 \\ \vdots & \vdots & \vdots & \ddots & \ddots & 0 \\ x_{k-1}' & x_{k-2} & \dots & \dots & x_1 & 1 \\ x_k' & x_{k-1} & \dots & \dots & x_2 & x_1 \end{vmatrix}.$$

Furthermore, $I_k' = k J_k$.

Proof. Since $\frac{\partial}{\partial \theta}$ and $\frac{\partial}{\partial z}$ commute, we have

$$\frac{\partial}{\partial z}\left(\frac{\frac{\partial C}{\partial \theta}}{C}\right) = \frac{\partial}{\partial \theta}\left(\frac{\frac{\partial C}{\partial z}}{C}\right)$$

so $I_k' = k J_k$. An easy computation using Cramer's Rule verifies the formulas for I_k and J_k. $\qquad \square$

Theorem 1. *Let $\mathcal{F} = \mathcal{D}_n$ be defined as above.*

(a) *Let $f \in \mathcal{F}$. One can determine in a finite number of steps whether there are v_i in $\mathcal{D}_{n-1}(\theta_n, f)$ and c_i in K, such that $f = v_0' + \sum_{i=1}^{m} c_i \frac{v_i'}{v_i}$ and find such elements if they exist.*

(b) *Let $f, g_i, i = 1, \dots, m$, be elements of $\mathcal{F}$. One can find in a finite number of steps $h_1, \dots, h_r$ in $\mathcal{D}_{n-1}(\theta_n, f, g_1, \dots, g_m)$ and a set S of linear algebraic equations in $m + r$ variables, with coefficients in K such that*

$$y' + fy = \sum_{i=1}^{m} c_i g_i$$

holds for $y \in \mathcal{D}_{n-1}(\theta_n, f, g_1, \dots, g_m)$ and elements c_i of K if and only if $y = \sum_{i=1}^{r} y_i h_i$, where y_i are elements of K and $c_1, \dots, c_m, y_1, \dots, y_r$ satisfy S.

Proof. We proceed by induction on n. Suppose $n = 0$, i.e., $\mathcal{F}$ is the algebraic closure of $K(z)$.

(a) Let $\mathcal{G} = K(z, f)$. For each place p of $\mathcal{G}$ we have a differential isomorphism of $\mathcal{G}$ into a field $K(((z-a)^{\frac{1}{t}}))$ or $K(((\frac{1}{z})^{\frac{1}{t}}))$ where $a \in K$, and t is a positive integer. Both a and t depend on p. We examine the expression $f = v_0 + \sum c_i \frac{v_i'}{v_i}$ at all finite places where f has a pole and also at the places above ∞.

We have at a finite p,

$$f = \frac{A_{-k}}{(z-a)^{\frac{k}{t}}} + \frac{A_{-k+1}}{(z-a)^{\frac{k-1}{t}}} + \cdots + \frac{A_{-t}}{z-a} + \text{terms in } (z-a)^{\frac{1}{t}} \text{ of degree} > -t$$

$$= \left[\frac{B_{-k+t}}{(z-a)^{\frac{k}{t}-1}} + \frac{B_{-k+t+1}}{(z-a)^{\frac{k-1}{t}-1}} + \cdots + \frac{B_{-1}}{(z-a)^{\frac{1}{t}}} + \text{terms of degree} > -1 \right]'$$

$$+ \sum_{i=1}^{m} c_i \frac{\text{ord}_p v_i}{t} \frac{1}{z-a} + \text{terms of degree} > -t.$$

The B's are determined by:

$$A_{-k} = -\left(\frac{k}{t} - 1\right) B_{-k+t}$$

$$\vdots$$

$$A_{-t+1} = -\frac{1}{t} B_{-1}$$

$$A_{-t} = \sum_{i=1}^{m} c_i \frac{\text{ord}_p v_i}{t}.$$

For each place p over ∞ we have:

$$f = A_k z^{\frac{k}{t}} + \cdots + A_0 + \cdots + A_{-t+1} z^{\frac{-t+1}{t}} + A_{-t} z^{-1}$$

$$+ \text{terms in } \left(\frac{1}{z}\right)^{\frac{1}{t}} \text{ of degree} > t$$

$$= \left[B_{k+t} z^{\frac{k}{t}+1} + \cdots + B_1 z^{\frac{1}{t}} + \text{terms of degree} > -1 \right]'$$

$$+ \sum_{i=1}^{m} c_i \frac{-\text{ord}_p v_i}{t} z^{-1} + \text{terms of degree} > t$$

so

$$A_k = \left(\frac{k}{t} + 1\right) B_{k+t}$$

$$\vdots$$

$$A_{-t+1} = \frac{1}{t} B_1$$

$$A_{-t} = -\sum_{i=1}^{m} c_i \frac{\text{ord}_p v_i}{t}.$$

Note that we need to look at these extra terms in the expression of f at ∞ because the derivation increases the order of an element with a pole above ∞.

Let Q be the divisor $\prod_i p_i^{-k_i}$ where v_0 has a pole of order k_i at p_i. We call a basis $\sigma_1, \ldots, \sigma_q$ for the field G a basis for Q which is *normal at* ∞ if every multiple[3] of Q is of the form $\sum_{i=1}^{q} g_i \sigma_i$ where $g_i \in K[z]$ and has a bounded degree which may be determined from Q. In [Bli66, p. 52], a basis normal at ∞ is defined under the condition that none of the places above ∞ are ramified. In Theorem 21.2 of [Bli66], it is shown how to construct such a basis and the property we use is proved. [Coa70] and [Tra84] also show how to construct such a basis.

However, even when there is ramification above ∞, one can construct a basis normal at ∞ in our sense. Consider the isomorphism φ of $G = K(z, f)$ onto $K(y, f)$ given by $\varphi(z) = \frac{1}{z-b} = y$ where none of the places lying over b are ramified. This induces a bijection $\bar{\varphi}$ between the places and divisors of $K(z, f)$ and those of $K(y, f)$ such that $\mathrm{ord}_p v = \mathrm{ord}_{\bar{\varphi}(p)} \varphi(v)$. If p lies above b, then $\bar{\varphi}(p)$ lies above ∞. We construct a basis $\varphi(\sigma_1), \ldots, \varphi(\sigma_q)$, normal at ∞ for $\bar{\varphi}(Q)$. A multiple of $\bar{\varphi}(Q)$ will be of the form $\sum_{i=1}^{q} g_i(y) \varphi(\sigma_i)$ where the bounds for the g_i are known. Thus a multiple of Q will be of the form $\sum_{i=1}^{q} g_i(z) \sigma_i$ with the same bounds for the g_i.

Now, by multiplying out the Puiseux expansions for the g_i and the σ_i at the various p we have examined, we get a simultaneous set of linear equations for the coefficients of the g_i which are the necessary and sufficient conditions for the existence of a $v_0 \in G$ having the principal parts we have found for it. If we find that these equations have a solution in K, we determine v_0 up to an additive element of K.

Let $r_1, \ldots, r_k$ be a basis for the vector space generated over the rationals by the various $\sum c_i \mathrm{ord}_p v_i$ we have found. We assert that if f has a representation $v_0' + \sum_{i=1}^{m} c_i \frac{v_i'}{v_i}$, it also has one of the form $v_0' + \sum_{i=1}^{k} \frac{r_i}{h} \frac{w_i'}{w_i}$ where h is a positive integer. Denote by $\langle c_i \rangle$ and $\langle r_i \rangle$ the vector space over the rationals spanned by the c_i and the r_i. We then have $\langle c_i \rangle = \langle r_i \rangle \oplus \langle s_i \rangle$, for some collection of elements s_i in $\langle c_i \rangle$. Thus we can write

$$ f = v_0' + \frac{1}{h} \sum_{i,j} n_{ij} r_j \frac{v_i'}{v_i} + \frac{1}{h} \sum_{i,j} m_{ij} s_j \frac{v_i'}{v_i} $$

where n_{ij}, m_{ij}, and h are integers, $h > 0$. Now

$$ f = v_0' + \frac{1}{h} \sum_j r_j \frac{w_j'}{w_j} + \frac{1}{h} \sum_j s_j \frac{y_j'}{y_j} \quad \text{where } w_j = \prod_i v_i^{n_{ij}} \text{ and } y_j = \prod_i v_i^{m_{ij}}. $$

It is clear that $\mathrm{ord}_p y_j = 0$ for each place p; thus the y_j are in K and the term involving them, in the equations for f, vanishes.

---

[3] Editors' comment: Risch is following the usage in [Bli66] where an element $f \in G$ is said to be a multiple of Q if its order at each p_i is greater than or equal to $-k_i$.

Let p_j, $j = 1, \ldots, s$, be the places where $\mathrm{ord}_{p_j} w_i \neq 0$ for some i. For each j, $\sum_i \frac{r_i}{h} \mathrm{ord}_{p_j} w_i$ is known and equals $\sum_i \frac{g_{ij}}{l_i} r_i$, where g_{ij} and l_i are integers that can be found. They are uniquely determined by the condition that for each i, l_i is the smallest positive integer such that these conditions hold. Let Q_i be the divisor $\prod_j p_j^{g_{ij}}$. It is clear that w_i has $Q_i^{\gamma_i}$ as its divisor, where γ_i is some positive integer. For any u_i which has $Q_i^{\delta_i}$ as its divisor, δ_i a positive integer, we have $\frac{1}{\gamma_i} \frac{w_i'}{w_i} = \frac{1}{\delta_i} \frac{u_i'}{u_i}$. We test to see if such w_i and γ_i exist for each i.

If $\int f$ is elementary, we have $f = v_0' + \sum_{i=1}^{k} \frac{r_i}{l_i \gamma_i} \frac{w_i'}{w_i}$. Of course, we must check whether this is so for the functions v_0, w_i which we have found.

As we did on p. 180 of [Ris69b], we remark that the analysis given serves to test whether an $f \in \mathcal{F}$ has an integral of a certain definite form. For example, to see whether there is a $v_0 \in \mathcal{G}$ such that $v_0' = f$; or if $v_1 \in \mathcal{G}$ is given, to test whether there is a $v_0 \in \mathcal{G}$ and $c \in K$ such that $f = v_0 + c \frac{v_1'}{v_1}$.

(b) In studying $y' + fy = \sum c_i g_i$ at a finite place p over a we set

$$y = \frac{A_\alpha}{(z-a)^{\frac{\alpha}{t}}} + \cdots, \quad f = \frac{B_\beta}{(z-a)^{\frac{\beta}{t}}} + \cdots, \quad \sum c_i g_i = \frac{C_\gamma}{(z-a)^{\frac{\gamma}{t}}} + \cdots,$$

where $-\alpha = \mathrm{ord}_p y$, $-\beta = \mathrm{ord}_p f$, and $-\gamma = \min_i \mathrm{ord}_p g_i$. β and γ are given and we want to find a bound for α.

$$\frac{-\frac{\alpha}{t} A_\alpha}{(z-a)^{\frac{\alpha}{t}+1}} + \cdots + \frac{A_\alpha B_\beta}{(z-a)^{\frac{\alpha+\beta}{t}}} + \cdots = \frac{C_\gamma}{(z-a)^{\frac{\gamma}{t}}} + \cdots$$

where the first nonzero term of the expansion of $\sum c_i g_i$ depends on the choice of the c_i.

Note that $\alpha > 0$ implies that $\beta > 0$ or $\gamma > 0$ and

$$\alpha + t \leq \gamma$$
$$\alpha + \beta \leq \gamma$$
$$\text{or } \alpha + t = \alpha + \beta > \gamma.$$

In the third case $\alpha = tB_\beta$. The largest of the three values determined will be our bound for α at p.

If p lies above ∞, we have

$$y = A_\alpha z^{\frac{\alpha}{t}} + A_{\alpha-1} z^{\frac{\alpha-1}{t}} + \cdots, \quad f = B_\beta z^\beta + \cdots, \quad \sum c_i g_i = C_\gamma z^\gamma + \cdots$$

so

$$\frac{\alpha}{t} A_\alpha z^{\frac{\alpha}{t}-1} + \cdots + A_\alpha B_\beta z^{\frac{\alpha+\beta}{t}} + \cdots = C_\gamma z^{\frac{\gamma}{t}} + \cdots.$$

When $\alpha \neq 0$,

$$\alpha - t \leq \gamma$$
$$\alpha + \beta \leq \gamma$$
$$\text{or } \alpha - t = \alpha + \beta > \gamma.$$

In the third case, $\alpha = -tB_\beta$.

Using the information found, we can determine an $R \in K[z]$ and S, T_i in $G = K(z, f, g_1, \ldots, g_m)$ which are integral over $K[z]$ (i.e. having no finite poles) such that $Y = Ry$ is integral over $K[z]$ and is of known bounded order at each place of G lying above ∞, and such that

$$RY' + SY = \sum_{i=1}^{m} c_i T_i \tag{1}$$

holds if and only if

$$y' + fy = \sum_{i=1}^{m} c_i g_i.$$

At an infinite p, let

$$Y = y_\alpha z^{\frac{\alpha}{t}} + y_{\alpha-1} z^{\frac{\alpha-1}{t}} + \cdots + y_\omega z^{\frac{\omega}{t}} + \cdots$$
$$R = r_\beta z^{\frac{\beta}{t}} + \cdots$$
$$S = s_\gamma z^{\frac{\gamma}{t}} + \cdots$$
$$\sum_{i=1}^{m} c_i T_i = t_\delta z^{\frac{\delta}{t}} + \cdots,$$

where $\delta = -\min_j \{\mathrm{ord}_p T_j\}$ and t_i is a linear homogeneous element of $K[c_1, \ldots, c_m]$. By plugging these expansions into (1) and equating powers of z on both sides, we get a set of linear equations S_p in $y_\alpha, \ldots, y_\omega, c_1, \ldots, c_m$, where ω is the smallest integer such that y_ω appears in a coefficient of a nonnegative power for z in (1). If Y is an element of G, integral over $K[z]$, and $c_1, \ldots, c_m$ are in K such that for each place p above ∞, the corresponding S_p is satisfied, then

$$RY' + SY - \sum_{i=1}^{m} c_i T_i$$

is an element of G that is integral over $K[z]$ and has no poles above ∞, so it must be in K. Since the coefficient of $\left(z^{\frac{1}{t}}\right)^0$ in one of these expansions is 0, we have that Y satisfies (1). Now we can find a basis $\sigma_1, \ldots, \sigma_q$ for G such that $Y = \sum_{s=1}^{q} g_s \sigma_s$, where $g_s = g_{k_s,s} z^{k_s} + \cdots + g_{0,s}$, k_s a known bound. For any place

210

p above ∞, the $y_i, i = 1, \ldots, \omega$, as above, can be expressed as a linear combination of the $g_{r,s}$. By substituting these expressions for the y_j into each S_p and taking the union of the S_p, we get a system S in the $g_{r,s}$ and $c_1, \ldots, c_m$. Let $h_{r,s} = \frac{z^r \sigma_s}{R}$. Then $y = \sum_{r,s} g_{r,s} h_{r,s}$ and S are equivalent to $y' + fy = \sum_{j=1}^m c_j g_j$ for a $y \in G$, as required.

Induction step. We assume both (a) and (b) hold for $\mathcal{D}_{n-1}$ and prove them for $\mathcal{F} = \mathcal{D}_n$ = the algebraic closure of $\mathcal{D}_{n-1}(\theta_n) = \mathcal{D}(\theta)$. Besides (a) as stated, we assume that the simpler variants, which occur when some c_i or v_i are given, have been established. See the discussion at the end of the proof of (a) for the case $n = 0$.

(a)   Case 1: $\theta = \log \eta$

At a place p centered at $a \in \mathcal{D}$, we have

$$f = \frac{A_{-k}}{(\theta - a)^{\frac{k}{t}}} + \cdots + \frac{A_{-t}}{\theta - a} + \text{terms in } (\theta - a)^{\frac{1}{t}} \text{ of degree} > -t$$

$$= \left[\frac{B_{-k+t}}{(\theta - a)^{\frac{k}{t} - 1}} + \cdots + \frac{B_{-1}}{(\theta - a)^{\frac{1}{t}}} + \text{terms of degree} > -1 \right]'$$

$$+ \sum_{i=1}^m c_i \left(\frac{\eta'}{\eta} - a' \right) \frac{\mathrm{ord}_p v_i}{t} \frac{1}{\theta - a} + \text{terms of degree} > -t.$$

If we differentiate the first term inside the brackets, we get

$$\frac{-(\frac{k}{t} - 1) B_{-k+t} (\frac{\eta'}{\eta} - a')}{(\theta - a)^{\frac{k}{t}}} + \frac{B'_{-k+t}}{(\theta - a)^{\frac{k}{t} - 1}}.$$

Thus we can determine the B's from the equations:

$$A_{-k} = - \left(\frac{k}{t} - 1 \right) \left(\frac{\eta'}{\eta} - a' \right) B_{-k+t}$$

$$\vdots$$

$$A_{-k+t} = - \left(\frac{k}{t} - 2 \right) \left(\frac{\eta'}{\eta} - a' \right) B_{-k+2t} + B'_{-k+t}$$

$$\vdots$$

$$A_{-t-1} = -\frac{1}{t} \left(\frac{\eta'}{\eta} - a' \right) B_{-1} + B'_{-t-1}$$

$$A_{-t} = \sum_{i=1}^m c_i \left(\frac{\eta'}{\eta} - a' \right) \frac{\mathrm{ord}_p v_i}{t} + B'_{-t}.$$

At a place p lying over ∞, we have

$$f = A_k\theta^{\frac{k}{t}} + A_{k-1}\theta^{\frac{k-1}{t}} + \cdots + A_0 + A_{-1}\theta^{-\frac{1}{t}} + \cdots + A_{-t+1}\theta^{-\frac{t-1}{t}}$$

$$+ A_{-t}\theta^{-1} + \text{terms of degree} > t \text{ in } \left(\frac{1}{\theta}\right)^{\frac{1}{t}}$$

$$= \left[B_{k+t}\theta^{\frac{k}{t}+1} + \cdots + B_1\theta^{\frac{1}{t}} + B_0 + B_{-1}\theta^{-\frac{1}{t}} + \cdots + B_{-t+1}\theta^{-\frac{t-1}{t}}\right.$$

$$\left. + B_{-t}\theta^{-1} + \text{terms of degree} > t\right]'$$

$$+ \sum_{i=1}^{m} C_i\frac{D_i'}{D_i} + C_{-1}'\theta^{-\frac{1}{t}} + \cdots + \left(C_{-t}' - \sum_{i=1}^{m} c_i\frac{\text{ord}_p v_i}{t}\frac{\eta'}{\eta}\right)\theta^{-1}$$

$$+ \text{terms of degree} > t$$

where A_i, B_i, C_i, D_i are elements of $\mathcal{D}$, and D_i is the leading coefficient of v_i.

We should justify our assertions about the C_i. We can, without loss of generality, assume that the leading coefficient of v_i for $i > 0$ is 1 at p since for $a \in \mathcal{D}$, $\frac{(av_i)'}{av_i} = \frac{a'}{a} + \frac{v_i'}{v_i}$. Let

$$v_i = \theta^{\frac{\ell}{t}} + x_1\theta^{\frac{\ell-1}{t}} + \cdots + x_t\theta^{\frac{\ell}{t}-1} + \ldots$$

$$v_i' = x_1'\theta^{\frac{\ell-1}{t}} + \cdots + \left(x_t' + \frac{\ell}{t}\frac{\eta'}{\eta}\right)\theta^{\frac{\ell}{t}-1} + \ldots$$

We then see that the coefficients of $\theta^{-\frac{1}{t}}, \ldots, \theta^{-\frac{t-1}{t}}$ in $\frac{v_i'}{v_i}$ differ only in sign from the $J_1, \ldots, J_{t-1}$ of Lemma 2 and so are derivatives of elements of $\mathcal{D}$. The coefficient of θ^{-1} equals

$$(-1)^{t-1}J_t + \begin{vmatrix} 1 & 0 & 0 & \ldots & 0 & 0 \\ x_1 & 1 & 0 & \ldots & 0 & 0 \\ x_2 & x_1 & 0 & \ldots & 0 & 0 \\ \vdots & \vdots & \vdots & \ddots & \vdots & \vdots \\ x_{t-2} & x_{t-3} & x_{t-4} & \ldots & 1 & 0 \\ x_{t-1} & x_{t-2} & x_{t-3} & \ldots & x_1 & \frac{\ell}{t}\frac{\eta'}{\eta} \end{vmatrix} = \text{a derivative} - \frac{\text{ord}_p v_i}{t}\frac{\eta'}{\eta}.$$

We determine the B's as follows:

$$0 = B_{k+t}', \quad \text{so } B_{k+t} \in K,$$

$$A_k = B_k' + \frac{\eta'}{\eta}\left(\frac{k}{t}+1\right)B_{k+t} \quad \text{or} \quad \int A_k = B_k + \left(\frac{k}{t}+1\right)B_{k+t}\log\eta.$$

By the induction assumption, we can determine whether there is a $B_k \in \mathcal{D}$ and a $B_{k+t} \in K$ such that the last relation holds. B_{k+t} will be uniquely determined since $\log \eta = \theta$ is a monomial over $\mathcal{D}$. B_k is uniquely determined up to an additive constant. We set $B_k = \bar{B}_k + b_k$ where $\bar{B}_k \in \mathcal{D}$ is fixed and $b_k \in K$ is undetermined. We continue

$$A_{k-t} = B'_{k-t} + k B_k \frac{\eta'}{\eta} \quad \text{or} \quad A_{k-t} - k \bar{B}_k \frac{\eta'}{\eta} = B'_{k+t} + k b_k \frac{\eta'}{\eta}.$$

As before, we determine b_k uniquely and B_{k-t} up to an additive constant. Continuing, we determine $\ldots B_{t-1}, \ldots, B_1$ up to an additive constant and, skipping the equation for A_0, we arrive at the equation,

$$A_{-1} = B'_{-1} + C'_{-1} + \frac{t-1}{t} B_{t-1} \frac{\eta'}{\eta}$$

yielding

$$A_{-1} - \frac{t-1}{t} \frac{\eta'}{\eta} \bar{B}_{t-1} = (B_{-1} + C_{-1})' + \frac{t-1}{t} b_{t-1} \frac{\eta'}{\eta}$$

which enables us to determine b_{t-1}. We obtain $b_{t-2}, \ldots, b_1$ similarly. Therefore we can determine $B_{k+t}, \ldots, B_{t+1}, B_{t-1}, \ldots, B_1$ uniquely. To determine B_t at each place above ∞, first consider a fixed place p_1 above ∞. In the equation for A_0, we have

$$A_0 = B'_0 + B_t \frac{\eta'}{\eta} + \sum_{j=1}^{m} c_j \frac{D'_j}{D_j} \quad \text{or} \quad A_0 - \bar{B}_t \frac{\eta'}{\eta} = B'_0 + b_t \frac{\eta'}{\eta} + \sum_{j=1}^{m} c_j \frac{D'_j}{D_j}$$

and we can assume that $b_t = 0$, i.e. that $b_t \frac{\eta'}{\eta}$ is absorbed by the term to its right, so $B_t = \bar{B}_t$. Thus the D_j that we determine here using the induction hypothesis are those v_i which are in $\mathcal{D}$. At the other places above ∞, we can determine b_t from an equation (derived from the equation for A_0) of the form:

$$\text{known element of } \mathcal{D} = B'_0 + b_t \frac{\eta'}{\eta}.$$

Therefore, at all places we can find the principal part of v_0. This information enables us, as in the $n = 0$ case, to determine v_0 up to an additive element of $\mathcal{D}$. To determine the v_i, $i > 0$, we note that at each place, the equation for A_{-t} is:

$$A_{-t} = \sum_{i=1}^{m} c_i \left(\frac{\eta'}{\eta} - a' \right) \frac{\text{ord}_p v_i}{t} + B'_{-t} \quad \text{(at finite places)}$$

$$A_{-t} = (B_{-t} + C_{-t})' - \sum_{i=1}^{m} c_i \frac{\text{ord}_p v_i}{t} \frac{\eta'}{\eta} \quad \text{(at infinite places).}$$

Using the induction hypothesis, these equations determine $\sum_{i=1}^{m} c_i \text{ord}_p v_i$ at each place. This data allows us to determine the v_i, $i > 0$, as we did in the case $n = 0$. The v_i are determined up to multiplication by an element of $\mathcal{D}$. We therefore are left to find elements $\bar{v}_0, \bar{v}_1, \ldots, \bar{v}_m$ in $\mathcal{D}$ and constants $\bar{c}_i$ such that

$$f - \left(v_0' + \sum_{i=1}^m c_i \frac{v_i'}{v_i} \right) = \bar{v}_0' + \sum_{i=1}^m \bar{c}_i \frac{\bar{v}_i'}{\bar{v}_i}$$

and that can be done by induction.

Case 2: $\theta = \exp \zeta$

We assume that there is no ramification at the places above 0 and ∞. This can be arranged by replacing $\theta = \exp(\zeta)$ by $\exp(\zeta/t)$ where t is the least common multiple of the ramification indices at these places.

At a place p above $a \in \mathcal{D}$, $a \neq 0$, we have

$$\left((\theta - a)^{\frac{k}{t}} \right)' = \frac{k}{t} (a\zeta' - a')(\theta - a)^{\frac{k}{t}-1} + \frac{k}{t} \zeta'(\theta - a)^{\frac{k}{t}}$$

where $(a\zeta' - a') \neq 0$. If p is a pole of f,

$$f = \frac{A_{-k}}{(\theta - a)^{\frac{k}{t}}} + \cdots + \frac{A_{-k+t}}{(\theta - a)^{\frac{k}{t}-1}} + \cdots + \frac{A_{-t}}{\theta - a}$$

$$+ \text{ terms in } (\theta - a)^{\frac{1}{t}} \text{ of degree } > -t$$

$$= \left[\frac{B_{-k+t}}{(\theta - a)^{\frac{k}{t}-1}} + \cdots + \frac{B_{-k+2t}}{(\theta - a)^{\frac{k}{t}-2}} + \cdots + \frac{B_{-1}}{(\theta - a)^{\frac{1}{t}}} \right.$$

$$\left. + \text{ terms of degree } > -1 \right]'$$

$$+ \sum_{i=1}^m c_i (a\zeta' - a') \frac{\text{ord}_p v_i}{t} \frac{1}{\theta - a} + \text{ terms of degree } > -t.$$

Therefore we can determine $B_{-k+t}, \ldots, B_{-1}$ and $\sum_{i=1}^m c_i \text{ord}_p v_i$ from the equations:

$$A_{-k} = -\left(\frac{k}{t} - 1 \right)(a\zeta' - a')B_{-k+t}$$

$$\vdots$$

$$A_{-k+t} = -\left(\frac{k}{t} - 2 \right)(a\zeta' - a')B_{-k+2t} + B_{-k+t}' - \left(\frac{k}{t} - 1 \right)\zeta' B_{-k+t}$$

$$\vdots$$

$$A_{-t} = \sum_{i=1}^m c_i (a\zeta' - a') \frac{\text{ord}_p v_i}{t} + B_{-t}' - \zeta' B_{-t}.$$

At the places p above 0,

$$f = A_{-k}\theta^{-k} + \cdots + A_{-1}\theta^{-1} + A_0 + \text{ terms in } \theta \text{ of degree } > 0$$

$$= \left[B_{-k}\theta^{-k} + \cdots + B_{-1}\theta^{-1} + B_0 + \text{ terms of degree } > 0 \right]'$$

$$+ \sum_{i=1}^{m} c_i \left[\frac{C_i'}{C_i} + \zeta' \text{ord}_p v_i \right] + \text{ terms of degree } > 0,$$

where C_i is the leading coefficient of v_i.

At a p above ∞, we have similarly,

$$f = A_k\theta^k + \cdots + A_1\theta + A_0 + \text{ terms in } \theta^{-1} \text{ of degree } > 0$$

$$= \left[B_k\theta^k + \cdots + B_1\theta + B_0 + \text{ terms of degree } > 0 \right]'$$

$$+ \sum_{i=1}^{m} c_i \left[\frac{C_i'}{C_i} - \zeta' \text{ord}_p v_i \right] + \text{ terms of degree } > 0.$$

Above 0 or ∞ we have for $j \neq 0$, $A_j = B_j' + j\zeta'B_j$, which is solvable by (b) of the induction hypothesis. B_j will be uniquely determined since $\exp \zeta$ is a monomial over $\mathcal{D}$.

We have determined the principal parts of v_0 and can now construct it explicitly in the same manner as described in the case $n = 0$. It is determined up to an additive element of $\mathcal{D}$.

Let p_1 be a fixed place above ∞. If $v_i = C_i\theta^k + \cdots, i > 0$, then $\frac{v_i'}{v_i} = k\zeta' + \frac{C_i'}{C_i} + \frac{w_i'}{w_i}$, where $w_i = 1 + E_1\theta^{-1} + \cdots$. Thus we can assume that at p_1, $\text{ord}_{p_1} v_i = 0$ and the leading coefficient of each v_i not in $\mathcal{D}$ is 1. Therefore,

$$A_0 = B_0' + \sum_{i=1}^{q} d_i \frac{D_i'}{D_i}.$$

By the induction hypothesis, we can determine B_0, d_i, D_i in $\mathcal{D}$ such that this holds. The D_i will be those $v_i, i > 0$, which lie in $\mathcal{D}$. The B_0 determined at p_1 determines v_0 exactly and so we know the B_0 term at the other places above 0 and ∞.

We now have

$$g = f - v_0' - \sum_{i=1}^{q} d_i \frac{D_i'}{D_i} = \sum_{i=1}^{l} c_i \frac{v_i'}{v_i}$$

where g is a known element of $\mathcal{G} = \mathcal{D}(\theta, f)$ and $\{v_i : i = 1, \ldots, l\}$ are those members of $\{v_i : i = 1, \ldots, m\}$ that do not lie in $\mathcal{D}$.

At p_1 we have $g = C_{-1}\theta^{-1} + C_{-2}\theta^{-2} + \cdots$ with $C_i \in \mathcal{D}$. Let us consider

$$G = \sum_{i=1}^{l} c_i \frac{\frac{\partial v_i}{\partial \theta}}{v_i}$$

at p_1. By Lemma 2, we have $G = \theta^{-2}\left(F_1 + F_2\theta^{-1} + F_3\theta^{-2} + \ldots\right)$ where $F_k = \sum_{i=1}^l c_i I_{i,k}$, where $I_{i,k}$ is the coefficient of θ^{-k-1} in $\frac{\frac{\partial v_i}{\partial \theta}}{v_i} = \theta^{-2}(I_{i,1} + I_{i,2}\theta^{-1} + \ldots)$. Using Lemma 2 and this notation, we have:

$$g = \sum_{i=1}^l c_i \frac{\frac{\partial v_i}{\partial z}}{v_i} + \zeta'\theta G$$

$$= \sum_{k=1}^\infty \left(-\frac{1}{k}F_k' + \zeta'F_k\right)\theta^{-k} = \sum_{k=1}^\infty C_{-k}\theta^{-k}.$$

Using the induction hypothesis for (b), each F_k can be determined uniquely by

$$-\frac{1}{k}F_k' + \zeta'F_k = C_{-k}.$$

At any p lying above 0, $\mathrm{ord}_p(G) \geq -1$. At any p lying above ∞, $\mathrm{ord}_p(G) \geq 0$. Since we already know the other places p where $\mathrm{ord}_p(G) < 0$ (they are those finite p where $\sum_{i=1}^m c_i \mathrm{ord}_p v_i \neq 0$), we can determine a divisor Q such that G is a multiple of Q. Using the fact that the number of zeros is equal to the number of poles, we can determine a bound h for the order of the zero that a multiple of Q can have at p_1.

In order to determine G explicitly, we take a basis $\sigma_1, \ldots, \sigma_q$ for Q that is normal at ∞. $G = \sum_{i=1}^q g_i\sigma_i$ where $g_i = g_{k_i,i}\theta^{k_i} + \cdots + g_{0,i}$, k_i a known bound. We express F_k, $k = 1, \ldots, h-1$ (the coefficients of $\theta^{-2}, \ldots, \theta^{-h}$ in G) as linear combinations of the $g_{j,i}$ with coefficients in $\mathcal{D}$. These equations have a unique solution, since otherwise there would be a multiple of Q that has a zero of order greater than h at p_1.

From the explicit expression $G = \sum_{i=1}^q g_i\sigma_i$ we determine $\sum_{i=1}^m c_i \mathrm{ord}_p v_i$ at the other p's lying above 0 and ∞ by finding the respective coefficients of θ^{-1} there. This information allows us to find the v_i, $i = 1, \ldots, l$, as in the $n = 0$ case, up to a multiplicative element of $\mathcal{D}$. They are then uniquely determined by the condition that their leading coefficient is 1 at p_1.

(b)   Case 1: $\theta = \log\eta$

The equations for determining a bound for $\mathrm{ord}_p y$, at a finite place p, are very similar to those in the case $n = 0$ and will be omitted. At a place above ∞ we have,

$$y = A_\alpha\theta^{\frac{\alpha}{t}} + \cdots, \quad f = B_\beta\theta^{\frac{\beta}{t}} + \cdots, \quad \sum_{i=1}^m c_i g_i = C_\gamma\theta^{\frac{\gamma}{t}} + \cdots.$$

Plugging these expressions into $y' + fy = \sum_{i=1}^m c_i g_i$, we get

$$A_\alpha'\theta^{\frac{\alpha}{t}} + A_{\alpha-1}'\theta^{\frac{\alpha-1}{t}} + \cdots + \left(A_{\alpha-t}' + \frac{\alpha}{t}\frac{\eta'}{\eta}A_\alpha\right)\theta^{\frac{\alpha-t}{t}} + \cdots$$

$$+A_\alpha B_\beta \theta^{\frac{\alpha+\beta}{t}} + \cdots = C_\gamma \theta^{\frac{\gamma}{t}} + \cdots .$$

When $\alpha \neq 0$,

$$\alpha - t \leq \gamma$$
$$\alpha + \beta \leq \gamma$$
$$\text{or } \alpha \geq \alpha + \beta \geq \alpha - t > \gamma.$$

In the third case, we have $\frac{\gamma - \alpha}{t} < -1$, so the coefficient of θ^{-1} in $\frac{y'}{y} + f$ will be 0. Using Lemma 2 as we did in Case 1 of part (a), we see that the coefficient of θ^{-1} is

$$\frac{\alpha}{t} \frac{\eta'}{\eta} + D' + B_{-t} = 0,$$

where $D \in \mathcal{D}$. Thus we find α uniquely from $\int B_{-t} = -D - \frac{\alpha}{t} \log \eta$.

As in the $n = 0$ case, we reduce the problem to solving the equivalent equation for $Y = Ry$:

$$RY' + SY = \sum_{i=1}^m c_i T_i \tag{2}$$

where $R \in \mathcal{D}[\theta]$ and S and T_i are elements of $\mathcal{G} = \mathcal{D}(\theta, f, g_1, \ldots, g_m)$ that are integral over $\mathcal{D}[\theta]$ and Y is an unknown element of $\mathcal{G}$ integral over $\mathcal{D}[\theta]$. For each place p above ∞, a bound μ for $\text{ord}_p y$ is known.

At a place p above ∞, (2) takes the following form:

$$\left[r_\lambda \theta^{\frac{\lambda}{t}} + r_{\lambda-1} \theta^{\frac{\lambda-1}{t}} + \cdots \right] \cdot \left[y'_\mu \theta^{\frac{\mu}{t}} + y'_{\mu-1} \theta^{\frac{\mu-1}{t}} + \cdots \right.$$
$$+ \left(y'_{\mu-t} + \frac{\mu}{t} \frac{\eta'}{\eta} y_\mu \right) \theta^{\frac{\mu-t}{t}} + \cdots \right] + \left[s_\lambda \theta^{\frac{\lambda}{t}} + s_{\lambda-1} \theta^{\frac{\lambda-1}{t}} + \cdots \right] \cdot \tag{3}$$
$$\left[y_\mu \theta^{\frac{\mu}{t}} + y_{\mu-1} \theta^{\frac{\mu-1}{t}} + \cdots \right] = t_{\lambda+\mu} \theta^{\frac{\lambda+\mu}{t}} + \cdots ,$$

where t_k are linear homogeneous elements of $\mathcal{D}[c_1, \ldots, c_m]$. At least one of r_λ, s_λ, and $t_{\lambda+\mu}$ is not equal to zero.

Let us look at the conditions for the coefficients of the non-negative powers of θ on both sides of (3) to cancel each other:

$$r_\lambda y'_\mu + s_\lambda y_\mu \qquad\qquad\qquad = t_{\lambda+\mu}$$
$$r_\lambda y'_{\mu-1} + s_\lambda y_{\mu-1} + r_{\lambda-1} y'_\mu + s_{\lambda-1} y_\mu \qquad = t_{\lambda+\mu-1}$$
$$\vdots$$
$$r_\lambda y'_{\mu-t} + s_\lambda y_{\mu-t} + \cdots + r_{\lambda-t} y'_\mu + \left(s_{\lambda-t} + \frac{\mu}{t} \frac{\eta'}{\eta} r_\lambda \right) y_\mu = t_{\lambda+\mu-t} \tag{4}$$
$$\vdots$$
$$r_\lambda y'_{-\lambda} + s_\lambda y_{-\lambda} + \cdots \qquad\qquad\qquad = t_0$$

Suppose $r_\lambda = s_\lambda = \ldots = r_{\lambda-k+1} = s_{\lambda-k+1} = 0$ while one of $r_{\lambda-k}, s_{\lambda-k} \neq 0$. $t_{\lambda+\mu} = \ldots = t_{\lambda+\mu-k+1} = 0$ is, by Lemma 1, equivalent to a linear system in

217

$c_1, \ldots, c_m$ with coefficients in K. We then proceed as below, with $r_{\lambda-k}, s_{\lambda-k}$ replacing r_λ, s_λ.

If one of $r_\lambda, s_\lambda \neq 0$, then $r_\lambda y'_\mu + s_\lambda y_\mu = t_{\lambda+\mu}$ is equivalent, by the induction assumption, to $y_\mu = \sum_{i=1}^{r} d_i h_i$, $d_i \in K$, $h_i \in \mathcal{D}$, and a linear system S_μ in $c_1, \ldots, c_m, d_1, \ldots, d_r$.

By substituting the expression for y_μ into the second equation of the triangular system (4), we get an equation for $y_{\mu-1}$ which is equivalent to $y_{\mu-1} = \sum_{i=1}^{s} e_i j_i$ and a system $S_{\mu-1}$ in $c_1, \ldots, c_m, d_1, \ldots, d_r, e_1, \ldots, e_s$. We continue in a like manner until we have determined $y_{-\lambda} = \sum_{i=1}^{w} q_i v_i$ and a linear system $S_{-\lambda}$ in $c_1, \ldots, c_m, \ldots, q_1, \ldots, q_w$.

We go through the above process for each place lying above ∞. As in the $n = 0$ case, we take a basis, $\sigma_1, \ldots, \sigma_q$, for the elements integral over $\mathcal{D}[\theta]$ and normal at ∞, $Y = \sum_{i=1}^{q} g_i \sigma_i$ where $g_i = g_{k_i,i}\theta^{k_i} + \cdots + g_{0,i}$, k_i a known bound. y_j, $j = \mu, \ldots, -\lambda$, is a homogeneous linear combination of the $g_{l,i}$ with coefficients in $\mathcal{D}$. For any choice of the y_j, the $g_{l,i}$ are uniquely determined, since otherwise there would be two distinct functions, integral over $\mathcal{D}[\theta]$, having the same principal parts and coefficient of θ^0 at each place above ∞; this is impossible. Thus, we can solve for the $g_{l,k}$, expressing them as homogeneous linear combinations of the y_j with coefficients in $\mathcal{D}$. Into this system, we substitute the expressions we have for y_j as a homogeneous linear combination of elements of $\mathcal{D}$ with constant coefficients. Now, plug these expressions for the $g_{l,i}$ into $y = \frac{\sum_{i=1}^{q} g_i \sigma_i}{R}$. This expression and the union of the systems $S_\mu, \ldots, S_{-\lambda}$ give us the conditions on y that are equivalent to $y' + fy = \sum_{i=1}^{m} c_i g_i$.

Case 2: $\theta = \exp \zeta$

At a place above $a \in \mathcal{D}$, $a \neq 0$, we have for $y' + fy = \sum_{i=1}^{m} c_i g_i$,

$$\frac{\frac{\alpha}{t}(a' - \zeta' a) A_\alpha}{(\theta - a)^{\frac{\alpha}{t}+1}} + \cdots + \frac{A_\alpha B_\beta}{(\theta - a)^{\frac{\alpha+\beta}{t}}} + \cdots = \frac{C_\gamma}{(\theta - a)^{\frac{\gamma}{t}}} + \cdots$$

Here also $\alpha > 0$ implies β or $\gamma > 0$. We have

$$\alpha + t \leq \gamma$$
$$\alpha + \beta \leq \gamma$$
$$\text{or } \alpha + t = \alpha + \beta > \gamma$$

In the third case, $\alpha = \frac{-t B_\beta}{a' - \zeta' a}$. Above 0, we get

$$\frac{A'_\alpha - \frac{\alpha}{t}\zeta' A_\alpha}{\theta^{\frac{\alpha}{t}}} + \cdots + \frac{A_\alpha B_\beta}{\theta^{\frac{\alpha+\beta}{t}}} + \cdots = \frac{C_\gamma}{\theta^{\frac{\gamma}{t}}} + \cdots \quad .$$

Thus,

$$\alpha \leq \gamma$$
$$\alpha + \beta \leq \gamma$$
$$\text{or } \alpha = \alpha + \beta > \gamma$$

In the third case we determine α by $\int B_0 = \frac{\alpha}{t}\zeta - \log A_\alpha$.
Above ∞, the analysis is the same as above 0.

We now proceed in the same manner as in the logarithmic case to find h_i and a linear system S. This completes the proof of the theorem. $\quad\square$

This theorem implies the two generalizations of the theorems in Section 3 of [Ris69b] mentioned at the beginning of this paper. Part (a) of the above theorem is just another way to look at statement (1). As for (2), recall that for an $f \in \mathcal{D}$, $\log f$ is not a monomial over $\mathcal{D}$ if and only if $\frac{f'}{f} = g'$ for some $g \in \mathcal{D}$; $\exp f$ is not a monomial over $\mathcal{D}$ if and only if $f' = \frac{1}{m}\frac{g'}{g}$ for an integer m and $g \in \mathcal{D}$. These are simple variants of part (a) of the above theorem, and our proof shows how to test for these.

References

[Bli66]   G. A. Bliss. *Algebraic Functions*. American Mathematical Society Collection Publications, Dover, 1966. First published as AMS Colloquium vol. XVI (1933).

[Che51]   C. Chevalley. *Algebraic Functions of One Variable*. Number VI in Mathematical Surveys and Monographs. American Mathematical Society, New York, 1951.

[Coa70]   J. Coates. Construction of rational functions on a curve. *Mathematical Proceedings of the Cambridge Philosophical Society*, 68:107–123, 1970.

[Dav81]   J. H. Davenport. *On the Integration of Algebraic Functions*, volume 102 of *Lecture Notes in Computer Science*. Springer Verlag, New York, 1981.

[Ris68]   R. H. Risch. *On the Integration of Elementary Functions Which are Built Up Using Algebraic Operations*. SDC Document SP - 2801/002/00, 26 June 1968.

[Ris69a]   R. H. Risch. *Further Results on Elementary Functions*. Technical report, IBM Publications, 1969.

[Ris69b]   R. H. Risch. The problem of integration in finite terms. *Transactions of the American Mathematical Society*, 139:167–189, 1969.

[Ris70]   R. H. Risch. The solution of the problem of integration in finite terms. *Bulletin of the American Mathematical Society*, 76:605–608, 1970.

[Tra84]   B. M. Trager. *Integration of Algebraic Functions*. PhD thesis, MIT, 1984.

[vdW53]   B. L. van der Waerden. *Modern Algebra*, volume 1. Frederick Ungar Publishing Company, New York, 1953. tr. Fred Blum.

Comments on Risch's
On the Integration of Elementary Functions which are Built Up Using Algebraic Operations

Clemens G. Raab[1]

Contents

In the two reports [Ris68, Ris69a], Risch devised a recursive algorithm that decides when a given elementary function has an elementary integral. At that time, only the complete algorithm for transcendental elementary integrands was published [Ris69b]. By abuse of language, "transcendental" elementary integrands is a short way of referring to the more restrictive class of elementary integrands that lie in differential fields whose generators are algebraically independent over the constants. For the general case, i.e. elementary integrands that also allow algebraic relations among their constituents, only the main ideas were published in a short note [Ris70]. In the years 1985/86, Michael Singer made some efforts on behalf of Risch to prepare a revised version of [Ris68] for publication, also incorporating unpublished results of [Ris69a]. Simultaneously, Bronstein in his dissertation [Bro87, Bro90b] was working on a generalization of Trager's algorithm [Tra84] to obtain a general algorithm for integrating elementary functions in finite terms. The paper appearing in this volume is the first version of Risch's complete decision procedure to be published. It reflects Singer's edition of [Ris68], incorporating parts of [Ris69a].

Risch's papers [Ris69b, Ris70] on the transcendental case and on the general case of his algorithm stimulated broader interest in exploring the algorithmic aspects of integration. Many variants of his algorithm have been proposed with the aim of making computations more efficient. His algorithm has also been extended and generalized in various directions and new algorithms have been developed for computing with integrals. Below, we give an overview of these developments, focusing on algorithms.

---

[1] The author was supported by the Austrian Science Fund (FWF): P 27229 and P 31952.

© Springer Nature Switzerland AG 2022

C. G. Raab und M. F. Singer (Hrsg.), *Integration in Finite Terms: Fundamental Sources*, Texts & Monographs in Symbolic Computation, https://doi.org/10.1007/978-3-030-98767-1_6

1 General Remarks on Integration in Finite Terms

Elementary functions and the problem of integration in finite terms can be formalized in different ways. In the context of differential fields, integration in finite terms usually refers to deciding the existence of and computing representations of antiderivatives in elementary extensions of given differential fields, which contain the integrand. All differential fields considered here have characteristic 0. Elementary extensions of a given differential field are differential fields generated by iteratively adjoining new elements that are logarithms and exponentials of other elements or satisfy polynomial equations. An antiderivative of an element of a differential field that lies in an elementary extension of that differential field is often called an *elementary integral* for short, even if it is not an elementary function. Elements of any elementary extension of the field of rational functions are elementary functions, which is taken as the definition of elementary functions in the context of differential fields.

For an introduction to general ideas and problems in integration in finite terms, we refer the reader to [ND79, Nor83, Raa13a], which also contain references to more detailed presentations. The notes of Bronstein [Bro98] discuss more algorithmic details and his book [Bro05] gives a self-contained presentation of many algorithms relevant for transcendental integrands along with detailed theoretical background.

Before discussing algorithms, we take a brief look at the question of decidability of integration in finite terms. Based on undecidability of certain types of Diophantine equations, some versions of integration in finite terms have been shown to be undecidable. A result of Richardson [Ric68] shows that, for integrands built (using addition, multiplication, and composition) from the real, single-valued functions $\exp(x)$, $\sin(x)$, and $|x|$, existence of an elementary integral is undecidable. However, working in a differential field, there cannot be additional functions satisfying $y^2 - x^2 = 0$ apart from x and $-x$. In [Ris69b], Risch pointed out that the use of multivalued functions in the definition of the integrand can be problematic, since the existence of an elementary integral may depend on the choice of branches. For example, this is the case for $(\log(\exp(x)) - x)\exp(x^2)$. He showed that, for elementary integrands, it is undecidable whether there is a choice of branches that admits an elementary integral. To still give a decision procedure, he restricts the definitions of integrands such that the existence of an elementary integral does not depend on the choice of the branches. His solution is adopted essentially in all variants and generalizations of the integration algorithm. Stated briefly, the differential field containing the integrand should be constructed as an extension of a computable field of constants such that no new constants are introduced.

The way functions are represented is an important issue in general, both for decidability and for concrete algorithms. Integration algorithms using the language of differential fields usually require that the differential field is given explicitly as a tower of nested extensions in terms of its generators and their differential algebraic properties. In addition to imposing the type of functions represented by these generators, many algorithms also depend on further conditions that the generators of the field have to satisfy. These conditions on the representation of the field depend on the algorithm and allow for a simpler treatment of the represented functions by explicitly exhibiting properties that otherwise might be hidden or by introducing additional properties that can be exploited.

2 Algorithmic Developments for Antiderivatives

The publication of Risch's decision procedure for finding elementary integrals of elementary functions that do not involve algebraic functions [Ris69b] attracted much attention. The procedure involves several subproblems. Very soon, parts of his algorithm for some of these subproblems were modified in order to reduce the computational effort needed. Even for rational integrands a lot of improvements were developed, such as the use of Hermite reduction [Her72] and variants of it [Ost45, Mac75], minimization of algebraic constants in the result via the Rothstein–Trager resultant [Rot76, Tra76], and computation of logarithms via subresultants [LR90, Mul97] or via Gröbner bases [Czi95]. In brief, Hermite reduction allows one to reduce higher order poles in the integrand to simple poles and to compute the rational part of the integral at the same time and the Rothstein–Trager resultant provides a way to compute all the residues of the integrand. Chapter 2 of [Bro05] contains a detailed discussion of algorithms for integrating rational functions.

Generalizations of these techniques to elementary and certain classes of non-elementary integrands were developed in parallel. Several improvements for transcendental elementary functions are presented in [Rot76]. Based on [Mac76], a generalization of Risch's decision procedure to elementary integrals of transcendental Liouvillian integrands is given in the appendix of [SSC85]. Liouvillian integrands may involve certain special functions such as polylogarithms, error functions, or the incomplete gamma function, for example. Bronstein realized that many techniques developed for Liouvillian integrands can be generalized to differential fields that are built up by what he calls monomials [Bro90a]. In short, monomials represent transcendental functions that satisfy an explicit first-order polynomial differential equation. The main part of his book [Bro05] presents such generalizations of algorithms for several subproblems arising in the computation of elementary integrals, like Hermite reduction and computation of new logarithms via subresultants, along with their theoretical foundation. Likewise, Czichowski's algorithm relying on Gröbner bases can be generalized to monomials [Raa12b]. These generalizations, however, do not yield a decision procedure for finding elementary integrals over differential fields involving non-elementary monomials. Complete decision algorithms along these lines were obtained only in [Raa12a] for Liouvillian monomials and under some technical restrictions also for non-Liouvillian monomials. By allowing non-Liouvillian monomials, integrands can be treated that may involve many common special functions, such as Bessel functions, complete elliptic integrals, or Legendre functions, for example. In [Raa13b], it was shown how to reduce the problem of finding elementary integrals over a differentially transcendental extension of a differential field to the problem of finding elementary integrals over that differential field itself.

As part of Risch's decision procedure and its variants, in addition to the subproblems mentioned above, inhomogeneous linear first-order differential equations need to be solved. As a result, this type of differential equation is called the Risch differential equation in this context by many authors. In fact, the inhomogeneous part of the equation depends linearly on constant parameters and it is required to find all parameter values such that the differential equation has a solution in the given differential field. In [Ris69b], not all details of this procedure are given and, aiming at increased efficiency and/or generality, many authors have given algorithms for this type of differential equation problem. Especially in the context of certain non-elementary integrands, other differential equation problems play a role as well. We do not discuss any of these algorithms here.

All the decision procedures for finding elementary integrals discussed above recursively follow the nested structure of the differential field. In addition, they involve several subalgorithms, some of which are recursive themselves, like the solution of the Risch differential equation. This complexity makes implementation in software rather difficult and, despite all improvements for efficiency, also may cause long runtimes. Therefore, a different and much simpler computational approach was proposed by Risch in 1976 for integrating transcendental elementary functions. The main idea is to treat all generators of the differential field at once in parallel. It was first implemented by Norman [NM77] and is also called the Risch–Norman algorithm. Basically, the Risch–Norman algorithm proceeds by making an ansatz for the integral with undetermined constant coefficients. A system of linear algebraic equations is then obtained by differentiating the ansatz and comparing it to the given integrand. Solving this linear system yields the values needed for the coefficients of the ansatz. Several improvements to this approach have been proposed, see e.g. [Fit81, Dav82b, GS89], and extensions to non-elementary integrands have been considered as well [Bro07, Boe10]. In contrast to the recursive versions of Risch's integration algorithm, the parallel approach of the Risch–Norman algorithm does not yield a decision procedure in general, i.e. an elementary integral may not be found even though it exists. Only some theoretical results regarding the denominator of the elementary integral and possible new logarithms arising in it could be obtained, first for integrands containing only logarithms [Dav82a, DT85] and later also for much more general integrands [Bro07]. Nevertheless, the Risch–Norman algorithm has turned out to be a rather powerful heuristic in practice. Moreover, it can be used for differential fields for which no other algorithm is available, see [Bro05, Bro07, Kra09, Boe10] and references therein.

Above, we focused on developments related to the transcendental case of Risch's decision procedure [Ris69b]. The general procedure sketched in [Ris70] also gave rise to algorithmic investigations. Relevant decision procedures for algebraic functions include [Dav81, Tra84]. For general elementary integrands, Bronstein gave a decision algorithm [Bro90b], see also [Bro98]. In [Tra79], an algorithm for computing the algebraic part of the integral was given for the case when radicals appear in the integrand. Heuristic approaches have also been considered for algebraic functions [Kau08] and for very general differential fields with algebraic relations among their generators [Boe10]. For further discussion of algorithms developed for integrands involving algebraic functions, see Trager's commentary in this volume.

All algorithms discussed so far dealt with the problem of finding elementary integrals over given differential fields. Even algorithms for computing non-elementary integrals have been developed, see e.g. [SSC85, Che85, Che86, Kno92, Kno93], where in addition to elementary functions also the error-function or the logarithmic integral are allowed in new expressions arising in the closed form of the integral. These algorithms depend on generalizations of Liouville's Theorem, like the one given in [SSC85]. For further discussion on various generalizations of Liouville's Theorem and related algorithms, see Singer's commentary in this volume.

All the recursive decision procedures for finding elementary integrals proceed by reducing the integrands step by step to a simpler form. Many other methods and algorithms for integration use some kind of reduction steps as well. For instance, there are closely related algorithms that can decide when given elements of certain types of differential fields have an integral in the same differential field, like Hermite reduction does for rational functions [Ost45, Her72] and for square-roots of rational functions [Her83]. In [CDL18, DGL+20], this has been achieved for certain differential fields built by primitive monomials and in [BLL+16] this has been worked out for differential fields built by differentially transcendental elements. More generally, parameterized reduction rules had already been considered for integration earlier. In [Nor90], an algorithmic way of constructing new reduction rules from given ones is presented. Reduction techniques are also used without representing the integrand as an element of a differential field. Reducing an integral to a form for which a closed form evaluation is already known is a very common method for computing integrals, especially before the advent of algebraic algorithms mentioned above. Many books with tables of explicit integrals have been compiled for this purpose. There are also efforts to make the use of precomputed integrals more systematic. Most notably, Rich [RJ09, RSA18] constructed a cleverly crafted collection of conditional reduction rules that transform a given integral into a simpler or even explicit form.

The principle of comparing the integrand to the derivative of an ansatz for the integral has also been applied in many other works apart from the Risch–Norman algorithm mentioned above. Without using the language of differential fields, Sonine [Son80] already computed integrals of functions involving cylinder functions in a certain way via an appropriate ansatz. The coefficients of his ansatz for the integral are not necessarily constant. As a consequence, the system obtained by differentiating the ansatz and comparing it to the given integrand is a differential system. For setting up the differential system, linear functional relations satisfied by the cylinder functions involved have to be exploited. Much later, this approach was generalized by Piquette to integrals involving products of certain special functions satisfying similar functional equations [PV84, Piq91]. Linear functional equations also play a crucial role in related algorithms for D-finite functions (i.e. functions satisfying a linear ODE with rational function coefficients). In these approaches, functions are represented by linear operators corresponding to the functional relations satisfied by those functions, like in [Zei90], and computations are done at the operator level. An algorithm for integrating D-finite functions as a linear combination with rational function coefficients of the integrand and its derivatives was given in [Chy00], for instance, which also proceeds by ansatz and finding rational solutions of linear differential systems. Many related algorithms using the operator viewpoint have been developed for D-finite and holonomic functions, see below. Another algorithm that uses the implicit representation of functions by linear operators is presented in [AH97], which relies on solving an adjoint equation.

3 Algorithms for Linear Relations of Definite Integrals

If an explicit antiderivative for the integrand can be found (e.g. by one of the approaches mentioned above), then by the fundamental theorem of calculus a definite integral can, at least in principle, be computed by evaluating the antiderivative at the bounds of the interval of integration. In practice, issues of singularities and branch cuts need to be taken into account, of course. For a discussion of these and related issues when computing definite integrals by computer, see e.g. [Dav03, Lic11].

If for the given integrand an antiderivative is not available in closed form, then other methods need to be used. One approach consists in relating the given definite integral to definite integrals for which an explicit evaluation is known, like integral representations of special functions (see e.g. [GGM+90]) or identities of general families of functions such as the Meijer G-functions (see e.g. [AM90]), or can be easily obtained otherwise. Similar to indefinite integrals, a lot of precomputed definite integrals are listed in integral tables. In applications, definite integrals often depend on additional parameters as they frequently arise from integral transforms or convolutions of functions, where it is sometimes possible to exploit properties of the integral transform in order to reduce the parameter integral to known cases.

For many parameter integrals, regardless of their origin, it is possible to systematically derive linear relations they satisfy. These can then be used to obtain more information on the parameter integral. Based on these linear relations and evaluations for fixed values of the parameters, it is often even possible to derive closed form evaluations of the parameter integral. A key technique for relating definite integrals is differentiation under the integral sign, since under mild conditions we have $\frac{d}{dy} \int_a^b f(y,x) \, dx = \int_a^b \frac{\partial f}{\partial y}(y,x) \, dx$. Abel already used this technique to obtain relations for certain parameter integrals [Abe27]. Development of computer algebra tools for the algorithmic evaluation of parameter integrals along these lines, related to differentiation under the integral sign, started around 1990 [AZ90, Zei90].

For obtaining linear relations of definite integrals as described in the following, the key idea is to construct new integrands for which an antiderivative can be found, which then yield relations of corresponding integrals. *Parametric integration*, sometimes also referred to as parameterized integration, generalizes indefinite integration in the sense that several integrands $f_1(x), \ldots, f_m(x)$ are given at once and the task is to find coefficients $c_1, \ldots, c_m$ (not depending on x) and a function $g(x)$ such that

$$\sum_{i=1}^m c_i f_i(x) = \frac{d}{dx} g(x). \tag{1}$$

One can also think of this as finding concrete values of the parameters c_i occurring linearly in the integrand $\sum_{i=1}^m c_i f_i(x)$ such that an antiderivative can be given in closed form. Integrating an identity of the form (1) from a to b yields the linear relation of definite integrals

$$\sum_{i=1}^{m} c_i \int_a^b f_i(x)\,dx = g(b) - g(a), \tag{2}$$

since the coefficients c_i do not depend on x.

For the closely related method of *creative telescoping*, usually only one integrand $f(\vec{y},x)$ depending on extra parameters $\vec{y}$ is considered and the goal is to find a linear operator L (acting on the parameters and commuting with $\frac{d}{dx}$ and with evaluation of x) and a function $g(\vec{y},x)$ such that $Lf(\vec{y},x) = \frac{d}{dx}g(\vec{y},x)$. From such a telescoping identity, the equation $LF(\vec{y}) = g(\vec{y},b) - g(\vec{y},a)$ satisfied by the parameter integral $F(\vec{y}) = \int_a^b f(\vec{y},x)\,dx$ is obtained by integration, since the operator L commutes with $\frac{d}{dx}$ and with evaluation of x. The operator L is often a differential operator or a recurrence operator. For example, if L is a differential operator, then the equation obtained is a differential equation for $F(\vec{y})$. This can be considered a generalization of differentiation under the integral sign. The relation between creative telescoping and parametric integration is that individual terms of $Lf(\vec{y},x)$ can be viewed as particularly chosen $f_i(x)$ in (1). For example, if L is a differential operator w.r.t. $\vec{y}$, then the integrands $f_i(x)$ for parametric integration can be chosen as partial derivatives of $f(\vec{y},x)$ w.r.t. $\vec{y}$. Then, the c_i obtained in solving (1) will be the corresponding coefficients in the differential operator L.

Originally, creative telescoping was a method to compute linear recurrences for sums by artificially creating a telescoping sum, see e.g. [Zei91], but the name was soon used for analogous problems in other contexts as well [CS98]. For surveys on the paradigm of creative telescoping, see e.g. [Kou13, Chy14], and for a short overview on computer algebra algorithms for parameter integrals, see also [Raa16]. In the following, we give only a brief summary.

Algorithms for parametric integration typically are designed to find a basis of all solutions $(c_0,\ldots,c_m,g(x))$ of (1) for which $g(x)$ lies in a given class of functions. In fact, Risch's differential equation problem in [Ris69b] covers parametric integration for the case of g lying in the same differential field as the integrands f_i. Algorithms that can deal with different classes of functions in the integrands were given in [Mac76, SSC85, Raa12a]. Most of these algorithms look for g in elementary extensions of the given differential field that contains the integrands and are variants of the recursive Risch algorithm. Similarly, the Risch–Norman algorithm can be used for parametric integration in a straightforward way. Based on the work of Zeilberger, algorithms for creative telescoping usually use holonomic systems of linear functional relations to describe the integrands [AZ90, Zei90, CS98, Chy00, AZ06, CKK14]. Holonomic systems can be thought of as systems of functional equations, which allow to uniquely specify a function with only finitely many initial values. As mentioned before in the context of D-finite functions, computations are mainly done at the operator level. Algorithmic results have also been obtained for non-holonomic systems [CKS09]. Recently, algorithms for creative telescoping based on reduction techniques generalizing Hermite reduction have emerged [CKS12, BCC+13, CHK+18, CDL18, BCL+18, Hoe21].

4 Summation in Finite Terms

The successful algorithmic developments for integration in finite terms also triggered interest in algorithmic approaches to summation in finite terms. We touch this topic only briefly here. For sequences that are given in terms of nested sums and products of rational expressions, Karr [Kar81, Kar85] presented a theory relying on difference fields and a corresponding algorithm for expressing the sequence of partial sums again in terms of nested sums and products of rational expressions. Later, the theory and algorithm were refined and generalized by Schneider, see [Sch15, Sch16, Sch17] and references therein. These generalizations also include algorithms for linear relations of definite sums in the spirit of [Zei91], which rely on the concept of parameterized telescoping, which is a summation analog of parametric integration discussed above. Both on the theoretical side and on the algorithmic side, summation in finite terms has striking similarities with integration in finite terms, but shows some fundamentally different phenomena as well. For instance, periodic sequences give rise to zero divisors, since the product of sequences $\prod_{i=1}^{k}(a-a_i)$ is zero if the sequence $a = (a_n)_{n \in \mathbb{N}}$ has period k.

References

[Abe27]   Niels H. Abel. Ueber einige bestimmte Integrale. *J. reine und ange-wandte Mathematik* 2, pp. 22–30, 1827.

[AH97]   Sergei A. Abramov and Mark van Hoeij. A Method for the Integration of Solutions of Ore Equations. *Proc. ISSAC'97*, pp. 172–175, 1997.

[AM90]   Victor S. Adamchik and Oleg I. Marichev. The algorithm for calculating integrals of hypergeometric type functions and its realization in REDUCE system. *Proc. ISSAC'90*, pp. 212–224, 1990.

[AZ90]   Gert E. T. Almkvist and Doron Zeilberger. The Method of Differentiating under the Integral Sign. *J. Symbolic Computation* 10, pp. 571–591, 1990.

[AZ06]   Moa Apagodu and Doron Zeilberger. Multi-variable Zeilberger and Almkvist–Zeilberger algorithms and the sharpening of Wilf–Zeilberger theory. *Advances in Applied Mathematics* 37, pp. 139–152, 2006.

[Boe10]   Stefan T. Boettner. *Mixed Transcendental and Algebraic Extensions for the Risch–Norman Algorithm*. PhD thesis, Tulane Univ., New Orleans, USA, 2010.

[BCC+13] Alin Bostan, Shaoshi Chen, Frédéric Chyzak, Ziming Li, and Guoce Xin. Hermite Reduction and Creative Telescoping for Hyperexponential Functions. *Proc. ISSAC'13*, pp. 77–84, 2013.

[BCL⁺18]   Alin Bostan, Frédéric Chyzak, Pierre Lairez, and Bruno Salvy. Generalized Hermite Reduction, Creative Telescoping, and Definite Integration of D-Finite Functions. *Proc. ISSAC'18*, pp. 95–102, 2018.

[BLL⁺16]   François Boullier, François Lemaire, Joseph Lallemand, Georg Regensburger, and Markus Rosenkranz. Additive normal forms and integration of differential fractions. *J. Symbolic Computation* 77, pp. 16–38, 2016.

[Bro87]   Manuel Bronstein. *Integration of Elementary Functions*. PhD thesis, Univ. of California, Berkeley, USA, 1987.

[Bro90a]   Manuel Bronstein. A Unification of Liouvillian Extensions. *Applicable Algebra in Engineering, Communication and Computing* 1, pp. 5–24, 1990.

[Bro90b]   Manuel Bronstein. Integration of Elementary Functions. *J. Symbolic Computation* 9, pp. 117–173, 1990.

[Bro98]   Manuel Bronstein. *Symbolic integration tutorial*. Course notes of an ISSAC'98 tutorial.
            http://www-sop.inria.fr/cafe/Manuel.Bronstein/
            publications/issac98.pdf

[Bro05]   Manuel Bronstein. *Symbolic Integration I – Transcendental Functions*. 2nd ed., Springer, 2005.

[Bro07]   Manuel Bronstein. Structure theorems for parallel integration. *J. Symbolic Computation* 42, pp. 757–769, 2007.

[CDL18]   Shaoshi Chen, Hao Du, and Ziming Li. Additive Decompositions in Primitive Extensions. *Proc. ISSAC'18*, pp. 135–142, 2018.

[CHK⁺18]   Shaoshi Chen, Mark van Hoeij, Manuel Kauers, and Christoph Koutschan. Reduction-based creative telescoping for fuchsian D-finite functions. *J. Symbolic Computation* 85, pp. 108–127, 2018.

[CKK14]   Shaoshi Chen, Manuel Kauers, and Christoph Koutschan. A Generalized Apagodu–Zeilberger Algorithm. *Proc. ISSAC'14*, pp. 107–114, 2014.

[CKS12]   Shaoshi Chen, Manuel Kauers, and Michael F. Singer. Telescopers for Rational and Algebraic Functions via Residues. *Proc. ISSAC'12*, pp. 130–137, 2012.

[Che85]   Guy W. Cherry. Integration in Finite Terms with Special Functions: the Error Function. *J. Symbolic Computation* 1, pp. 283–302, 1985.

[Che86]   Guy W. Cherry. Integration in finite terms with special functions: the logarithmic integral. *SIAM J. Computing* 15, pp. 1–21, 1986.

[Chy00]   Frédéric Chyzak. An extension of Zeilberger's fast algorithm to general holonomic functions. *Discrete Mathematics* 217, pp. 115–134, 2000.

[Chy14]   Frédéric Chyzak. *The ABC of Creative Telescoping – Algorithms, Bounds, Complexity*. Habilitation à diriger des recherches (HDR), Univ. Paris-Sud 11, France, 2014.
            https://specfun.inria.fr/chyzak/Publications/
            Chyzak-2014-ACT.pdf

[CKS09]   Frédéric Chyzak, Manuel Kauers, and Bruno Salvy. A Non-Holonomic Systems Approach to Special Function Identities. *Proc. ISSAC'09*, pp. 111–118, 2009.

[CS98]   Frédéric Chyzak and Bruno Salvy. Non-commutative elimination in Ore algebras proves multivariate identities. *J. Symbolic Computation* 26, pp. 187–227, 1998.

[Czi95]   Günter Czichowski. A Note on Gröbner Bases and Integration of Rational Functions. *J. Symbolic Computation* 20, pp. 163–167, 1995.

[Dav81]   James H. Davenport. *On the Integration of Algebraic Functions*. Springer, 1981.

[Dav82a]   James H. Davenport. The Parallel Risch Algorithm (I). *Proc. EURO-CAM'82*, pp. 144–157, 1982.

[Dav82b]   James H. Davenport. On the Parallel Risch Algorithm (III): Use of tangents. *SIGSAM Bull.* 16, pp. 3–6, 1982.

[Dav03]   James H. Davenport. *The Difficulties of Definite Integration*. 2003. http://staff.bath.ac.uk/masjhd/Calculemus2003-paper.pdf

[DT85]   James H. Davenport and Barry M. Trager. On the Parallel Risch Algorithm (II). *ACM Trans. Mathematical Software* 11, pp. 356–362, 1985.

[DGL⁺20]   Hao Du, Jing Guo, Ziming Li, and Elaine Wong. An Additive Decomposition in Logarithmic Towers and Beyond. *Proc. ISSAC'20*, pp. 146–153, 2020.

[Fit81]   John P. Fitch. User-based Integration Software. *Proc. SYMSAC'81*, pp. 245–248, 1981.

[GGM⁺90]   Keith O. Geddes, M. Lawrence Glasser, R. A. Moore, and Tony C. Scott. Evaluation of Classes of Definite Integrals Involving Elementary Functions via Differentiation of Special Functions. *Applicable Algebra in Engineering, Communication and Computing* 1, pp. 149–165, 1990.

[GS89]   Keith O. Geddes and L. Yohanes Stefanus. On the Risch–Norman Integration Method and Its Implementation in MAPLE. *Proc. ISSAC'89*, pp. 212–217, 1989.

[Her72]   Charles Hermite. Sur l'intégration des fractions rationnelles. *Annales scientifiques de l'École Normale Supérieure* (2ᵉ série) 1, pp. 215–218, 1872.

[Her83]   Charles Hermite. Sur la réduction des intégrales hyperelliptiques aux fonctions de première, de seconde et de troisième espèce. *Bull. sciences mathématiques et astronomiques* (2ᵉ série) 7, pp. 36–42, 1883.

[Hoe21]   Joris van der Hoeven. Constructing reductions for creative telescoping – The general differentially finite case. *Applicable Algebra in Engineering, Communication and Computing* 32, pp. 575–602, 2021.

[Kar81]   Michael Karr. Summation in Finite Terms. *J. Association of Computing Machinery* 28, pp. 305–350, 1981.

[Kar85]   Michael Karr. Theory of Summation in Finite Terms. *J. Symbolic Computation* 1, pp. 303–315, 1985.

[Kau08]   Manuel Kauers. Integration of Algebraic Functions: A Simple Heuristic for Finding the Logarithmic Part. *Proc. ISSAC'08*, pp. 133–140, 2008.

[Kno92]   Paul H. Knowles. Integration of a Class of Transcendental Liouvillian
            Functions with Error-Functions, Part I. *J. Symbolic Computation* 13,
            pp. 525–543, 1992.

[Kno93]   Paul H. Knowles. Integration of a Class of Transcendental Liouvillian
            Functions with Error-Functions, Part II. *J. Symbolic Computation* 16,
            pp. 227–241, 1993.

[Kou13]   Christoph Koutschan. Creative Telescoping for Holonomic Functions.
            In *Computer Algebra in Quantum Field Theory*, pp. 171–194, Springer,
            2013.

[Kra09]   Robert Kragler. On Mathematica Program for "Poor Man's Integrator"
            Implementing Risch–Norman Algorithm. *Programming and Computer
            Software* 35, pp. 63–78, 2009.
            Russian original in *Programmirovanie* 2009, no. 2, pp. 10–27.

[LR90]   Daniel Lazard and Renaud Rioboo. Integration of Rational Func-
            tions: Rational Computation of the Logarithmic Part. *J. Symbolic
            Computation* 9, pp. 113–115, 1990.

[Lic11]   Daniel Lichtblau. Symbolic definite (and indefinite) integration: meth-
            ods and open issues. *ACM Communications in Computer Algebra* 45,
            pp. 1–16, 2011.

[Mac76]   Carola Mack. *Integration of affine forms over elementary functions.*
            Computational Physics Group Report UCP-39, Univ. of Utah,
            1976.

[Mac75]   Dieter Mack. *On Rational Integration.* Computational Physics Group
            Report UCP-38, Univ. of Utah, 1975.

[Mul97]   Thom Mulders. A Note on Subresultants and the Lazard/Rioboo/Trager
            Formula in Rational Function Integration. *J. Symbolic Computation* 24,
            pp. 45–50, 1997.

[Nor83]   Arthur C. Norman. Integration in Finite Terms. In *Computer Algebra:
            Symbolic and Algebraic Computation*, pp. 57–69, Springer, 1983.

[Nor90]   Arthur C. Norman. A Critical-Pair/Completion based Integration
            Algorithm. *Proc. ISSAC'90*, pp. 201–205, 1990.

[ND79]   Arthur C. Norman and James H. Davenport. Symbolic integration —
            The dust settles?. *Proc. EUROSAM'79*, pp. 398–407, 1979.

[NM77]   Arthur C. Norman and P. M. A. Moore. Implementing the new Risch
            Integration algorithm. *Proc. 4th International Colloquium on Advanced
            Computing Methods in Theoretical Physics*, pp. 99–110, 1977.

[Ost45]   Mikhail V. Ostrogradsky. De l'intégration des fractions rationnelles.
            *Bull. classe physico-mathématique de l'Académie Impériale des Sciences
            Saint-Pétersbourg* 4, pp. 145–167 and 286–300, 1845.

[Piq91]   Jean C. Piquette. A Method for Symbolic Evaluation of Indefinite
            Integrals Containing Special Functions or their Products. *J. Symbolic
            Computation* 11, pp. 231–249, 1991.

[PV84]   Jean C. Piquette and Arnie L. Van Buren. Technique for evaluating
            indefinite integrals involving products of certain special functions. *SIAM
            J. Mathematical Analysis* 15, pp. 845–855, 1984.

[Raa12a]   Clemens G. Raab. *Definite Integration in Differential Fields*. PhD thesis, Johannes Kepler Univ. Linz, Austria, 2012.
            http://www.risc.jku.at/publications/download/risc_4583/PhD_CGR.pdf

[Raa12b]   Clemens G. Raab. Using Gröbner bases for finding the logarithmic part of the integral of transcendental functions. *J. Symbolic Computation* 47, pp. 1290–1296, 2012.

[Raa13a]   Clemens G. Raab. Generalization of Risch's Algorithm to Special Functions. In *Computer Algebra in Quantum Field Theory*, pp. 285–304, Springer, 2013.

[Raa13b]   Clemens G. Raab. Integration of Unspecified Functions and Families of Iterated Integrals. *Proc. ISSAC'13*, pp. 323–330, 2013.

[Raa16]   Clemens G. Raab. Symbolic Computation of Parameter Integrals. Tutorial Abstract, *Proc. ISSAC'16*, pp. 13–15, 2016.

[RJ09]   Albert D. Rich and David J. Jeffrey. A Knowledge Repository for Indefinite Integration Based on Transformation Rules. *Proc. Calculemus/MKM 2009*, pp. 480–485, 2009.

[RSA18]   Albert D. Rich, Patrick Scheibe, and Nasser M. Abbasi. Rule-based integration: An extensive system of symbolic integration rules. *J. Open Source Software* 3, Article 1073, 2018.

[Ric68]   Daniel Richardson. Some undecidable problems involving elementary functions of a real variable. *J. Symbolic Logic* 33, pp. 514–520, 1968.

[Ris68]   Robert H. Risch. *On the Integration of Elementary Functions which are Built Up Using Algebraic Operations*. SDC report SP-2801/002/00, 1968.

[Ris69a]   Robert H. Risch. *Further Results on Elementary Functions*. IBM report RC-2402, 1969.

[Ris69b]   Robert H. Risch. The problem of integration in finite terms. *Trans. American Mathematical Society* 139, pp. 167–189, 1969.

[Ris70]   Robert H. Risch. The solution of the problem of integration in finite terms. *Bull. American Mathematical Society* 76, pp. 605–608, 1970.

[Rot76]   Michael Rothstein. *Aspects of Symbolic Integration and Simplification of Exponential and Primitive Functions*. PhD thesis, Univ. of Wisconsin-Madison, USA, 1976.
            http://www.cs.kent.edu/~rothstei/dis.pdf

[Sch15]   Carsten Schneider. Fast Algorithms for Refined Parameterized Telescoping in Difference Fields. In *Computer Algebra and Polynomials*, pp. 157–191, Springer, 2015.

[Sch16]   Carsten Schneider. A difference ring theory for symbolic summation. *J. Symbolic Computation* 72, pp. 82–127, 2016.

[Sch17]   Carsten Schneider. Summation Theory II: Characterizations of $R\Pi\Sigma^*$-extensions and algorithmic aspects. *J. Symbolic Computation* 80, pp. 616–664, 2017.

[SSC85]   Michael F. Singer, B. David Saunders, and Bob F. Caviness. An Extension of Liouville's Theorem on Integration in Finite Terms. *SIAM J. Computing* 14, pp. 966–990, 1985.

[Son80]   Nikolay Y. Sonine. Recherches sur les fonctions cylindriques et le développement des fonctions continues en séries. *Mathematische Annalen* 16, pp. 1–80, 1880.

[Tra76]   Barry M. Trager. Algebraic Factoring and Rational Function Integration. *Proc. SYMSAC'76*, pp. 219–226, 1976.

[Tra79]   Barry M. Trager. Integration of simple radical extensions. *Proc. EUROSAM'79*, pp. 408–414, 1979.

[Tra84]   Barry M. Trager. *Integration of algebraic functions*. PhD thesis, MIT, USA, 1984. Reprinted in this volume

[Zei90]   Doron Zeilberger. A holonomic systems approach to special functions identities. *J. Computational and Applied Mathematics* 32, pp. 321–368, 1990.

[Zei91]   Doron Zeilberger. The Method of Creative Telescoping. *J. Symbolic Computation* 11, pp. 195–204, 1991.

Integration of Algebraic Functions

Barry M. Trager

PhD. Thesis, Massachusetts Institute of Technology (1984)

ACKNOWLEDGEMENTS

First of all I wish to thank my parents who have always been very supportive and always hoped that someday I would finish this thesis and finally become a "doctor".

Next I would like to thank Professor Joel Moses for introducing me to the field of Computer Algebra in general and symbolic integration in particular, Professors Steven Kleiman and Mike Artin for several illuminating discussions, Dr. James Davenport for constructive criticism and conversations about topics of common interest and Dr. Richard Zippel for his helpful comments.

I would also like to thank the many friends who encouraged, cajoled, and browbeat me into finally finishing this dissertation, especially Patrizia who actually got me to finish, David and Joel who tried everything they could think of, Anni who supplied me with a continuing source of interesting integrals to solve, and Dick and the others at IBM research who provided the necessary time, tools, and financial support.

© Springer Nature Switzerland AG 2022
C. G. Raab und M. F. Singer (Hrsg.), *Integration in Finite Terms: Fundamental Sources*, Texts & Monographs in Symbolic Computation,
https://doi.org/10.1007/978-3-030-98767-1_7

Contents

Chapter 1
Introduction

This thesis will provide a practical decision procedure for the indefinite integration of algebraic functions. Unless explicitly noted, we will always assume x to be our distinguished variable of integration. An algebraic function y of x is defined as a solution of a monic polynomial in y with coefficients that are rational functions in x. Each of these rational functions in x can be written as a quotient of polynomials in x whose coefficients are constants (i.e. not dependent on x). If we adjoin each of these constant coefficients to our base field $\mathbf{Q}$ (rational numbers), then we have constructed a finitely generated extension of $\mathbf{Q}$ that we will call our coefficient field, $\mathbf{K}$. We can assume the integral has the form $\int y \, dx$ where $f(x, y)$ is the unique monic polynomial of least degree that y satisfies. An elementary function is one that can be expressed using combinations of algebraic functions, exponentials and logarithms. We propose either to express the integral as an elementary function or guarantee that this cannot be done.

1.1 Review of Previous Work

During the 18th and 19th centuries this problem attracted much interest [1]. Euler (1748) and others studied elliptic functions and discovered they were not integrable in terms of elementary functions. Abel (1826) first studied the integrals of general algebraic functions, which later became known as abelian integrals. Liouville [48] proved a key theorem that forms the basis for the decision procedure. He showed that $\int y \, dx$, if integrable in terms of elementary functions, could be expressed as

$$\int y \, dx = v_0 + \sum c_i \log v_i \tag{1}$$

where the c_i are constants and the v_i are rational combinations of x and y. The basic idea in his proof is that integration cannot introduce any new functions other than constant multiples of logs since differentiation would not remove them. Liouville (1833) also gave an integration algorithm [37] for the special case of algebraic integrals that could be expressed without logarithms, but he found no method for solving the general problem. This problem was so difficult that Hardy (1916) pessimistically stated [25] "there is reason to suppose that no such method can be given".

The next major steps in this area were taken by Risch (1968). First he gave a complete decision procedure for integrating elementary functions that were free of algebraic functions [45]. Then he turned his attention to the algebraic problem, and in [46] he sketched a procedure that reduced algebraic integration to a problem in algebraic geometry, determining a bound for the torsion subgroup of the divisor class group of an algebraic function field. Finally in [47] by referring to some recent results in algebraic geometry he outlined a theoretical solution to the problem. While this indeed disproved Hardy's undecidability conjecture, it did not really present a practical algorithm that could be used to actually solve integration problems.

Risch refined equation (1) by showing one could assume $c_i \in \overline{K}$ and $v_i \in \overline{K}(x, y)$ where $\overline{K}$ is the algebraic closure of K (see also [49]). He also showed that $v_0 \in K(x, y)$. The integration problem is now reduced to finding the c_i and the v_i. The basic approach that Risch used and that we will use is to construct these functions by analyzing their singularities. By considering algebraic functions on their Riemann surface, they are no longer multi-valued and have only a finite number of poles as singularities. At each point of the Riemann surface a function can be expressed locally as a Laurent series in terms of some uniformizing parameter. When the parameter is expressed as $(x - a)^{1/n}$ for some $a \in K$ and $n \in N$ the series is called a Puiseux expansion. The finite initial segment composed of the terms with negative exponents is called the principal part of the series expansion. If we compute the principal parts of the integrand at all its poles, then we can integrate these sums termwise. If we discard the log terms that arise by integrating terms of the form $r_j(x - a)^{-1}$, then the remaining terms form the principal parts of the expansions of v_0 at all of its poles. Thus we make the key observation that the singular places of v_0 are a subset of the singular places of the integrand. Now we are in the position of trying to find a function with a given set of principal parts. Since an algebraic function with no poles is constant, this function is only determined up to an additive constant, as expected from an integration problem. The least degree term of each principal part specifies the order n_j of the pole at that place p_j on the Riemann surface. This allows us to associate an integer with each place if we agree to associate the integer 0 with places where the integrand was non-singular. The formal sum over all places $\sum n_j p_j$ is called a *divisor*, and by the Riemann–Roch theorem the set of algebraic functions with orders not less than those specified by the divisor form a finite dimensional vector space over $\overline{K}$. Bliss [7] gives a technique for finding a basis for this vector space. A general member of this vector space can be expressed as a linear combination of basis elements with indeterminate coefficients. By equating the principal parts of a general member of the vector space with the principal parts of the integral we get a system of linear equations that give a necessary and sufficient condition for the existence of the algebraic part of the integral.

To find the logarithmic part we now examine the terms of the form $r_j(x - a)^{-1}$ that we ignored earlier. The coefficients r_j are precisely the residues of the integrand at the singular places p_j of the integrand. If the integral exists then it can be presented in a form where the c_i's form a basis for the Z-module generated by the residues. In fact as we shall show in chapter 6 the c_i generate the minimal algebraic extension of the coefficient field required to express the integral. The algorithms we present will avoid Puiseux expansions and never introduce algebraic quantities not required for the final answer.

To simplify our presentation of Risch's approach, we will now assume all the residues are rational numbers. The general case will be treated in chapter 6. Let m be a least common denominator of all the residues. We will now construct a divisor $\mathbf{D}$ from our set of residues, in general we would have k divisors where k is the rank of the $\mathbf{Z}$-module generated by the residues. We define the order of $\mathbf{D}$ at each place p to be m times the residue of the integrand at p. Thus $\mathbf{D} = \sum (mr_i)p_i$. Since there are only a finite number of places where the integrand has non-zero residue $\mathbf{D}$ is a well defined divisor. Since the sum of the residues of the integrand is zero, $\deg \mathbf{D} = \sum (mr_i) = 0$. For any function $f \in \mathbf{K}(x, y)$ we have a divisor $\text{div}(f) = \sum (\text{ord}_p f)p$ where $\text{ord}_p f$ is the order of f at p. A divisor is called principal if it is the divisor of some algebraic function. If the divisor associated with the residues is principal describing a function f then we have determined the log term $\frac{1}{m} \log f$. However, if the divisor is not principal it is possible for some integer multiple of it to become principal. If k times a divisor $\mathbf{D}$ is principal with associated function v, then we have a log term $\frac{1}{mk} \log v$. This is the fundamental theoretical obstruction to the integration of algebraic functions. The integrand contains complete information about the location and orders of the poles of the algebraic part of the integral, but for the log parts the residues enable us to find reduced divisors only (the coefficients of the places are relatively prime), not necessarily the actual divisors of the logands. The "points of finite order problem" is to find an integer bound $\mathbf{B}$ such that if $k\mathbf{D}$ is not principal for some divisor $\mathbf{D}$ and all integers $1 \le k \le \mathbf{B}$ then no integer multiple of $\mathbf{D}$ is principal. Under componentwise addition and subtraction of the order coefficients the set of divisors becomes an abelian group. The quotient of the group of divisors of degree zero by the subgroup of principal divisors is called the divisor class group. The divisor classes for which some multiple is principal form the torsion subgroup of the divisor class group. Thus we can reformulate our question as finding a bound for the torsion subgroup. In order for such a bound to exist it is critical that our constant field be a finitely generated extension of the rationals and not an algebraically closed field.

The approach that Risch outlined for determining the bound uses a technique that has lately come into vogue in many areas of algebraic manipulation. We take a difficult problem and homomorphically map it into a simpler domain, hoping to find some technique for lifting the solution back to the original domain. Although the original *points of finite order* problem seems quite difficult, Weil (1948) showed that the problem is easily solvable for function fields in one variable over finite fields [59]. For these fields the divisor class group is finite and Weil's rationality formula for the zeta function shows that the order of the class group can be computed by counting points on a non-singular model. Using Weil's proof of the extended Riemann hypothesis for such fields we can explicitly give bounds for the divisor class group over a finite field. An upper bound is $(\sqrt{q} + 1)^{2g}$ where q is the order of the finite field and g is the genus, and we obtain a lower bound of $(\sqrt{q} - 1)^{2g}$.

The natural approach to reducing our problem is "reduction mod p," where p is the prime ideal for some discrete valuation of our constant field, e.g. if the constant field is $\mathbf{Q}$ then p will be a prime ideal in $\mathbf{Z}$. But we must verify when such a reduction is "good", i.e. gives us useful information for determining our bound. Risch observed [47] that if we get a projective non-singular model for our function field and if its reduction mod p is still non-singular then the reduction is good. This means that the induced homomorphism of divisor class groups is injective for all divisors whose order is relatively prime to the characteristic of the finite field ([51] or [54]). Since the torsion subgroup is a finite abelian group, it can be decomposed into a product of the group of elements whose orders are relatively prime to p and the p-sylow subgroup. Thus if we find two distinct rational primes that give us good reduction, we can multiply the bounds for the two divisor class groups and get our desired bounds. Risch claimed that all of these steps were known to be effective, but he provided no explicit algorithms. Dwork and Baldassarri [4] independently duplicated Risch's approach to this problem in the process of finding algebraic solutions for second order linear differential equations.

James Davenport has independently investigated this problem. His algorithms are constructed along the lines suggested by Risch. He uses Puiseux expansions and Coate's [14] algorithm to construct a basis for the multiples of a divisor. This construction is used for both the algebraic and transcendental part of the answer. To bound the divisor torsion he uses special purpose algorithms depending on whether the curve is genus 1 and whether there are parameters present. In the case there are parameters present, he produces an explicit test for torsion without computing the bound. In the case of genus 1, he uses arithmetic on the curve to compute the torsion. While these special purpose tests can be reasonably efficient, in the general case, he reverts to the Weil bound given previously. His algorithms are general, but his implementation is currently limited to algebraic functions that can be expressed with nested square roots.

Risch and Davenport depend on Puiseux expansions to unravel the singularities of the function field defined by the integrand. Some such mechanism is necessary to distinguish apparent poles from actual ones. For example the function y/x has an apparent pole at the origin, but if $y^2 = x^2(x + 1)$ then the function is actually holomorphic there. We will use integral bases to determine the nature of the poles of an algebraic function. An integral basis for an algebraic function field of degree n is a set of n functions such that an element of the function field can be expressed as a linear combination of basis elements with polynomial coefficients if and only if that element has no singularities in the finite plane, i.e. the element is an integral algebraic function. Good algorithms for computing integral bases are a subject of ongoing research [63]. Bliss [7] shows that finding integral bases is no harder than computing Puiseux expansions, and in the very important case of function fields defined by a single radical, they are immediate.

In this thesis we present a new algorithm for integration of algebraic functions. we rely on the construction of an integral basis to provide an affine non-singular model for the curve. This will allow us to construct the algebraic part of the answer by a generalization of Hermite's algorithm for integrating rational functions. This integral basis can also be used to test whether divisors are principal and to test for good reduction.

A very serious problem in symbolic calculations is that intermediate expressions are frequently much larger than the final answer and can thus be very time and space consuming. One of the counterparts to intermediate expression swell that we are forced to deal with in this problem is intermediate constant field extensions. Risch and Davenport make the simplifying assumption of an algebraically closed constant field. Making unnecessary extensions of the constant field greatly increases the cost of arithmetic. We will present an algorithm that will perform all its operations in the minimum extension field in which the answer can be expressed. Risch's more extensive use of Puiseux expansions forced him to operate in an extension field of much higher degree for his intermediate computations.

We will present a new algorithm for algebraic function integration that is strongly analogous to recent efficient algorithms for rational function integration [56]. Using integral bases to normalize the problem, we will be able to reduce the finding of the algebraic part of the integral to solving a set of linear equations. Finding the logarithmic part is indeed more difficult and does involve determining whether a given divisor is of finite order to guarantee termination of the algorithm. In addition to obtaining bounds that guarantee termination, we will present a novel algorithm for actually obtaining the logand associated with a divisor. Unlike earlier approaches that constructed this function from the divisor alone, we will use the integrand to create the ideal of functions that are multiples of the divisor in all finite places. Generators for this ideal can be derived almost by inspection, and there remains only to determine whether there is a principal generator.

Algebraic function integration is significantly more complicated than rational function integration since one can no longer depend on unique factorization. Ideals were created by Dedekind to restore unique factorization to algebraic number fields. Since that time they have been studied in increasingly abstract settings, to the point that their origins are almost forgotten. We intend to actually use ideals, as Dedekind intended 100 years ago, to combat the lack of unique factorization in algebraic function fields. In addition, our explicit generators for the ideal associated with a divisor will permit us to reduce the generators and compute the exact order of this divisor mod p. Thus instead of working with bounds for the torsion, we can compute the exact order of torsion divisors. We then need only test whether or not this particular power is principal, instead of testing all powers up to some calculated bound. This will be shown to lead to a very practical decision procedure for algebraic function integration.

Finally we will investigate the possibility of extending this procedure to include exponentials and logarithms enabling one to determine whether the integral of any elementary function can be expressed as an elementary function.

1.2 Outline of Thesis

Chapter Two will present an algorithm for computing an integral basis for our function field. This is essentially the same algorithm presented by Ford [23] for algebraic number fields, with proofs of validity in this more general situation and extended to normalize the basis elements at infinity. This fundamental construction will be used throughout the thesis and effectively provides us with an affine non-singular model for our function field. It enables us to determine the poles of our integrand and

241

characterize the form of the answer. We will also use this integral basis to help find principal generators for divisors and test for "good reduction".

Since the running time of many of our algorithms depends critically on the degree of the defining relationship for our function field, it is very useful to guarantee that our defining polynomial is irreducible over the algebraic closure of our coefficient domain, i.e. is absolutely irreducible. In Chapter Three we present a new algorithm for finding an absolutely irreducible factor of a multivariate polynomial that seems to be significantly better than other known approaches. As pointed out by Duval [21], the number of absolutely irreducible factors of the defining polynomial is the same as the dimension of the vector space of functions that have no poles. We can compute this simply using our normalized integral basis, and then we only need to perform our factorization algorithm when it is known to yield a lower degree factor. Since an irreducible polynomial will usually be absolutely irreducible, this can prevent a lot of wasted effort.

Armed with a minimal defining polynomial and an integral basis we are ready to find the purely algebraic part of the integral. Chapter Four proceeds by analogy with the standard approaches to rational function integration. It shows how Hermite's algorithm can be generalized to deal with algebraic functions. This approach will always succeed in reducing the integral to one with only simple finite poles and perhaps poles at infinity. In fact we are able to show that if the original problem had no poles at infinity, and after removing the algebraic part we introduce poles at infinity, then the original problem was not integrable. This approach has the advantage of allowing one to obtain canonical reduced forms even for problems that are not integrable. In this stage only linear congruence equations are solved and no new algebraic numbers are generated. It is difficult to remove poles at infinity by a Hermite-like method, so the original integral is transformed by a simple change of variables so that there are no poles at infinity. This simplifying transformation is one of the problems to be overcome in trying to generalize this algorithm to handle mixed transcendental as well as algebraic extensions.

If the original problem was integrable, the remaining simple finite poles must be canceled by the derivative of a linear combination of logarithmic terms. In the rational function case these log terms can be found by factorization or by computing gcd's of polynomials. Unfortunately algebraic function fields are not unique factorization domains so this approach cannot be used. As described earlier we will use the residues of the integrand to construct divisors associated with each log term. In Chapter Five we present an algorithm that computes a polynomial whose roots are all the residues using resultants. This is an extension of the idea we used in [55] for rational functions and does not use power series expansions or extensions of the coefficient domain. Given this polynomial we finally need to extend our coefficient domain to include its splitting field. We show that this is the minimal extension required for expressing the integral. After computing the $\mathbf{Z}$-linear basis for the residues, we construct a model for the divisor associated with each basis element. Our construction of the model appears new, and provides us with a simple construction to find a principal generator if one exists.

Finally in Chapter Six we are ready to address the "points of finite order problem". We need to know for each of the divisors we have constructed, whether there is some power of it that is principal. As discussed in Risch [47] and Davenport [16], we will use the technique of "good reductions" to solve this problem. However our explicit representation of a divisor will allow us to compute the order of individual divisors exactly instead of merely computing a bound on the orders of all divisors. While we do perform a sequence of tests for principality on powers of divisors, these tests are all performed over finite constant fields and thus much less expensive then testing successive powers of a divisor over our original coefficient field. After computing what should be the order of our original divisor if it were finite, we merely perform a single test for that power of our divisor over the original function field. Both of these improvements should make a substantial difference in running time.

In Chapter Seven we summarize our contributions and suggest ways to extend the work done here. In an appendix we present one step toward a complete algorithm for handling both transcendental and algebraic extensions.

Chapter 2
Integral Basis

In later chapters we will be very concerned with the problem of creating functions with prescribed singularities. It will be very useful to be able to recognize and generate functions whose only poles lie at places over infinity. Such functions are called integral algebraic functions. In the field $\mathbf{K}(x)$ these are simply polynomials in x, i.e. a rational function has a finite pole if and only if it has a nontrivial denominator. Any function that is algebraic over $\mathbf{K}(x)$ satisfies a unique monic irreducible polynomial with coefficients in $\mathbf{K}(x)$

$$Z^m + a_1 Z^{m-1} + \cdots + a_m. \tag{1}$$

Such a function is integral over $\mathbf{K}[x]$ if and only if the coefficients are in fact in $\mathbf{K}[x]$, i.e. polynomials in x. In the rest of this thesis we will abbreviate "integral over $\mathbf{K}[x]$" to "integral".

Let $\mathbf{K}(x, y)$ be a finite algebraic extension of degree n over $\mathbf{K}(x)$, then the integral functions form a free module of rank n over $\mathbf{K}[x]$, i.e. any such function can be written as linear combination of n basis functions with coefficients that are polynomials in x. Such a basis is called an *integral basis*. If we allow the coefficients to be rational functions in x then these same n functions comprise a vector space basis for $\mathbf{K}(x, y)$ over $\mathbf{K}(x)$. Thus each element of $\mathbf{K}(x, y)$ has a unique representation in terms of a particular integral basis, and has no finite poles if and only if each coefficient is a polynomial, i.e. no denominators. In this chapter we present an algorithm for computing an integral basis.

One technique for finding such a basis is given in [7]. What we call an integral basis, he would term *multiples except at infinity of the divisor 1*. His basic technique involves Puiseux expansions. We wish to avoid performing such expansions for two reasons: (1) A large amount of code is required (2) Many algebraic numbers need to be introduced to compute Puiseux expansions even though none of them are actually required to express the final basis elements. While Puiseux expansions are very useful in their own right, we don't really need them and choose to avoid the time cost of doing unnecessary algebraic number computations and the space cost of all that additional code.

The algorithm we present here is based on work by Zassenhaus and Ford [23]. They were primarily interested in the case of algebraic number fields, but their algorithm also applies to function fields in one variable. In fact since we will always assume the characteristic of $\mathbf{K}$ is zero or greater than n, the algorithm can be somewhat simplified. The algorithm presented here is a generalization to function fields of the first of the two algorithms Ford presents. We choose to use this one since it is much simpler and it avoids fully factoring the discriminant.

We are given $\mathbf{K}(x, y)$ where $\mathbf{K}$ is a computable field, x is a distinguished transcendental element, and $f(x, y)$ is an irreducible separable polynomial of degree n over $\mathbf{K}[x]$. Without loss of generality we can also assume f monic. If not then let $\hat{y} = ay$ where a is the leading coefficient, then $\hat{y}$ satisfies a monic polynomial and generates the same function field. The elements of $\mathbf{K}(x, y)$ that are integral over $\mathbf{K}[x]$ form a ring called the integral closure of $\mathbf{K}[x]$ in $\mathbf{K}(x, y)$. As noted above this ring is also a free module of rank n. Since y is integral over $\mathbf{K}[x]$, and the sum or product of integral elements are integral, $[1, y, \ldots, y^{n-1}]$ constitutes a basis for an integral $\mathbf{K}[x]$ module. This is our first "approximation" to an integral basis. Each iteration of the algorithm will produce a basis for a strictly larger integral $\mathbf{K}[x]$ module until the integral closure is reached.

One important measure of the relative sizes of full (i.e. rank n) sub-modules of the integral closure is given by the discriminant. Let $[w_1, \cdots, w_n]$ be n elements of $\mathbf{K}(x, y)$. Since $\mathbf{K}(x, y)$ is a separable algebraic extension of $\mathbf{K}(x)$ of degree n, there are n distinct embeddings σ_i into a given algebraic closure. The images of an element under these mappings are called the conjugates of that element. The conjugate matrix of $\overline{w} = [w_1, \ldots, w_n]$ is defined by

$$
M_{\overline{w}} = \begin{bmatrix} \sigma_1(w_1) & \cdots & \sigma_n(w_1) \\ \vdots & & \vdots \\ \sigma_1(w_n) & \cdots & \sigma_n(w_n) \end{bmatrix}.
\tag{2}
$$

The *discriminant* of $\overline{w}$ is the square of the determinant of the conjugate matrix. The discriminant is non-zero if and only if the w_i generates a full module (i.e. they are linearly independent). We can define the trace (sp) of an element $w \in \mathbf{K}(x, y)$ as $\mathrm{sp}(w) = \sum \sigma_i(w)$. Since this is a symmetric function of the conjugate, it is an element of $\mathbf{K}(x)$. If we re-express the discriminant as the determinant of the product of the conjugate matrix and its transpose, we see that the product entries are traces of products of the original matrix entries. Thus the discriminant could be defined as determinant($\mathrm{sp}(w_i w_j)$).

$$
SP_{\overline{w}} = \begin{bmatrix} \sigma_1(w_1) & \cdots & \sigma_n(w_1) \\ \vdots & & \vdots \\ \sigma_1(w_n) & \cdots & \sigma_n(w_n) \end{bmatrix} \begin{bmatrix} \sigma_1(w_1) & \cdots & \sigma_1(w_n) \\ \vdots & & \vdots \\ \sigma_n(w_1) & \cdots & \sigma_n(w_n) \end{bmatrix}
$$

$$
= \begin{bmatrix} \mathrm{sp}(w_1^2) & \cdots & \mathrm{sp}(w_1 w_n) \\ \vdots & & \vdots \\ \mathrm{sp}(w_n w_1) & \cdots & \mathrm{sp}(w_n^2) \end{bmatrix}.
\tag{3}
$$

If the w_i's are integral functions then their traces are polynomials, and thus the discriminant is a polynomial. If $\overline{v} = [v_1, \ldots, v_n]$ is a basis for a full module that contains $\overline{w}$ then each w_i can be written as a polynomial combination of the v_i, i.e. $\overline{w} = A\overline{v}$ where the change of basis matrix A is an $n \times n$ matrix of polynomials. Thus the conjugate matrix $M_{\overline{w}} = A.M_{\overline{v}}$ and $\mathrm{Disc}(\overline{w}) = \det(A)^2\mathrm{Disc}(\overline{v})$. $\overline{w}$ and $\overline{v}$ generate the same module if and only if A is invertible as a matrix over $\mathbf{K}[x]$, i.e. $\det(A) \in \mathbf{K}$. If $\overline{v}$ strictly contains $\overline{w}$ then $\det(A)$ is a polynomial p of nonzero degree and $\mathrm{Disc}(\overline{v}) = \mathrm{Disc}(\overline{w})/p^2$ thus each time we are able to produce a strictly larger $\mathbf{K}[x]$ modules, we eliminate a squared factor from the discriminant and the process can only continue for finitely many steps.

We will now state and prove the key algebraic result on which the algorithm is based. Let $\mathbf{R}$ be a *principal ideal domain*, in our case $\mathbf{K}[x]$ while Ford and Zassenhous assume $\mathbf{R} = \mathbf{Z}$. Let $\mathbf{V}$ be a domain that is a finite integral extension of $\mathbf{R}$. Then $\mathbf{V}$ is also a free module of rank equal to the degree of $\mathbf{QF}(\mathbf{V})$ (the quotient field of $\mathbf{V}$) over $\mathbf{QF}(\mathbf{R})$. Let $\overline{v} = [v_1, \ldots, v_n]$ be a basis for $\mathbf{V}$ over $\mathbf{R}$. The discriminant of $\overline{v}$ generates an ideal in $\mathbf{R}$ that we will call the discriminant of $\mathbf{V}$ over $\mathbf{R}$. The discriminant of any other basis of $\mathbf{V}$ over $\mathbf{R}$ differs by the square of a unit and thus generates the same ideal. If m is an ideal in a ring $\mathbf{S}$, we define the idealizer $\mathrm{Id}(m)$ to be the set of all $u \in \mathbf{QF}(\mathbf{S})$ such that $um \subseteq m$. $\mathrm{Id}(m)$ clearly contains $\mathbf{S}$ and is the largest ring in which m is still an ideal.

Theorem 2.1. *Under the above conditions $\mathbf{V}$ is integrally closed if only if the idealizer of every prime ideal containing the discriminant equals $\mathbf{V}$.*

We will break this into a succession of small lemmas.

Lemma 2.2. *If $\mathbf{m}$ is a nonzero ideal in $\mathbf{V}$ the idealizer of $\mathbf{m}$ is integral over $\mathbf{R}$.*

Proof. Since $\mathbf{V}$ is a finite $\mathbf{R}$-algebra $\mathbf{m}$ is finitely generated over $\mathbf{R}$. Let $m_1, \ldots, m_k$ span $\mathbf{m}$ over $\mathbf{R}$ and let $u \in \mathbf{QF}(\mathbf{V})$ such that $u\mathbf{m} \subseteq \mathbf{m}$. Then $um_i = \sum_j r_{ij}m_j$ with $r_{ij} \in \mathbf{R}$. Let $\mathbf{M}$ be the matrix $r_{ij} - \delta_{ij}u$ where δ_{ij} is the Kronecker index. Then $\mathbf{M}$ annihilates the vector $[m_1, \ldots, m_k]$ and if we multiply by the adjoint of $\mathbf{M}$ we see that $\det\mathbf{M}$ kills each of m_i and thus annihilates $\mathbf{m}$. Since $\mathbf{V}$ is an integral domain, $\det(\mathbf{M})$ must be zero, but this gives a monic polynomial over $\mathbf{R}$ that u satisfies, and hence u is integral over $\mathbf{R}$. □

Next we define the *inverse* of an ideal. If $\mathbf{m}$ is an ideal in $\mathbf{V}$ then $\mathbf{m}^{-1}$ is the set of all $u \in \mathbf{QF}(\mathbf{V})$ such that $u\mathbf{m} \subseteq \mathbf{V}$. This notion is similar to the idealizer. The idealizer of an ideal is the subset of the quotient field that sends an ideal into itself, while the inverse of an ideal sends it into the ring $\mathbf{V}$. Having defined the inverse of an ideal we will say that an ideal $\mathbf{m}$ is *invertible* if $\mathbf{mm}^{-1} = \mathbf{V}$. By definition we have $\mathbf{mm}^{-1} \subseteq \mathbf{V}$. If $\mathbf{V}$ is integrally closed then it is a Dedekind domain and has the property that every non-zero ideal is invertible. By proposition 9.7 and 9.8 in [2, p. 97] we have

Lemma 2.3. *If all the non-zero prime ideals in $\mathbf{V}$ are invertible the $\mathbf{V}$ is integrally closed.*

Lemma 2.4. *Any prime ideal not containing the discriminant of $\mathbf{V}$ over $\mathbf{R}$ is invertible.*

Proof. Let $\mathbf{m}$ be a prime ideal not containing the discriminant of $\mathbf{V}$ over $\mathbf{R}$. Then if we localize at $\mathbf{m}$ the discriminant becomes a unit and thus the local ring is integrally closed and locally $\mathbf{m}$ is invertible, in fact principal. If we localize at any other maximal ideal of $\mathbf{V}$ $\mathbf{m}$ contains units and is its own inverse. Since the localization of $\mathbf{m}$ at each prime ideal is invertible then by [2, proposition 9.6] $\mathbf{m}$ is invertible. □

Corollary 2.5. *If $\mathbf{V}$ is not integrally closed there is some prime ideal containing the discriminant that is not invertible.*

Proof. By lemma 2.3 there is some prime ideal that is not invertible and by lemma 2.4 it must contain the discriminant. □

The following lemma is proved in [31, p. 607].

Lemma 2.6. *If $\mathbf{V}$ properly contains an ideal $\mathbf{m}$ then $\mathbf{m}^{-1}$ properly contains $\mathbf{V}$.*

Now we are ready to finish the proof of theorem 2.1.

Proof. If $\mathbf{V}$ is integrally closed the idealizer of any non-zero ideal equals $\mathbf{V}$ by lemma 2.2. If $\mathbf{V}$ is not integrally closed there is some prime ideal $\mathbf{m}$ that contains the discriminant but is not invertible. Thus $\mathbf{m}^{-1}\mathbf{m}$ is an ideal containing $\mathbf{m}$ but properly contained in $\mathbf{V}$. Since $\mathbf{m}$ is maximal we must have $\mathbf{m}^{-1}\mathbf{m} = \mathbf{m}$. Thus in this case $\mathbf{m}^{-1} = \mathrm{Id}(\mathbf{m})$. By lemma 2.6 we have the idealizer of $\mathbf{m}$ properly contains $\mathbf{V}$. □

Using theorem 2.1 we could compute the integral closure by computing the idealizer of the finitely many ideals that contain the discriminant. Either the result in each case will be $\mathbf{V}$ in which case $\mathbf{V}$ must be integrally closed, or we will find a ring strictly larger than $\mathbf{V}$ that is integral over $\mathbf{V}$. This can happen only a finite number of times since each such ring will remove a squared non-unit factor from the discriminant. We will improve this idea by dealing with all the ideals dividing the discriminant at the same time. First we observe the following property of idealizers:

Lemma 2.7. *If $\mathbf{m}$ and $\mathbf{n}$ are ideals then the idealizer of the product contains the idealizer of either ideal.*

Proof. Any element of $\mathbf{mn}$ is of the form $\sum m_i n_i$ where $m_i \in \mathbf{m}$ and $n_i \in \mathbf{n}$. If $u \in \mathrm{Id}(\mathbf{m})$ then $um_i \in \mathbf{m}$ and thus $um_i n_i \in \mathbf{mn}$. □

Next we must introduce the notion of the radical of an ideal. The radical of an ideal $\mathbf{m} \subseteq \mathbf{V}$ is the set of all $u \in \mathbf{V}$ such that some power of u is in $\mathbf{m}$ and can also be characterized as the intersection of all prime ideals containing $\mathbf{m}$. Our algorithm is based on the following corollary to theorem 2.1.

Corollary 2.8. *$\mathbf{V}$ is integrally closed if and only if the idealizer of the radical of the discriminant equals $\mathbf{V}$.*

Proof. Since all the non-zero prime ideals of $\mathbf{V}$ are maximal, the radical of the discriminant is also the product of all prime ideals containing the discriminant. By theorem 2.1 if $\mathbf{V}$ is not integrally closed the idealizer of one of these primes must be strictly larger than $\mathbf{V}$. By lemma 2.7 the idealizer of the radical contains the idealizer of that prime and thus must also strictly contain $\mathbf{V}$. If $\mathbf{V}$ is integrally closed again the idealizer of any ideal must equal $\mathbf{V}$. □

248

Thus our algorithm for computing the integral closure of $\mathbf{V}$ is:

1. Find the radical of the discriminant of $\mathbf{V}$ over $\mathbf{R}$.
2. Compute the idealizer $\widehat{\mathbf{V}}$ of that radical.
3. If $\widehat{\mathbf{V}}$ is strictly larger than $\mathbf{V}$ then set $\mathbf{V}$ to $\widehat{\mathbf{V}}$ and go to step (1).
4. Return $\mathbf{V}$.

Although this algorithm works, there are a few optimizations possible. The first time through $\mathbf{V} = \mathbf{R}[\alpha]$ where α satisfies equation (1). To compute the discriminant in this case, one simply computes the resultant of equation (1) and its derivative. Prime factors of the discriminant that appear only to the first power can be ignored. When returning from step 3 to step 1 one only needs to concentrate on the factors of the discriminant that have actually been reduced in the previous iteration. Since if there is some p whose p-radical is invertible, it will stay that way throughout the rest of the computation. This leads us to the following improved version of the algorithm for computing the integral closure of $\mathbf{V} = \mathbf{R}[\alpha]$ over $\mathbf{R}$ assuming f(X) is the minimal equation of integral dependence for α.

1. Let $d = \text{Resultant}(f, f')$ and $k = d$.
2. Let $q = \prod p_i$ such that p_i is prime, $p_i \mid k$ and $p_i^2 \mid d$. If q is a unit the return $\mathbf{V}$.
3. Find $\mathbf{J}_q(\mathbf{V})$, the radical of (q) in $\mathbf{V}$.
4. Find $\widehat{\mathbf{V}}$, the idealizer of $\mathbf{J}_q(\mathbf{V})$ along with $\mathbf{M}$ the change of basis matrix from $\widehat{\mathbf{V}}$ to $\mathbf{V}$.
5. Let k be the determinant of $\mathbf{M}$. If k is a unit then return $\mathbf{V}$.
6. Set $d = d/k^2$ and $\mathbf{V} = \widehat{\mathbf{V}}$ and go to 1.

2.1 Radical of the Discriminant

The discriminant is a principal ideal generated by some element d of $\mathbf{R}$. We wish to compute the radical of the ideal d generates in $\mathbf{V}$. Since $\mathbf{R}$ is a principal ideal domain it is also a unique factorization domain. Let $(p_1, \ldots, p_k)$ be the distinct prime factors of d in $\mathbf{R}$. Since the radical of (d) is the intersection of the prime ideals containing d, it is also the intersection of the radicals of the p_i. Let us therefore consider how to compute the radical in $\mathbf{V}$ of a principal ideal generated by a prime element p of $\mathbf{R}$. Following Ford we call such an ideal the *p-radical* of $\mathbf{V}$.

u is in the p-radical if and only if the coefficients a_i in equation (1) of the monic minimal polynomial for u over $\mathbf{R}$ are divisible by p. ([2, Proposition 5.14 and 5.15]). This gives us a membership test, but we are looking for ideal generators. Zassenhaus and Ford observed that under appropriate conditions the trace map from $\mathbf{V}$ to $\mathbf{R}$ provides us with linear constraints on the members of the p-radical. For any u in $\mathbf{V}$ the degree of u over $\mathbf{R}$ must divide the rank of $\mathbf{V}$ over $\mathbf{R}$ that is also the degree of $\mathbf{QF}(\mathbf{V})$ over $\mathbf{QF}(\mathbf{R})$. If m is the degree of u over $\mathbf{R}$ then $\text{sp}(u) = -(n/m)a_1$. Thus if $u \in$ p-radical then p divides $\text{sp}(u)$. Again following Ford we define the *p-trace-radical* as the set of $u \in \mathbf{V}$ such that for all $w \in \mathbf{V}$, $p \mid \text{sp}(uw)$. This leads us to the following lemma:

Lemma 2.9. *p-radical $\subseteq$ p-trace-radical.*

Proof. If u is in the p-radical then for any $w \in \mathbf{V}$, uw is in the p-radical. But the previous argument shows that $\mathrm{sp}(uw)$ is divisible by p. $\qquad\qquad\qquad\qquad\square$

Now we find conditions under which the two sets of lemma 2.9 are the same. If w is in the p-trace-radical the $\mathrm{sp}(w^k) \equiv 0 \bmod p$ for all $k > 0$. Let m be the degree of w over $\mathbf{R}$. Then $\mathbf{R}[w]$ is a free R-module of rank m dividing n, the rank of $\mathbf{V}$ over $\mathbf{R}$. There is a reduced trace map from $\mathbf{R}[w]$ to $\mathbf{R}$ denoted by sp_w satisfying $(n/m)\mathrm{sp}_w(u) = \mathrm{sp}(u)$ for any $u \in \mathbf{R}[w]$. If $n/m \notin (p)$ then p dividing $\mathrm{sp}(u)$ implies p divides $\mathrm{sp}_w(u)$ and thus the p-trace-radical of $\mathbf{R}[w]$ equals the intersection of the p-trace-radical of $\mathbf{V}$ with $\mathbf{R}[w]$. If n/m is zero in $\mathbf{R}/(p)$ then the characteristic of $\mathbf{R}/(p)$ must divide n. To avoid this problem, we now make the assumption that the characteristic of $\mathbf{R}/(p)$ is greater than n the rank of $\mathbf{V}$ over $\mathbf{R}$. In our application where $\mathbf{R} = \mathbf{K}[x]$ the characteristic of $\mathbf{R}/(p)$ is the same as the characteristic of $\mathbf{K}$, and we shall see that the restriction will not cause any problems. In Ford and Zassenhaus' case where $\mathbf{R} = \mathbf{Z}$, $p \in \mathbf{Z}$ and the characteristic of $\mathbf{R}/(p)$ equals p. Thus when the discriminant has small prime factors, our assumption would be invalid, so they compute the p-radical for small values of p using the kernel of powers of the Frobenius automorphism instead of the trace map.

Assume w is in the p-trace-radical of $\mathbf{R}[w]$. Then $\mathrm{sp}_w(w^k)$ is divisible by p for all $k > 0$. We wish to relate the traces of powers of w to the coefficients of the minimal polynomial for w in equation (1). This is provided by Newton's identities ([29, p. 208]). If we let $s_k = \mathrm{sp}_w(w^k)$ then for $1 \le k \le n$ we have:

$$s_k + a_1 s_{k-1} + \cdots + a_{k-1}s_1 = -k a_k. \qquad (4)$$

Thus $a_1 = -s_1$ is divisible by p and by induction assume a_i is divisible by p for all $i < k \le n$. Thus the left-hand side of equation (4) is divisible by p and as long as the characteristic of $\mathbf{R}/(p)$ is greater than n we can divide by k and a_k must also be divisible by p. Since the coefficients of its minimal polynomial are divisible by p, w must be in the p-radical by $\mathbf{V}$, which proves the following partial converse to lemma 2.9.

Theorem 2.10. *If the characteristic of $\mathbf{R}/(p)$ is greater than the rank of $\mathbf{V}$ over $\mathbf{R}$, then the p-trace-radical of $\mathbf{V}$ equals the p-radical of $\mathbf{V}$.*

2.1.1 Computing the p-trace-radical

Let $[w_1, \ldots, w_n]$ be a basis for $\mathbf{V}$ over $\mathbf{R}$. The p-trace-radical $\mathbf{J}_p(\mathbf{V})$ was defined as the set of $u \in \mathbf{V}$ such that $p \mid \mathrm{sp}(uw)$ for all $w \in \mathbf{V}$.

$$\mathrm{sp}(uw) \equiv 0 \bmod p \text{ for all } w \in \mathbf{V} \iff$$

$$\mathrm{sp}(uw_i) \equiv 0 \bmod p \text{ for } 1 \le i \le n \iff$$

$$\sum_{j=1}^{n} u_j \mathrm{sp}(w_j w_i) \equiv 0 \bmod p \text{ for } 1 \le i \le n.$$

Using the trace matrix $SP_{\overline{w}}$ defined by equation (3) we can write this as:

$$SP_{\overline{w}} \bullet \overline{u} = \begin{bmatrix} sp(w_1^2) & \cdots & sp(w_1 w_n) \\ \vdots & & \vdots \\ sp(w_n w_1) & \cdots & sp(w_n^2) \end{bmatrix} \begin{bmatrix} u_1 \\ \vdots \\ u_n \end{bmatrix} \in p\mathbf{R}^n \qquad (5)$$

where $p\mathbf{R}^n$ represents the set of vectors of length n whose elements are divisible by p. $J_p(V)$ is determined by the solutions to (5) with $u_i \in \mathbf{R}$. We actually wish to compute $J_q(V)$ where q is a certain product of distinct primes dividing the discriminant

$$J_q(V) = \bigcap_{p_i | q} J_{p_i}(V).$$

Thus we want to find all $\overline{u} \in \mathbf{R}^n$ such that the left side of equation (5) is in fact in $q\mathbf{R}^n$. We need to add some equations to (5) to guarantee that each element u_i of our solution vectors must lie in $\mathbf{R}$ as opposed to $\mathbf{QF(R)}$. Let $\mathbf{I}_n$ be the $n \times n$ identity matrix. Then $\overline{u} \in \mathbf{R}$ if and only if $q\mathbf{I}_n \bullet \overline{u} \in q\mathbf{R}^n$. Thus if we let $\mathbf{M}_q$ be the vertical concatenation of $SP_{\overline{w}}$ and $q\mathbf{I}_n$, $J_q(V)$ is the set of all $u \in \mathbf{QF}(V)$ such that $\mathbf{M}_q \bullet \overline{u} \in q\mathbf{R}^{2n}$. If we left multiply $\mathbf{M}_q$ by an invertible $\mathbf{R}$-matrix, the $\mathbf{R}$-module of solutions remains unchanged. Invertible $\mathbf{R}$-matrices are called unimodular, and are characterized by having determinants that are units in $\mathbf{R}$. Since $\mathbf{R}$ is a principal ideal domain, there is some unimodular matrix that converts $\mathbf{M}_q$ into an upper triangular matrix. ([44, Theorem II.2] gives a constructive proof of this). This process allows us to reduce the $2n$ relations imposed by $\mathbf{M}_q$ and equation (5) to an equivalent set of n independent relations. Once this is done, we invert the square matrix determined by the first n rows of the reduced $\mathbf{M}_q$. The columns of this inverse matrix provide a basis for the solutions to equation (5).

In our application we have the even stronger restriction that $\mathbf{R}$ is a euclidean domain. This allows us to triangularize $\mathbf{M}_q$ by elementary row operations only. This process is called Hermitian row reduction and is somewhat analogous to gaussian elimination that is used for matrices over fields. With gaussian elimination any nonzero element can be used to zero out its entire column. With hermitian row reduction one can only multiply by elements of $\mathbf{R}$ and a nonzero element can reduce the other members of its column to be smaller than it. Since $\mathbf{R}$ is a euclidean domain this process can only continue for finitely many steps before we find an element that divides everyone else in its column. Finally this element can be used to clear the column. Let d be the size function associated with $\mathbf{R}$, for $\mathbf{R} = \mathbf{K}[x]$ we use the degree, and for $\mathbf{R} = \mathbf{Z}$ we use absolute value. To simplify the presentation of the following algorithm we define $d(0) = \infty$. Let $nrows$ and $ncols$ be respectively the numbers of rows and columns of a given matrix $\mathbf{M}$. We assume that $nrows \geq ncols$ and $\mathbf{M}$ has rank $ncols$ in the following algorithm for hermitian row reduction.

1. loop for $j = 1$ thru $ncols$
2.   choose k such that $d(\mathbf{M}_{kj})$ is minimal for $j \leq k \leq nrows$.
3.   exchange rows j and k
4.   loop for $i = j + 1$ thru $nrows$

5.   let q be the polynomial part of $\mathbf{M}_{ij}/\mathbf{M}_{jj}$
6.   replace $\mathbf{M}_i$ with $\mathbf{M}_i - q\mathbf{M}_j$
7.   if there is some $\mathbf{M}_{ij} \neq 0$ for $j < i \leq nrows$ then go to 2
8. return $\mathbf{M}$

2.2 Computing the Idealizer

Given $(m_1, \ldots, m_n)$ that form an $\mathbf{R}$-basis for an ideal $\mathbf{m}$ in $\mathbf{V}$, we wish to compute an $\mathbf{R}$-basis for the idealizer of $\mathbf{m}$. The idealizer of $\mathbf{m}$ was defined as the set of $u \in \mathbf{QF}(\mathbf{V})$ such that $u\mathbf{m} \subseteq \mathbf{m}$. This concept is very similar to the inverse ideal of $\mathbf{m}$, $\mathbf{m}^{-1}$, which was defined as the set of $u \in \mathbf{QF}(\mathbf{V})$ such that $u\mathbf{m} \subseteq \mathbf{V}$. Although we don't need to compute inverses to find an integral basis, we will need them in chapter 5, and we present both algorithms here to display their similarities. The inverse procedure will be given first since it is slightly simpler.

We assume $\bar{\mathbf{v}} = (v_1, \ldots, v_n)$ forms a basis for $\mathbf{V}$ over $\mathbf{R}$ that we hold fixed throughout this section. $u \in \mathbf{m}^{-1}$ if and only if $um_i = \sum r_{ij}v_j$ with $r_{ij} \in \mathbf{R}$ for $1 \leq i \leq n$. Multiplication by m_i is a linear transformation on $\mathbf{V}$. Let $\mathbf{M}_i$ represent multiplication by m_i with respect to our fixed choice of basis $\bar{\mathbf{v}}$. (for details on constructing $\mathbf{M}_i$ see [55]) Since $\bar{\mathbf{v}}$ also forms a basis for $\mathbf{QF}(\mathbf{V})$ over $\mathbf{QF}(\mathbf{R})$, $\mathbf{u}$ can be represented as $\sum u_i v_i$ where $u_i \in \mathbf{QF}(\mathbf{R})$. Then $\mathbf{M}_i[u_1, \ldots, u_n]^t$ yields the vector of coefficients of um_i. $\mathbf{u} \in \mathbf{m}^{-1}$ if and only if the product coefficients lie in $\mathbf{R}$ for all i. Thus we are left with the linear algebra problem of finding a basis for $\mathbf{R}$-module of vectors $[u_1, \ldots, u_n]$ such that $\mathbf{M}_i([u_1, \ldots, u_n]^t)$ is a vector of elements of $\mathbf{R}$ for all i. Let $\mathbf{M}$ be the $n^2 \times n$ matrix that is the vertical concatenation of the $\mathbf{M}_i$. Then we are looking for all vectors $\bar{\mathbf{u}}$ over $\mathbf{QF}(\mathbf{R})$ such that $\mathbf{M}\bar{\mathbf{u}}$ is an n^2 vector of elements of $\mathbf{R}$. By Hermitian row reduction we can zero out the last $n^2 - n$ rows of $\mathbf{M}$, so we have reduced to an $n \times n$ matrix $\widehat{\mathbf{M}}$ such that $\mathbf{M}\bar{\mathbf{u}} \in \mathbf{R}^{n^2}$ if and only if $\widehat{\mathbf{M}}\bar{\mathbf{u}} \in \mathbf{R}^n$. The columns of $\widehat{\mathbf{M}}^{-1}$ form a basis for $\mathbf{m}^{-1}$.

Essentially the same approach will allow us to find the idealizer of $\mathbf{m}$. Now we require that $um_i = \sum r_{ij}m_j$ with $r_{ij} \in \mathbf{R}$. The $\mathbf{M}_i$ still represent multiplication by m_i but now the input and output bases are different. The $\mathbf{M}_i$ necessary to compute the idealizer again take inputs expressed in terms of $\bar{\mathbf{v}}$ but give output vectors expressed in terms of the basis for $\mathbf{m}$. Other than this one change the algorithm for idealizers is identical to the previous one for inverses. A summary of both algorithms follows:

1. Let $\mathbf{M}_i$ be the matrix representing multiplication by m_i with input base $\bar{\mathbf{v}}$ and output basis $\bar{\mathbf{v}}$ for computing inverses or $\bar{\mathbf{m}}$ for calculating the idealizer.
2. Let $\widehat{\mathbf{M}}$ be the first n rows of the Hermitian row reduction of the vertical concatenation of the $\mathbf{M}_i$.
3. Return the columns of $\widehat{\mathbf{M}}^{-1}$ as the result expressed with respect to $\bar{\mathbf{v}}$.

Note the transpose of $\widehat{\mathbf{M}}^{-1}$ is the change of basis matrix required in step 3 of the integral basis algorithm.

2.3 Normalize at Infinity

In this section we return to the special case $\mathbf{R} = \mathbf{K}[x]$. Having constructed an integral basis we can recognize finite poles of functions, but we also need to deal with singularities at "infinity". Given an arbitrary basis for a $\mathbf{K}[x]$ module (an integral basis is a special case), we wish to minimize the sum of the orders of the basis elements at infinity. A characteristic property of an integral basis $[w_1, \ldots, w_n]$ is that $\sum a_i(x)w_i$ is integral over $\mathbf{K}[x]$ if and only if each $a_i(x)w_i$ is. In other words there is no cancellation of singularities from different summands. We would like our basis to have the same property with respect to the local ring of $\mathbf{K}(x)$ at infinity and we will characterize this by saying that such a basis is *normal at infinity*.

We will need the concept of a *local ring* at a place p of the function field $\mathbf{K}(x)$. The local ring at p is defined as the set of functions in $\mathbf{K}(x)$ that have no pole at p. If p is a finite place centered at $x = a$ then the local ring at p consists of rational functions whose denominators are not divisible by $x - a$. The order of a rational function at *infinity* is the degree of its denominator less the degree of its numerator. Thus the local ring of $\mathbf{K}(x)$ at ∞ consists of those rational functions whose numerator degree does not exceed their denominator degree. A function in $\mathbf{K}(x, y)$ is said to be integral over the local ring at p. (for brevity integral at p) if it satisfies a monic polynomial with coefficients in the local ring at p. Analogous to the global integral basis, there exists a local integral basis at each place p of $\mathbf{K}(x)$ such that all functions in $\mathbf{K}(x, y)$ that are integral at p can be written as a linear combination of basis elements with coefficients in the local ring at p of $\mathbf{K}(x)$. We will find it convenient to introduce the slightly weaker concept of a normal basis. $[w_1, \ldots, w_n]$ is a normal basis at p if there exist rational functions $r_j \in \mathbf{K}(x)$ such that $r_i w_i$ form a local integral basis at p. In other words there exists rational scaling factors that convert a normal basis into an integral basis.

Armed with this terminology we see that a basis is normal at infinity if and only if some rational multiple of the basis elements is a local integral basis at infinity. Let $[w_1, \ldots, w_n]$ be a basis for a $\mathbf{K}[x]$ module. Assume we are given a local integral basis at infinity $[v_1, \ldots, v_n]$. We wish to modify the original basis to make it normal at infinity without disturbing its basis properties everywhere else. We first represent the w_i in terms of the v_j:

$$w_i = \sum_{j=1}^{n} M_{ij} v_j \text{ where } M_{ij} \in \mathbf{K}(x).$$

If w_i were a normal basis at infinity there would be rational functions r_i such that the change of basis matrix $r_i m_{ij}$ would have a determinant that was a unit in the local ring of $\mathbf{K}(x)$ at infinity, i.e. a rational function whose numerator has the same degree as its denominator.

Define the representation order, $k(w_i)$ of w_i at infinity as the $\min_j \mathrm{ord}_\infty M_{ij}$ for $1 \le j \le n$. We will initially choose $r_i = x^{-k(w_i)}$. This guarantees that $r_i w_i$ is integral at infinity and its representation order is 0. Note that the order at infinity of the determinant of $\mathbf{M}$ is always $\ge \sum k(w_i)$. We will show that our basis becomes normal when these numbers are equal. Let $\widehat{\mathbf{M}}$ be the change of basis matrix for $r_i w_i$. Each row of $\widehat{\mathbf{M}}$ is just r_i times the corresponding row of $\mathbf{M}$. $r_i w_i$ is an integral basis if and only if the determinant of $\widehat{\mathbf{M}}$ has order zero at infinity. Since this determinant is integral at infinity, this is equivalent to it having a non-zero value at infinity. Let $\mathbf{N}$ be a matrix where N_{ij} is the value of $\widehat{M}_{ij}$ at infinity. Since taking determinants commutes with evaluation, the determinant of $\mathbf{N}$ equals the value of the determinant of $\widehat{\mathbf{M}}$ at infinity. Thus $r_i w_i$ is a local integral basis at infinity if and only if $\mathbf{N}$ has nonzero determinant. If the determinant of $\mathbf{N}$ is zero then there are a set of constants $c_i \in \mathbf{K}$ such that $\sum_{i=1}^n c_i N_{ij} = 0$ for $1 \le j \le n$. Let $i0 = i$ such that $c_i \ne 0$ and $k(w_i)$ is minimal. Define

$$\hat{w}_{i0} = \sum_{i=1}^n c_i x^{k(w_{i0}) - k(w_i)} w_i.$$

Then replacing w_{i0} by $\hat{w}_{i0}$ still yields a global integral basis. Similarly replacing $\widehat{M}_{i0j}$ by $\sum c_i \widehat{M}_{ij}$ yields a row whose orders are strictly positive. Thus the representation order $k(\hat{w}_{i0})$ is strictly greater than $k(w_{i0})$. The order of the determinant of the change of basis matrix is preserved by our new basis. After a finite number of such steps this order will be equal to the sum of the representation orders of our basis elements and the basis will be normal.

We have shown how to make an arbitrary basis normal at infinity given a local integral basis at infinity. We next show how to compute the later. If we replace $x = 1/z$ then infinity gets transformed to zero in the z-space. In order to compute a local integral basis at zero, we can use the algorithm of the previous section after making one optimization . In the local ring at 0 any polynomial not divisible by z is a unit. Thus we can replace the discriminant by the maximal power of z dividing it. After computing this local integral basis, we merely substitute $1/x$ for z and obtain a local integral basis at infinity.

2.4 Genus Computation

Our integral basis algorithm will also afford us an easy way to compute the genus of a function field. The global discriminant divisor of $\mathbf{K}(x, y)$ over $\mathbf{K}(x)$ is the product of the local discriminant divisors over each place of $\mathbf{K}(x)$. As a by-product of our integral basis computation, we have computed two discriminants, a polynomial $\text{disc}_{\text{finite}}(x)$ that is the product of the discriminants over all finite places of $\mathbf{K}(x)$ and a monomial in $z = 1/x$, $\text{disc}_\infty(1/x)$ that is the local discriminant at infinity. The degree of the discriminant divisor of our function field $\mathbf{K}(x, y)$ over $\mathbf{K}(x)$ is the sum of degree of $\text{disc}_{\text{finite}}(x)$ and the order at infinity of $\text{disc}_\infty(1/x)$. If we call this total discriminant degree d and we assume that $\mathbf{K}$ is the exact constant field of $\mathbf{K}(x, y)$, then the genus of our function field $\mathbf{K}(x, y)$ can be computed by the following formula given in [22, p. 134]:

$$g = d/2 - [\mathbf{K}(x, y) : \mathbf{K}(x)] + 1.$$

2.5 Simple Radical Extensions

If $\mathbf{F}$ is a field with y algebraic over $\mathbf{F}$ of degree n such that $y^n \in \mathbf{F}$, then we will call $\mathbf{F}(y)$ a simple radical extension of $\mathbf{F}$. If the characteristic of $\mathbf{F}$ is relatively prime to n, then extending $\mathbf{F}$ if necessary we can assume that $\mathbf{F}$ contains ω a primitive n^{th} root of unity. There is a unique differential automorphism of $\mathbf{F}(y)$ over $\mathbf{F}$ such that $\sigma(y) = \omega(y)$. Define the operator

$$T_i = \frac{1}{n} \sum_{j=0}^{n-1} \frac{\sigma^j}{\omega^{ij}}.$$

Note that $T_i(y^j) = y^j$ if $i = j$ else 0. Thus letting $g = \sum g_i y^i$ with $g^i \in \mathbf{F}$, we have $T_i(g) = g_i y^i$. Since σ sends integral functions to integral functions, and sums and products of integral functions are integral, we have that the operators T_i map integral functions to integral functions. If g is an integral function this shows that each $g_i y^i$ must also be integral, which means that the basis y^i is normal everywhere proving the following:

Proposition 2.11. *If* $\mathbf{K}(x,y)$ *is a simple radical extension of* $\mathbf{K}(x)$ *of degree* n *relatively prime to the characteristic, then the natural basis,* $1, y, \ldots, y^{n-1}$ *is normal everywhere.*

Without loss of generality we can assume that y satisfies the following equation:

$$y^n = \prod_{i=0}^{n-1} p_i^i$$

where $p_i \in \mathbf{K}[x]$ and has no repeated factors. Thus to convert our natural basis into an integral basis we have to find polynomials $d_i(x)$ of maximal degree such that $y^i/d_i(x)$ is integral. Raising this expression to the nth power implies that $\prod p_j^{ij}/d_i^n \in \mathbf{K}[x]$. It is easy to show that the maximal $d_i(x)$ is the following:

$$d_i = \prod_{j=0}^{n-1} p_j^{[ij/n]}.$$

Thus the following functions provide an integral basis for our simple radical extension:

$$\frac{y^i}{d_i(x)}.$$

2.5 Simple Radical Extensions

If we hold with a primitive over k of degree n such that $\ldots \in k$, then we will call...

$$z = \int \frac{dx}{\ldots}$$

Assume that ... Such is integral functions ...

$$z = \prod \ldots$$

Chapter 3
Absolute Irreducibility

We have assumed that the defining polynomial $f(x, y)$ for the integrand y is irreducible over $\mathbf{K}(x)$, but during the integration process we may need to extend our coefficient field $\mathbf{K}$, and over the extended field f may no longer be irreducible. In fact, in order to compute our bound for the "points of finite order", we need to be able to guarantee that f will remain irreducible. Another difficulty is the precise determination of our coefficient field. Initially we defined our coefficient field $\mathbf{K}$ to be the minimal extension of $\mathbf{Q}$ necessary to express the defining polynomial. Any element of our function field that is algebraic over k could also have been considered part of the coefficient field. For example, if $f(x, y) = y^4 - 2x^2$ then y^2/x is a $\sqrt{2}$ and thus algebraic over $\mathbf{Q}$. Note that once we adjoin $\sqrt{2}$ to $\mathbf{K}$, f is no longer irreducible and y satisfies a polynomial of degree 2 over this extended coefficient field. It will be advantageous to make our coefficient field as large as possible since that will decrease the degree of our function field and thus speed up our computation time that is strongly dependent on this degree. We now define the *exact coefficient field* $\mathbf{K}^0$ of $\mathbf{K}(x, y)$ to be the set of all elements of $\mathbf{K}(x, y)$ that are algebraic over $\mathbf{K}$, also called the relative algebraic closure of $\mathbf{K}$ in $\mathbf{K}(x, y)$. From the previous example the existence of elements of $\mathbf{K}(x, y)$ that are algebraic over $\mathbf{K}$ seems connected with the question of the absolute irreducibility of $f(x, y)$. In the next section we will prove that this is indeed the case and in fact the process for finding an absolutely irreducible polynomial for y will lead us to discover the true coefficient field of $\mathbf{K}(x, y)$.

As the previous example seems rather contrived, one might be led to suspect that defining polynomials that are irreducible but not absolutely irreducible are quite rare in practice. This is indeed the case, however the integral basis computation from the previous chapter can be used to perform a quick test for absolute irreducibility. The integral basis for that example is $1, y, y^2/x, y^3/x$. Note that the first and third of these functions have no poles anywhere and thus have divisor (1). As observed by Duval in [21] the dimension of the multiples of the divisor (1) is the number of absolutely irreducible components of the function field. Since we have already computed an integral basis that is normal at infinity, we merely need to test how many of our basis elements have no poles at infinity. If the answer is one, we can skip the algorithm derived in this chapter, since we are guaranteed that our defining polynomial is absolutely irreducible.

In this chapter we will permit $\mathbf{K}$ to be a perfect field of arbitrary characteristic. As a consequence if $\mathbf{F}$ is any finite algebraic extension of $\mathbf{K}$, it can be generated by a single element and is thus a simple extension of $\mathbf{K}$, $\mathbf{F} = \mathbf{K}(\alpha)$. Assuming that $f(x, y)$

is irreducible over $\mathbf{K}$, we will present a new algorithm for performing absolutely irreducible factorizations, i.e. factorization over the algebraic closure of $\mathbf{K}$, $\overline{\mathbf{K}}$. The initial difficulty is that all current algebraic factoring algorithms only operate over a finitely generated field and the algebraic closure of $\mathbf{K}$ is not finitely generated. Thus we must find some subfield of $\overline{\mathbf{K}}$ that is finitely generated and is sufficient for performing the factorization. Risch showed the problem was decidable in [45, p. 178]. His approach was to convert a multivariate polynomial to a univariate one by the Kronecker substitution ([57, p. 135])

$$f(x_1, x_2, \ldots, x_v) \mapsto f(t, t^d, \ldots, t^{v-1}).$$

The key property of this substitution is that different power products of the x_i go into different powers of t assuming d is chosen larger than the degree of any variable appearing in f. Risch argued that the splitting field of this univariate polynomial suffices for the factorization. Assuming the original polynomial had v variables each with maximum degree d then his splitting field can be an algebraic extension of degree $d^v!$, whereas the algorithm presented below can test for absolute irreducibility or find an absolutely irreducible factor by operating over an extension of degree d_{min}, the minimum of the degrees of all the variables. By examining the algebraic structure of the fields involved, we will also discover some quick tests for absolute irreducibility.

3.1 Exact Coefficient Fields and Regular Extensions

First we will need some purely algebraic results (see also [52, pp. 194–198]). All fields are assumed to be contained in some universal field. If $\mathbf{E}$ and $\mathbf{F}$ are fields then $\mathbf{EF}$ is the composition that is the field generated by $\mathbf{E} \cup \mathbf{F}$. If $\mathbf{K}$ is a finite algebraic extension of k then $[\mathbf{K} : k]$ denotes the degree of this extension. From the previous section we see that determining the exact coefficient field requires us to find the relative algebraic closure of one field in another. The next lemmas give some properties of such fields.

Lemma 3.1. *If x is transcendental over k then k is algebraically closed in $k(x)$.*

Proof. Let y be an element of $k(x)$ that is not in k. y can be written as $u(x)/v(x)$ with $u, v \in k[x]$. Then x satisfies the polynomial $P(X) = u(X) - v(X)y$. If P is not identically zero, this implies x is algebraic over $k(y)$. Let $u(X) = \sum u_i X^i$ and $v(X) = \sum v_i X^i$ and choose j such that $v_j \neq 0$. If P were identically zero then $u_j - y v_j = 0$, but since $u_j, v_j \in k$ this implies $y \in k$ contrary to the assumptions. Thus x is algebraic over $k(y)$ and y cannot be algebraic over k. $\square$

Lemma 3.2. *Let $\mathbf{K} \supseteq k$ be fields with k algebraically closed in $\mathbf{K}$ and $k(\alpha)$ a simple algebraic extension of k. Then $[\mathbf{K}(\alpha) : \mathbf{K}] = [k(\alpha) : k]$.*

Proof. Any factor of the monic minimal polynomial for α over k has coefficients that are polynomials in the conjugates of α and thus algebraic over k. If these coefficients were in $\mathbf{K}$ they would also be in k since the latter is algebraically closed in the former. $\square$

Corollary 3.3. *Let x be transcendental over k and $\mathbf{F}$ be a simple algebraic extension of k, then $[\mathbf{F}: k] = [\mathbf{F}(x) : k(x)]$.*

We are looking for a defining polynomial for y that is irreducible over $\overline{\mathbf{K}}$. Factoring the bivariate polynomial f over $\overline{\mathbf{K}}$ is exactly the same as factoring it over $\overline{\mathbf{K}}(x)$ by Gauss's lemma. If a perfect field k is algebraically closed in $\mathbf{K}$, then $\mathbf{K}$ is said to be a *regular* extension of k. Thus the search for an absolutely irreducible defining polynomial is equivalent to finding a presentation of our function field as a regular extension of the coefficient field. We have the following key theorem that connects our twin problems of absolute irreducibility and exact coefficient fields.

Theorem 3.4. *If $f(x, y)$ is irreducible over a perfect field k then it is absolutely irreducible if and only if k is algebraically closed in $k(x, y)$, i.e. $k = k^0$*

Proof. f is absolutely irreducible if and only if it is irreducible over any finite algebraic extension $\mathbf{F}$ of k. We wish to prove that f is irreducible over any such $\mathbf{F}$ if and only if k is algebraically closed in $k(x, y)$.

$$[\mathbf{F}(x, y) : k(x)] = [\mathbf{F}(x, y) : \mathbf{F}(x)][\mathbf{F}(x) : k(x)] = [\mathbf{F}(x, y) : k(x, y)][k(x, y) : k(x)]. \tag{1}$$

By corollary 3.3 $[\mathbf{F}(x) : k(x)] = [\mathbf{F} : k]$ and using equation (1) we have:

$$[\mathbf{F}(x, y) : \mathbf{F}(x)] = [k(x, y) : k(x)] \iff [\mathbf{F} : k] = [\mathbf{F}(x, y) : k(x, y)]. \tag{2}$$

$f(x, y)$ is irreducible over $\mathbf{F}$ if and only if $[\mathbf{F}(x, y) : \mathbf{F}(x)] = [k(x, y) : k(x)]$, and using equation (2) we have f is absolutely irreducible if and only if $[\mathbf{F} : k] = [\mathbf{F}(x, y) : k(x, y)]$ for any finite algebraic extension $\mathbf{F}$ of k.

If k is algebraically closed in $k(x, y)$ then by lemma 3.1 $[\mathbf{F}(x, y) : k(x, y)] = [\mathbf{F} : k]$ for any such $\mathbf{F}$. Conversely if f is absolutely irreducible then choose $\mathbf{F}$ to be the algebraic closure of k in $k(x, y)$. Thus $[\mathbf{F}(x, y) : k(x, y)] = 1$ since $\mathbf{F} \subseteq k(x, y)$. By equation (2) we have $[\mathbf{F} : k] = 1$ showing that k is in fact algebraically closed in $k(x, y)$. $\qquad\square$

Corollary 3.5. *If $f(x, y)$ is irreducible over k^0, the algebraic closure of k in $k(x, y)$, then f is absolutely irreducible.*

Proof. k^0 is algebraically closed in $k^0(x, y) = k(x, y)$. Thus f is absolutely irreducible by the theorem. $\qquad\square$

Corollary 3.5 gives us a finitely generated field to factor over.

Note that we have also demonstrated that $[\mathbf{K}^0 : \mathbf{K}]$ divides $[\mathbf{K}(x, y) : \mathbf{K}(x)]$. Our choice of x as the independent variable and y as algebraic over x was somewhat arbitrary. If we reverse the roles we see that $[\mathbf{K}^0 : \mathbf{K}]$ must divide both $\deg_y f$ and $\deg_x f$. So if these two numbers are relatively prime then f must be irreducible. More generally for an irreducible multivariate polynomial, if the gcd of all the degrees appearing in the polynomial is 1, then it must be absolutely irreducible.

259

3.2 Algorithmic Considerations

Armed with these algebraic results, we return to the question of finding an absolutely irreducible equation for y. We now realize that $\mathbf{K}^0$ is the field we want to factor over, but we have no explicit presentation of $\mathbf{K}^0$. We know that $\mathbf{K}^0 \subseteq \mathbf{K}(x, y)$ with each element of $\mathbf{K}^0$ algebraic over $\mathbf{K}$. Since each element of $\mathbf{K}^0$ is independent of x, we can also view $\mathbf{K}^0$ as a subfield of the field $\mathbf{K}(u, w)$ where w satisfies $f(u, w) = 0$. Again as in chapter 2 without loss of generality we will assume that $f(X, Y) \in \mathbf{K}[X, Y]$ is monic as a polynomial in Y.

The minimal polynomial of each element of $\mathbf{K}^0$ is monic with coefficients in $\mathbf{K}$. Thus $\mathbf{K}^0[u]$ is certainly an integral algebraic extension of $\mathbf{K}[u]$. Let $\mathbf{A}$ be the integral closure of $\mathbf{K}[u]$ in $\mathbf{K}(u, w)$. Any factorization of $f(x, y)$ with coefficients in $\mathbf{K}^0$ yields a factorization with coefficients in $\mathbf{A}$. By the results of the last chapter any element of $\mathbf{A}$ can be written as a polynomial in u and w divided by the discriminant of $\mathbf{K}[u, w]$ over $\mathbf{K}[u]$. This discriminant is a polynomial in u and is the same as the discriminant of $f(u, w)$ viewed as a polynomial in w. The discriminant of a monic polynomial is non-zero if and only if the polynomial is square-free, i.e. has no repeated factors. Since we have assumed that $\mathbf{K}$ is perfect, any irreducible polynomial over $\mathbf{K}$ must be square-free and hence the discriminant of f is non-zero.

If $\mathbf{K}$ is sufficiently large then we can find a $u_0 \in \mathbf{K}$ such that $\mathrm{disc}(f)$ doesn't vanish at $u = u_0$. Then the map sending u to u_0 is a well defined homomorphism from $\mathbf{A}$ onto a ring $\mathbf{B}$. $\mathbf{B}$ can be presented as $\mathbf{K}[w]$ modulo $f(u_0, w)$. Let $f_1(w)$ be an irreducible factor of $f(u_0, w)$ over $\mathbf{K}$. There is a natural homomorphism from $\mathbf{B}$ onto the field $\mathbf{B}_1 = \mathbf{K}[w]/(f_1(w))$. If we have a non-trivial factorization of f with coefficients in $\mathbf{A}$, then applying both of the above homomorphisms we arrive at a non-trivial factorization with coefficients in $\mathbf{B}_1$. Our choice of an irreducible factor of $f(u_0, w)$ was arbitrary, so we may as well choose the factor of least degree. In particular, if there are any linear factors, then $f(x, y)$ must already be absolutely irreducible. In any case, we have found a presentation for an algebraic extension of $\mathbf{K}$ that contains $\mathbf{K}^0$. Note that it was unnecessary to compute $\mathrm{disc}(f)$ to verify our choice of u_0. $\mathrm{Disc}(f)$ is zero if and only if f has a repeated factor. Thus we pick successive values of u_0 until we find one such that $f(u_0, y)$ remains square-free. If m is the degree of f in u, then we must test at worst $2mn$ values until we find one that works.

We have justified the following algorithm for finding an absolutely irreducible factor of a polynomial $f(x, y)$ whose discriminant with respect to y is nonzero and irreducible over a sufficiently large field $\mathbf{K}$.

1. If $\gcd(\deg_x f, \deg_y f) = 1$ then return $f(x, y)$.
2. Find an x_0 in $\mathbf{K}$ such that $f(x_0, y)$ is square free. (May fail if $\mathbf{K}$ is finite.)
3. Factor $f(x_0, y)$ over $\mathbf{K}$ and let $f_1(y)$ be a factor of least degree.
4. If $f_1(y)$ is linear return $f(x, y)$.
5. Else factor $f(x, y)$ over $\mathbf{K}[w]/(f_1(w))$ and return a factor of minimal degree.

For the more general problem of finding absolutely irreducible factors of multivariate polynomials, the same approach works. Step 0 is changed to take the gcd of the degrees of all variables present, and in step 1 we must find values for all variables but one such that the resulting univariate polynomial is square free and the leading coefficient doesn't vanish. We have assumed that $\mathbf{K}$ is large enough so that we can find substitution points that leave f square free. This may fail if $\mathbf{K}$ is a finite field. In that case if we let m be the gcd of the degrees of all variables present, then we do know that $[\mathbf{K}^0 : \mathbf{K}]$ divides m. Thus $\mathbf{K}^0$ is a subfield of the unique extension of $\mathbf{K}$ of degree m. It therefore suffices to factor over that extension.

Once we have found an absolutely irreducible defining polynomial for y, then we adjoin the coefficients of all the monomials to $\mathbf{K}$ and this generates $\mathbf{K}^0$ the exact coefficient field.

3.3 Binomial Polynomials

If $f(x,y)$ is of the form $y^n - g(x)$ then a much simpler algorithm exists for obtaining an absolutely irreducible factorization. This is based on the following theorem proven in [33, p. 221].

Theorem 3.6. *Let k be a field and n an integer ≥ 2. Let $a \in k$, $a \neq 0$. Assume that for all prime numbers p such that $p|n$ we have $a \notin k^p$, and if $4|n$ then $a \notin -4k^4$. Then X^{n-a} is irreducible in $k[X]$.*

We define square-free factorization of a polynomial $g(x) \in \mathbf{K}[x]$ to be

$$g(x) = c \prod_i g_i^{e_i} \tag{3}$$

where $c \in \mathbf{K}$, $\gcd(g_i, g_j) = 1$ for $i \neq j$, and each g_i has no repeated factors. If we assume that $\mathbf{K}$ is perfect, then $g_i(x) \in \mathbf{K}[x]$ for all i. This leads us to the following simple algorithm for finding an absolutely irreducible factor of $y^n = g(x)$.

1. Compute a square-free factorization of $g(x)$ as in equation (3).
2. Let d be the gcd of all the e_i and n.
3. If $d = 1$ then f is absolutely irreducible and return it.
4. An absolutely irreducible factor of $y_n - g(x)$ is

$$y^{\frac{n}{d}} - c^{\frac{1}{d}} \prod_i g_i^{\frac{e_i}{d}}.$$

Chapter 4
The Rational Part of Integral

By Liouville's theorem if the integral of an algebraic function is expressible in terms of elementary functions, it can also be written as the sum of an algebraic function and constant multiples of logs of algebraic functions. We will label the sum of logs the *transcendental part* of the integral and the rest the *rational part*. This terminology is carried over from rational function integration. We use it since we want to draw strong parallels between our algorithms for algebraic function integration and the well known ones for rational function integration.

4.1 Rational Function Integration

There are primarily three basic algorithms for finding the rational part of the integral of a rational function, however they all share a common first stage. Polynomial division is employed to convert the integrand into the sum of a polynomial and a proper rational function. (Proper means the degree of the numerator is less than that of the denominator). The polynomial part can be trivially integrated termwise. At this point the three algorithms diverge.

4.1.1 Full factorization

The simplest algorithm conceptually completely factors the denominator of the reduced integrand over the algebraic closure of the constant field. If only approximations to the roots were used this algorithm would be acceptable, but we want exact solutions and this requires constructing the splitting field of the denominator. This construction involves algebraic factoring and can be very expensive. Then a complete partial fraction decomposition is performed. Each term in the result is easy to integrate, but the result contains many algebraic quantities. They are all unnecessary since the rational part of the integral can always be expressed without extending the constant field, but the task of trying to eliminate the algebraics can be costly. This algorithm is close in spirit to the one proposed by Risch in [46] and partially implemented by Davenport in [16]. The similarity becomes more evident if we reexpress the algorithm in terms of power series. The portion of the partial fraction expansion involving a particular root of the denominator is identical to the principal part (negative degree terms) of the power series expansion of the integrand at that root. Thus the algorithm can be expressed as power series computation and termwise integration. The additional step in Risch's algorithm involves reconstruction of the rational part of the integral from the principal parts of its expansion at all its poles. This is a fairly complex process for an algebraic function, however a rational function is simply determined up to an additive constant as the sum of its principal parts.

4.1.2 Linear equations

Another algorithm for rational function integration was proposed by Horowitz [30]. This approach is global as contrasted with the local techniques used above. First a square-free factorization of the denominator is performed

$$D = \prod D_i^i, \gcd(D_i, D_j) = 1 \text{ for } i \neq j \tag{1}$$

and each D_i is square-free (has no multiple factors). By observing that the integral of $(x-c)^{-k-1}$ is $-(x-c)^{-k}/k$, we see that integration reduces the order of a pole by one. Thus

$$\int \frac{A}{\prod D_i^i} = \frac{B}{\prod D_i^{i-1}} + \int \frac{C}{\prod D_i}$$

where the last integral produces the transcendental part. The degrees of B and C are constrained since both are numerators of proper rational functions. By letting B and C be polynomials with undetermined coefficients, differentiating both sides of the above equation and equating coefficients of the same powers of x we get a system of linear equations for the coefficients of B and C. Since the rational part of the integral is uniquely determined up to an additive constant and we have constrained the constant by requiring a proper rational function, the above system has a unique solution. This method is relatively insensitive to sparseness in the input or output.

4.1.3 Hermite's algorithm

The third technique is due to Hermite [28]. It attempts to reduce the "complexity" of the problem using a succession of linear first degree congruence equations. Again we start by performing a square-free factorization of the denominator. But now instead of trying to find the entire rational part in a single step, we will repeatedly find pieces of the rational part that can be used to reduce the multiplicity of the denominator of the integrand. The algorithm we present here treats the factors of the denominator one at a time. Mack [39] presents a variant that treats all the factors at once, but his algorithm would only make our formulas more complicated, and the underlying theory is essentially the same.

Again we assume a square-free factorization of the denominator of the integrand as in (1). Let $V = D_{j+1}$ for some $j > 0$, i.e. V is a multiple factor of the denominator. Let U be the cofactor of V, $U = D/V$. By (1) U and V are relatively prime and we can write the integral as

$$\int \frac{A}{UV^{j+1}} dz.$$

We will attempt to repeatedly reduce the multiplicity of the denominator while constructing portions of the final answer. We claim that there exist polynomials B and C such that

$$\int \frac{A}{UV^{j+1}} = \frac{B}{V^j} + \int \frac{C}{UV^j}.$$

After differentiating both sides and multiplying by UV^{j+1} we get

$$A = UVB' - jBUV' + VC.$$

This is a differential equation with two unknowns so it would seem that we have made the problem more complicated. We will additionally claim, however, that there is a unique solution B such that $\deg(B) < \deg(V)$. We then reduce the equation modulo V. This eliminates the B' and C terms from the above equation and we are left with

$$A \equiv -jBUV' \bmod V. \tag{2}$$

This equation will indeed have a unique solution as long as $j \neq 0$ and $\gcd(V, V') = 1$. But V is square-free by construction so the latter requirement is satisfied. As long as $j > 0$ we can find a unique B solving (2) and then subtracting $(B/V^j)'$ from the integrand will reduce the multiplicity of the denominator. By repeating this process with all multiple factors of the denominator, we see that the integral of any rational function can always be expressed as the sum of a rational function and an integral whose denominator has multiplicity one. The latter integral has no rational part, i.e. it is expressible exclusively as a sum of logs.

4.2 Algebraic Functions

We choose to base our algorithm for finding the rational part of the integral of an algebraic function on Hermite's method for rational function integration. In fact all three of the algorithms presented in the previous section can be generalized to handle algebraic functions. The generalization of the first approach requires Puiseux expansions and algebraic number computations that we wish to avoid. The advantage of the third approach over the second is that we are provided with insight into how an integral may fail to be elementary and its step by step reductive nature will often allow us to return partial results instead of returning "not integrable".

To simplify matters we will transform the integral so that there are no poles or branch points at *infinity*. We assume an integral of the form $\int \sum R_i(x)y^i/Q(x)\,dx$ where R_i and Q are polynomials and y satisfies $f(y,x) = 0$. Let a be an integer that is neither a root of Q nor a root of the discriminant of f. We can "move" the point a to ∞ by defining $z = 1/(x-a)$ or $x = a + 1/z$. Our new integral is

$$\int \sum \frac{R_i(a+1/z)y^i}{Q(a+1/z)}(-z^{-2})dz.$$

265

After performing the integration we can apply the inverse transformation to express the answer in terms of x instead of z. If we let m be the degree of f in x, then the transformed minimal polynomial for y is $g(y,z) = z^m f(y, a + 1/z)$. Using the results of the last chapter, we can find a basis, $[w_1, \ldots, w_n]$ for the integral closure of $\mathbf{K}[z]$ in $\mathbf{K}(z, y)$. In terms of this basis the integrand can be expressed as $\sum A_i(z) w_i / D(z)$ where D and A_i are polynomials. Since this integrand has no poles at ∞, $\deg(A_i(z)) < \deg(D(z))$ for all i.

We now attempt to imitate Hermite's algorithm for rational functions. Again we start by performing a square-free factorization of the denominator $D(z)$ yielding $D = \prod D_i^i$. Let $V = D_{k+1}$ for some $k > 0$ and $U = D/V$. Following Hermite's algorithm we might now look for polynomials B_i and C_i such that

$$\int \sum A_i \frac{w_i}{UV^{k+1}} dz = \sum B_i \frac{w_i}{V^k} + \int \sum C_i \frac{w_i}{UV^k} dz. \tag{3}$$

Unfortunately we won't be able to find them in general without additional restrictions on U. The difficulty is caused by the fact that y' has a non-trivial denominator. Hermite depended on the fact that the derivative of a polynomial is a polynomial, but this is not necessarily true for algebraic functions. If y satisfies $f(y, z) = 0$ then we can find $y' = dy/dz$ by taking the total derivative of the defining polynomial

$$\frac{\partial f}{\partial y} dy + \frac{\partial f}{\partial z} dz = 0$$

$$\frac{dy}{dz} = -\frac{\partial f/\partial z}{\partial f/\partial y}.$$

Thus y' is a rational function in y and z. Similarly w_i' in general has a nontrivial denominator. Let E be the least common denominator of w_i' for all i. Then the derivative of the second term in (3) will in general have E as a factor of its denominator. This will force us to permit E to be a factor in the denominator of the third term. We want to use (3) iteratively; this means the third term on one iteration will become the first term for the next iteration. Therefore we may as well assume the denominator of the first term is also divisible by E. In the last iteration $k = 1$ making the denominator of the third term UV, so our additional restriction on equation (3) is that $E \mid UV$. We can always guarantee this by multiplying the A_i and U by a suitable factor of E. Note that there is then an uncanceled gcd between the numerator and denominator of the first term in (3). We need to be sure that the new U is still relatively prime to V. This can only be guaranteed if we know E to be square-free. Fortunately that is always the case.

Lemma 4.1. *If $v_p w \geq 0$ and $v_p(z-a) > 0$ then $v_p(z-a)w' > 0$.*

Proof. $v_p w \geq 0$ implies $v_p dw \geq 0$, and $v_p(z-a) > 0$ implies $v_p(z-a) = v_p dz + 1$ ([10, IV.8 Lemma 1]). But $dw = w' dz$ and thus $0 \leq v_p w' dz < v_p(z-a)w'$. □

We therefore begin with equation (3) with the following conditions:

$$E \mid UV, \quad \gcd(U, V) = 1, \quad \gcd(V, V') = 1. \tag{4}$$

Now as before we perform the differentiation and multiply through by UV^{k+1} yielding

$$\sum A_i w_i = U \sum \left(V B_i' + B_i V^{k+1} \left(\frac{w_i}{V^k} \right)' \right) + V \sum C_i w_i. \tag{5}$$

We then reduce the equation modulo V

$$\sum A_i w_i = \sum B_i U V^{k+1} \left(\frac{w_i}{V^k} \right)' \bmod V. \tag{6}$$

We must show that this equation always has a unique solution. This is equivalent to showing that the $S_i = U V^{k+1} \left(\frac{w_i}{V^k} \right)'$ is a local integral basis, i.e. any integral function can be expressed as a linear combination of them with rational function coefficients and denominator relatively prime to V. There are two ways this can fail, either the S_i are not linearly independent over $\mathbf{K}(z)$ or there exists an integral function whose representation requires a factor of V as a denominator. Since the former case implies that zero is a nontrivial linear combination of the S_i, both cases may be summarized by saying there exists an integral function that can be represented as

$$\frac{1}{V} \sum T_i U V^{k+1} \left(\frac{w_i}{V^k} \right)' \text{ where } V \text{ does not divide } UT_i \text{ for some } i. \tag{7}$$

For purposes of generating a contradiction we assume that the S_i do not form a local integral basis and thus there exists an integral function (7). We can add $\sum (UT_i)' w_i$ to equation (7) yielding a new integral function G such that

$$G = \sum V^k \left((UT_i)' \frac{w_i}{V^k} + UT_i \left(\frac{w_i}{V^k} \right)' \right) = V^k \sum \left(UT_i \frac{w_i}{V^k} \right)'. \tag{8}$$

Assuming that such an integral function G exists implies there exists a function for whom differentiation doesn't increase the order of its poles; this is the source of the contradiction. Let $F = \sum UT_i w_i / V^k$. The restriction on the T_i in (7) says that a smaller value of k would be insufficient, i.e. there is some place p such that $v_p F < v_p (1/V^{k-1})$ and $v_p V > 0$ where v_p is the order function at p.

Lemma 4.2. *If u is a nonzero function such that $v_p u \neq 0$ and p is a finite place (not over ∞) then $v_p u' = v_p u - r$ where r is the ramification index of p with respect to* $\mathbf{K}(z)$.

Proof. We embed our function field in its p-adic completion and let t be a uniformizing parameter at p. We can then write

$$u = \sum_{i=j}^{\infty} c_i t^i.$$

The coefficients in the series are algebraic over our constant field and we can extend our derivation d/dt uniquely to the completion yielding:

$$u' = \sum_{i=j}^{\infty} i c_i t^{i-1} t'.$$

Since $j \neq 0$, $v_p u' - j - 1 + v_p t'$. Since p is a finite place, the p-adic expansion of z begins $z - a_0 + a_1 t' + \cdots$ where $a_r \neq 0$ so $v_p(\mathrm{d}z/\mathrm{d}t) = r - 1$. Thus $v_p t' = v_p(\mathrm{d}t/\mathrm{d}z) = 1 - r$ and $v_p u' = (j-1) + (1-r) = j - r$. □

Since V is a square-free polynomial in z and $v_p V > 0$ we have $v_p V = r$ and thus $v_p(1/V^k) = v_p(1/V^{k-1}) - r$. If $k > 0$ then $v_p F < v_p(1/V^{k-1})$ implies $v_p F < 0$. Using lemma 4.2 we see that $v_p F' < v_p(1/V^{k-1}) - r = v_p(1/V^k)$. But according to equation (8) F' can be written as G/V^k where G is an integral function. This contradicts the fact that $v_p F' < v_p(1/V^k)$. Thus the S_i do indeed form a local integral basis, and equation (6) will have a unique solution modulo V as long as $k > 0$.

By our choice of E there must exist polynomials M_{ij} such that

$$E w_i' = \sum_j M_{ij} w_j.$$

Since $E \mid UV$ let $TE = UV$ for some polynomial T.

$$UV w_i' = TE w_i' = T \sum_j M_{ij} w_j. \tag{9}$$

Substituting (9) into equation (6) yields:

$$\sum_i A_i w_i \equiv \sum_i (-kUV'B_i) w_i + \sum_i B_i T \sum_j M_{ij} w_j \bmod V. \tag{10}$$

If we now equate the coefficients of w_i on both sides of (10) we get a set of linear congruence equations with B_i as unknowns

$$A_i \equiv -kUV'B_i + T \sum_j B_j M_{ji} \bmod V. \tag{11}$$

We have shown that this system will always have a unique solution for $k > 0$. Thus the determinant of the coefficient matrix must be relatively prime to V and the system can always be solved, e.g. using Cramer's rule.

For efficiency reasons it is important to recognize when the system (11) decouples, i.e. each equation involves only one unknown. When $V \mid T$ the summation in (11) vanishes and we are reduced to solving a succession of equations of the form

$$A_i \equiv -kUV'B_i \bmod V.$$

This case of algebraic integration is almost exactly the same as in rational function integration. $V \mid T$ if and only if $\gcd(V, E) = 1$. Thus it can become worthwhile to split V into two factors, one that divides E and one relatively prime to E and treat each case separately.

The other situation in which the system decouples is when the matrix M_{ij} in (9) is diagonal. This means that $w_i' = R_i w_i$ where R_i is a rational function in z. We can solve this differential equation, and the solution will be an algebraic function if and only if $m_i^m \in \mathbf{K}(z)$ [49]. Thus the matrix can be diagonal if and only if $\mathbf{K}(z, y)$ is a compositum of single rational extensions.

4.3 Poles at Infinity

Repeated application of the reductions presented in the previous section will leave us with an integral whose denominator is square-free. One might hope that we have removed all singularities from the integrand except those that should be cancelled by log terms as in the rational function case. Unfortunately while we have been reducing the order of the finite poles, we may have introduced poles at ∞. In this section we will prove that if our basis is normal at infinity the presence of such poles will prove that the original problem was non-integrable.

The problem is caused by the second term in equation (3), $B = \sum B_i w_i / V^k$. By construction we have the restriction that $\deg(B_i) < \deg(V)$. Thus at each iteration of the previous algorithm, the coefficients of the answer produced will be proper rational functions, i.e. the degree of the numerator is less than the degree of the denominator. This guarantees that the rational function coefficients of the portion of the answer we have produced will have positive order at ∞, but the w_i themselves will in general have poles at ∞. We will see that if this algebraic portion of the answer has poles at infinity, then the original problem was not integrable.

Lemma 4.3. *If u is a nonzero function such that $v_p u \neq 0$ with p a place over ∞, then $v_p u' = v_p u + r$ where r is the ramification index of p.*

Proof. The proof is identical to Lemma 4.2 except that $v_p t'$ is different. The p-adic expansion of z is $z = a_{-r} t^{-r} + \cdots$ so $v_p(\mathrm{d}z/\mathrm{d}t) = -r - 1$. Thus $v_p t' = v_p(\mathrm{d}t/\mathrm{d}z) = r + 1$ and $v_p u' = (v_p u - 1) + (r + 1) = v_p u + r$. $\qquad\square$

Lemma 4.4. *If u is a nonzero function such that $v_p u \neq 0$ at some place p, then $v_p \mathrm{d}u = v_p u - 1$.*

Let us assume that the third term in equation (3) has poles at infinity. Since the original integrand had zero residue at infinity, and the derivative of any algebraic function has zero residue everywhere this third term must also have zero residue at infinity. Thus if it has poles there, they must be at least double poles. If this term is integrable it must be expressible as an algebraic function and a sum of constant multiples of logs. This algebraic function can not have any finite poles, since the integrand has only simple poles in the finite plane. Thus this algebraic function must be expressible as a polynomial multiple of our basis element w_i. Since this function has a pole at infinity, either one of the coefficients is a polynomial of positive degree, or some non-constant basis element has a non-zero coefficient. Although the second term in equation (3) could also have poles at infinity, when we add these functions those poles do not cancel. The coefficients of the basis elements in the second term are all proper rational functions, i.e. the degree of the numerator is less than the degree of the denominator. This is a consequence of the fact that $\deg(B_i) < \deg(V)$ since each B_i is computed modulo V. When we add this polynomial multiple of our basis elements to the second term, we arrive at a function with at least one coefficient that is an improper rational function. If this coefficient has a pole at infinity then the function itself has a pole at infinity since the basis is normal at infinity and no basis elements have zeros at all places over infinity. If none of the coefficients have poles at infinity, then some non-constant basis function must have a coefficient of order zero at infinity. But since any non-constant basis function must have a pole at infinity, both possibilities imply the complete algebraic part of the integral must have a pole at infinity. But by lemma 4.4 this would imply that the integrand would have a pole at infinity that contradicts our initial assumption. Thus we see that if in the process of finding the algebraic part, we introduce poles at infinity then the original problem was not integrable and there is no need to try to remove these poles. We have demonstrated the following theorem and corollary:

Theorem 4.5. *If $g\mathrm{d}x$ is a differential with zero residues and no poles at infinity, then the algorithm of the previous section computes a function h such that $(g - h')\mathrm{d}x$ has no finite poles and is zero if $g\mathrm{d}x$ was integrable.*

Corollary 4.6. *If $g\mathrm{d}x$ also has non-zero residues then $(g - h')\mathrm{d}x$ has only simple poles in the finite plane and has no poles at infinity if $g\mathrm{d}x$ was integrable.*

Chapter 5
Log Terms and Divisors

In this chapter we will present algorithms for finding the logarithmic or transcendental part of the integral. Using the results of the previous chapter we can assume that the rational part has been removed and the integrand has been reduced to a differential with only simple poles. Using Liouville's theorem we know that if the integral is expressible as an elementary function, then it can be written as

$$\int R(x,y)\mathrm{d}x = \sum c_i \log v_i(x,y) \qquad (1)$$

where $c_i \in \mathbf{K}'$ and $v_i \in \mathbf{K}'(x,y)$ with $\mathbf{K}'$ a finite algebraic extension of $\mathbf{K}$. To find the rational part of the integral no extension of the ground field was necessary, but the same is not true for the logarithmic part. We will show that the residues of the integrand generate the unique $\mathbf{K}'$ of minimal degree over $\mathbf{K}$ sufficient to express the answer. Additionally these residues provide us with clues about the orders of the poles and zeros of the v_i.

5.1 Properties of Logarithmic Differentials

We will start by examining the relationship between $f \in \mathbf{K}(x,y)$ and the differential $\frac{\mathrm{d}f}{f}$. We are interested in the local behavior at a place p with t as a uniformizing parameter. If a function f has order k at p then $f = t^k g$ where g is a function whose value at p is finite and nonzero. Any differential $f\mathrm{d}g$ can be written as $h\mathrm{d}t$ where $h = f\frac{\mathrm{d}g}{\mathrm{d}t}$. The order at p of the differential $f\mathrm{d}g$ is defined to be the order of the function h above. The residue at p of the differential $h\mathrm{d}t$ is the coefficient of the t^{-1} term in the series expansion of h in powers of t. Both the residue and the order of a differential are independent of the choice of uniformizing parameter.

Next we examine the properties of differentials of the form $\frac{\mathrm{d}f}{f}$ at a place p with local uniformizing parameter t. If f has order k at p it can be written as $f = t^k g$ for some function g of order 0 at p and similarly the differential can be rewritten as

$$\frac{\mathrm{d}f}{f} = k\frac{\mathrm{d}t}{t} + \frac{\mathrm{d}g}{g}.$$

As shown in chapter four, since g has no pole at p, dg must have non-negative order at p also. By construction g has order 0 at p and thus $\frac{dg}{g}$ has non-negative order at p. Since t is a local uniformizing parameter at p the order of $k\frac{dt}{t}$ is the same as the order of $\frac{k}{t}$ that is precisely -1 as long as k is nonzero. Similarly the residue of $\frac{df}{f}$ is always greater than or equal to -1, and its residue at any place p is the same as the order of f.

This leads to a solution to a special case of our original problem. When can a differential be expressed in the form $\frac{df}{f}$ for some function f in our function field? The following two necessary properties give quick failure tests:

1. The order of the differential must be greater than or equal to -1 everywhere.
2. The residues must all be integers since they correspond to the orders of the desired function f.

If a differential passes those tests, then we try to determine if there exists a function whose order at every place is equal to the residue of the differential at that place. Thus the residues of the differential provide us with a formal specification of the location and orders of the poles and zeros of the desired function. Since the differential only has a finite number of poles, it can only have nonzero residue at a finite number of places. A divisor is a formal integer linear combination of places that has a finite number of non-zero coefficients. Divisors that correspond to the orders of the poles and zeros of actual functions are called principal divisors. A differential with integer residues everywhere immediately provides us with a divisor and we are left with the problem of determining whether or not it is principal. An algorithm for answering this question will be presented later in this chapter.

The solution to the problem discussed in the previous paragraph leads to some computational difficulties, but no great theoretical problems. But if we generalize things slightly we arrive at a problem that many mathematicians around the turn of the century worked on but were unable to solve. In fact Hardy even went so far as to state "there is reason to believe that no solution exists". The generalization involves introducing one additional constant coefficient. Instead of asking whether a given differential can be expressed as $\frac{df}{f}$, we allow one more degree of freedom and try to express it as $1/m\frac{df}{f}$ for some integer m. By the same reasoning as above we see that if we multiply all the residues of the original differential my m we arrive at the desired divisor for f. But we don't know the value of m. This is the source of the theoretical difficulty. We derive a divisor from the residues of the integrand and we can test whether or not it is principal. If it is then we are done. If it is not, however, perhaps by scaling all the orders specified by two, we arrive at a principal divisor. If not try scaling by three, ... etc. The real difficulty is knowing when to stop. More formally given a divisor D we need to be able to determine a bound M such that if for all positive integers $j < M$, jD is not principal, then we are guaranteed that there exists no multiple of D that is principal. It is only in the last thirty years that a solution to the problem has been discovered using the technique of good reduction from algebraic geometry. Basically the original coefficient field is reduced to a finite field. There each divisor has finite order, i.e. there is some finite value of j such that

jD is principal. This information is used to limit the potential set of j's that needs to be examined. We will present this construction in the next chapter.

Armed with some intuition from the previous special cases, we will now investigate the fully general problem indicated by equation (1). We wish to write the given differential as

$$R(x,y)\mathrm{d}x = \sum_i c_i \frac{\mathrm{d}v_i}{v_i} \qquad (2)$$

for some constants c_i and some rational functions $v_i(x,y)$. As indicated earlier the constants c_i and the coefficients of the v_i will in general lie in a finite algebraic extension $\mathbf{K}'$ of the field of constants of the original function field. Our first step involves the construction of the minimal extension of $\mathbf{K}$ sufficient to express the answer.

The decomposition of the $R(x,y)\mathrm{d}x$ indicated by equation (2) is certainly not unique. In fact let b_j be another set of constants such that each c_i can be expressed as integer multiples of the b_j, $c_i = \sum n_{ij}b_j$. Define new functions $w_j = \prod_i v_i^{n_{ij}}$ then:

$$\sum_i c_i \frac{\mathrm{d}v_i}{v_i} = \sum_j b_j \frac{\mathrm{d}w_j}{w_j}.$$

Since a linear dependence among the coefficients implies we can express the sum with fewer terms, the answer with the smallest number of summands will have coefficients that are linearly independent over the rationals.

We next investigate the relationship between the coefficients of the log terms and the residues of the integrand. Since the residues of $\frac{\mathrm{d}v}{v}$ is always an integer for any function v, we see that the residues of the integrand are always integer linear combinations of the c_i's. Thus the coefficients of the log terms generate a $\mathbf{Z}$-module that contains all the residues of the integrand. We will show that the c_i's can be chosen so that they generate the same vector space over the rationals that the residues of the integrand do. Let a_j form a basis for the vector space generated by the residues. Let b_k extend this basis to include the coefficients of the log terms. Thus each c_i can be written uniquely as $\sum r_{ij}a_j + \sum s_{ik}b_k$ where the r_{ij} and s_{ik} are rational numbers. We will in fact assume the r_{ij} and s_{ik} are integers, which can be accomplished by suitably scaling the basis elements. Then as shown in the previous paragraph, we can construct functions w_j and u_k such that:

$$\sum_i c_i \frac{\mathrm{d}v_i}{v_i} = \sum_i a_i \frac{\mathrm{d}w_i}{w_i} + \sum_k b_k \frac{\mathrm{d}u_k}{u_k}. \qquad (3)$$

Since however the a_j's form a basis for the vector space containing all the residues of the entire sum and the b_k's and the a_j's form a linearly independent set by construction, the residues of each $\frac{\mathrm{d}u_k}{u_k}$ must be zero everywhere. This implies that each u_k has order zero everywhere, is thus a constant. Since the differential of any constant is zero, the second sum in equation (3) is identically zero.

We have shown that if a differential has a decomposition as in equation (2), it has one where the coefficients form a basis for the vector space spanned by the residues of $R(x, y)dx$. If we compute such a basis r_i, then by lemma 1 [*ambiguous in original text*], we are guaranteed that there exists integers n_i such that $\frac{r_i}{n_i}$ can be chosen as the coefficients of the log terms. Let $\mathbf{K}'$ be $\mathbf{K}$ extended by all the residues of the integrand. Assume we have found a representation for the integrand fdx as a sum of logarithmic differentials, $\sum c_i \frac{dv_i}{v_i}$. We have shown that we can assume the coefficients of the log terms c_i lie in $\mathbf{K}'$. We wish to show that we can also assume that the $v_i \in \mathbf{K}'(x, y)$. If instead $v_i \in \mathbf{E}(x, y)$ where $\mathbf{E}$ is a finite algebraic extension of $\mathbf{K}'$ of degree j then by applying a trace from $\mathbf{E}$ to $\mathbf{K}'$, $\text{tr}_{\frac{E}{K'}}$ we arrive at a solution whose constant field is exactly $\mathbf{K}'$

$$f dx = \sum c_i \frac{dv_i}{v_i}$$

$$\text{tr}(f dx) = \sum \text{tr}(c_i \frac{dv_i}{v_i}).$$

Since $f \in \mathbf{K}(x, y)$, $\text{tr}(f dx) = j f dx$ and $\text{tr}(d/v)$ is the same as $d(Nv)/(Nv)$ where N is the norm from $\mathbf{E}$ to $\mathbf{K}'$

$$j f dx = \sum c_i \frac{d N v_i}{N v_i}.$$

But Nv is a rational function with coefficients in $\mathbf{K}'$, so we have shown if the integral can be expressed over some algebraic extension of $\mathbf{K}$ then in fact it can be expressed over the extension of $\mathbf{K}$ generated by the residues of the integrand, and no smaller extension will suffice.

5.2 Computing the Residues

Now that we have demonstrated the importance of the residues of the integrand, we need an efficient way to compute these residues. If the differential is expressed in the form $f dt$ where t is the local uniformizing parameter for some place p, then the residue was defined to be the coefficients of t^{-1} in the series expansion of f at p. If f is known to have order greater than or equal to -1 at p, then we can just compute the value of tf at p. Thus we need to be able to find local uniformizing parameters for places, and be able to compute the value of functions at places. We can view the Riemann surface associated with the differential as a multisheeted covering of the complex $\mathbf{X}$-plane. Thus each finite place p_0 can be associated with some x-value x_0 by projection. The order of the line of projection at a place defines the branch index of that place. Branch places are those where the line of projection is tangent to the Riemann surface and thus have branch index greater than one. Since $x = x_0$ is the equation of the line of projection from p to the $\mathbf{X}$-plane, the order of the function $x - x_0$ at p is equal to the branch index of p. If p is not a branch place, $x - x_0$ has order 1 and can thus be used as a local uniformizing parameter.

Theorem 5.1. *Let $f dx$ be a differential with order greater than or equal to -1 at some place p with branching index r centered at x_0. The residue of $f dx$ at p is equal to the value of the function $r(x - x_0)f$ at p.*

Proof. Let t be a uniformizing parameter at p. Since $x - x_0$ has order r at p, it can be written as

$$x - x_0 = t'g \tag{5}$$

where g has order zero at p

$$dx = (rt^{r-1}g + t^r \frac{dg}{dt})dt.$$

Since dg/dt has non-negative order at p, dx has order $r - 1$ at p and f must have order greater than or equal to $-r$ at p

$$f dx = rt^{r-1}f g dt + t^r f(\frac{dg}{dt})dt \tag{6}$$

the second term on the right side of equation (6) is holomorphic at p so the residue of $f dx$ at p is the same as the residue of the first term on the right side of (6). By the argument of the preceding paragraph, this residue can be computed by evaluating the function $rt'fg$ at p. Using equation (5) this function can be written as $r(x - x_0)f$. $\square$

Theorem 5.2. *Let $f dx$ be a differential with at most simple poles in the finite plane. Let $D(x)$ be a polynomial whose roots include the x-projections of the poles at f. Let $g = fD$, then $g dx$ is holomorphic in the finite plane and the residue of $f dx$ at any place p with branch index r centered over a root of $D(x)$ is equal to the value of rg/D' at that place.*

Proof. Since $f dx$ has only simple poles in the finite plane, $g dx$ has no finite poles. Let p be a place over a root x_0 of $D(x)$. From the previous theorem we only need to show that at p $(x - x_0)f = \frac{g}{D'}$. Let $D(x) = (x - x_0)C(x)$. Then since $f = \frac{g}{D}$, $(x - x_0)f = \frac{g}{C}$. But $D' = C + (x - x_0)C'$ and thus $D'(x_0) = C(x_0)$ at p. $\square$

We could use the previous theorem to compute separately each residue of the integrand, but it will be more convenient to find them all at once. We will construct a polynomial whose roots are rational multiples of the residues of the differential $f dx$. f is in general a rational function of x and y. After rationalizing the denominator we can assume $f(x, y) = g(x, y)/D(x)$ where g and D are polynomials. Let Z be a new indeterminate and define $R(Z) = \prod ZD' - g$ where the product is taken over all places centered above the roots of D. The roots of R are the residues of $f dx$ divided by the branch orders. Since the branch orders are always positive integers, the splitting field of R is precisely the minimum extension of $\mathbf{K}$ containing all the residues of the integrand, and thus the smallest extension of the coefficient field in which the integration can be performed. Similarly a $\mathbf{Q}$-basis for the roots of R provides us with a $\mathbf{Q}$-basis for the residues. A key observation is that R can be computed without extending the coefficient field. Let $F(x, y) = 0$ be the defining polynomial of our function field. For a particular value of x, x_0, $\prod ZD'(x_0) - g(x_0, y)$ taken over all places sitting over x_0 is just the resultant$_Y$ $(ZD'(x_0) - g(x_0, Y), F(x_0, Y))$ both viewed as polynomials in Y. Extending that product over all roots x_0 of D is just another resultant of the previous result with $D(x)$, both viewed as polynomials in X. Thus we can compute

$$R(Z) = \text{resultant}_X(\text{resultant}_Y(ZD'(X) - g(X,Y), F(X,Y)), D(X)). \qquad (7)$$

Since the roots of R are nonzero rational multiples of the residues, the splitting field of R is the minimal extension containing all the residues. We have found R using only rational operations over $\mathbf{K}$ but to actually compute the log terms we will have to work in a coefficient field that contains all the roots of R. [55] gives an algorithm for computing the splitting field of R using algebraic factoring. This step can be very expensive, since if R has degree n the splitting field may be of degree n factorial. However there is no escaping this expense since this is the smallest extension in which the answer can be expressed.

We next need to compute a basis for the vector space spanned by the roots of R over the rationals. If the coefficients of R are all rational numbers, then we can view the splitting field of R as a vector space over the rationals. Each root is thus a vector with rational coefficients, and we are interested in finding a basis for the space spanned by these vectors. This can be done using standard techniques from linear algebra. In general however the coefficients of R will come from some finitely generated extension of $\mathbf{Q}$. This occurs when the integrand contains additional parameters or algebraic numbers. Let $\mathbf{K}$ be the field generated by the coefficients of R. The splitting field of R is a vector space over $\mathbf{K}$ and thus each root of R is representable as a vector of elements of $\mathbf{K}$ with respect to some chosen basis of the splitting field over $\mathbf{K}$. We need to view these roots as elements of a finite dimensional vector space over $\mathbf{Q}$. We will do this by replacing each coefficient from $\mathbf{K}$ by a finite dimensional vector over $\mathbf{Q}$. Let b_i be a basis for the splitting field of R over $\mathbf{K}$. Then the roots of R can be represented as $r_j = \sum c_{ij} b_j$ where $c_{ij} \in \mathbf{K}$. The c_{ij} will in general be rational quantities but by a different choice of basis elements we can guarantee they are polynomials in the generators of $\mathbf{K}'$ over $\mathbf{K}$. Let d_i be a common denominator for the ith components of all the vectors, i.e. for c_{ij} with fixed i. If we choose as a new basis for $\mathbf{K}'$ over $\mathbf{K}$, b_i/d_i then with respect to this basis all the coefficients will be polynomials. Now we have represented the roots as vectors of multivariate polynomials with rational coefficients. If we choose a basis for the set of monomials appearing in these polynomials, then with respect to the tensor product of this monomial basis and the original basis of $\mathbf{K}'$ over $\mathbf{K}$, our roots are expressible as vectors of rational numbers and can thus be viewed as elements of a finite dimensional vector space over $\mathbf{Q}$. At this point we apply gaussian elimination to this collection of vectors adjoined to an identity matrix, to find a minimal basis over $\mathbf{Q}$, and the rational linear equations expressing all the roots in terms of this basis set.

5.3 Constructing Divisors

The basis for the residues over $\mathbf{Q}$ become our candidates for coefficients of the log terms. They are only candidates since they may be integer multiples of the correct coefficients. For each candidate we next proceed to compute an associated divisor. The minimum multiple of this divisor that is principal will provide us with the appropriate scaling of the candidate coefficient. We first need to construct a set of building blocks from which we will construct our divisors. For each root of R we will construct a divisor that has order one at each place where the integrand has that root as residue, and order zero elsewhere.

We assume the integrand is of the form $(G(x,y)/D(x))dx$ as above. In order to simplify matters we will also assume that all places centered above roots of D are not branch places. In the next section we will show how to convert the general problem to satisfy this restriction. Under this assumption the roots of R are precisely the residues of the integrand, and at any place above the root x_0 of D, $x - x_0$ is a local uniformizing parameter. By theorem 5.2 G/D' is a function whose value over all places centered above roots of D is the residue of the integrand. Let r be a root of R, then $G/D' - r$ is a function that vanishes wherever the integrand has residue equal to r and at other places above roots of D, has a nonzero value. Define $B(x) = D(x)/\gcd(D, D')$. Then B contains all factors of D taken with multiplicity one.

Proposition 5.3. *The minimum of the orders of $B(x)$ and $G - rD'$ at any finite place p is equal to one if the integrand has residue r at p and zero otherwise.*

Proof. If the integrand has residue r at p then $G - rD'$ vanishes at p and $B(x)$ vanishes to order one at p. If p is any place over a root of D where the integrand has residue different from r then $G - rD'$ has order zero at p. If p is a finite place not centered over a root of D then $B(x)$ has order zero at p. □

If our function field had unique factorization we would proceed by computing the greatest common divisor of $B(x)$ and $G - rD'$ and use these to build our divisors. But unless the function field happens to have genus 0, we are not guaranteed that the notion of gcd is well defined. We will work instead with the ideal generated by these two functions over the ring of integral algebraic functions. In a gcd-domain this ideal would have a single generator but not in general.

Now that we have created these ideal building blocks, we need algorithms for multiplication and division. In general a set of generators for the product of two ideals can be computed simply as the set cross products of generators from one times generators from the other. This means that the product of two ideals with m and n generators respectively will require mn generators. Our primitive building blocks have only two generators, and we will first show how this property can be maintained under multiplication and division.

First we will modify our building blocks slightly. We define the support of a divisor as the set of places in the divisor with nonzero order. Our building blocks are of the form $[h(x,y,r), B(x)]$. We know that the zeros of h at places over roots of B coincide with the zeros of the desired divisor, but h can have zeros of multiplicity greater than one, which is compensated in the ideal by the fact that B has only zeros of order one. Our first step is to modify h so that it has only simple zeros. This can be done by adding a random integer multiple of B to h. For all but a small finite number of choices, this will produce an h with only simple zeros at places over the roots of B. To complete this step we only need a test to guarantee our random integer was not "unlucky". The norm of h is a polynomial in X whose degree is the sum of the order of the finite zeros of h. For any chosen integer j we can compute $N(x) = \text{Norm}(h + jB)$. We can split write $N = N_1 N_2$ where N_2 is the part of N that is relatively prime to B.

We wish to choose j so that N_1 has as small a degree as possible. When this is done, the degree of N_1 will be the same as the number of places where the integrand had residue r that is the same as the multiplicity of r in $R(x)$. Thus j is "lucky" as long as degree N_1 is the same as the multiplicity of r in $R(x)$.

Now we have divisor descriptions of the form $(h(x, y, r), A(x))$ for divisor D and satisfying the following properties:

1. $\text{order}(h) = \text{order}(D)$ at all places over the roots of the $A(x)$.
2. $\text{order}(D) = 0$ at all other places.
3. Both h and A are multiples of D except at infinity.

(h, A) can be viewed as a locally principal model of the divisor D since at any place either the $\text{order}(D) = \text{order}(h)$ or $\text{order}(D) = \text{order}(A)$. Given two such descriptions $(h_1(x, y), A_1(x))$ for divisor D_1 and $(h_2(x, y), A_2(x))$ for divisor D_2, where the support of A_1 equals the support of A_2 we claim the description (h_1, h_2, A_1, A_2) is of the same form for divisor $D_1 D_2$.

Quotients of divisors are slightly more complicated. h_1/h_2 satisfies property 1 but not necessarily property 3. It may have extraneous poles outside the support of D_1 and D_2. These can be removed by rationalizing its denominator and removing any factors that are relatively prime to A_1 or A_2. If we let h_3 be this normalization of h_1/h_2 whose finite poles are all above A_1 or A_2, then (h_3, A_1, A_2) satisfies all three properties for the divisor D_1/D_2.

Now that we have the basic building blocks, and the simple algorithms for multiplying and dividing them, we are ready to construct the divisors associated with each of the residue basis elements. Let b_1 be the ith basis element. Each root r_j of $R(x)$ can be represented uniquely as an integer linear combination of the basis elements $r_j = \sum c_{ij} b_i$. Also each r_j can be associated with the set of places P_j where the integrand had residue equal to r_j. The divisor D_i associated with basis element b_i is defined at $\prod P_j^{c_{ij}}$. We have seen how to represent P_j as the basic building block associated with r_j, and we now construct D_i as appropriate products and quotients of the P_j. This gives us a computable representation for the divisors associated with our candidate logands. In the next chapter we will see how to determine if there is some multiple of these divisors that are principal and thus whose generators will furnish us with the desired log terms.

5.4 Dealing with Branch Places

In the previous section we assumed that the branch places of the function field did not lie above any of the poles of the integrand. We now will show how to bring the integrand into this form. If the integrand involves only a single unnested radical, then we claim this assumption is guaranteed. In this case the defining polynomial for our function field is of the form $Y^n = F(x)$. The integrand can be written as $(\sum G_i(x) Y^i) dx$. As shown in the appendix, this is integrable if and only if each summand is. Thus we are reduced to integrands of the form $G(x) Y dx$ with possibly a different choice of Y. We can also assume without loss of generality that $F(x)$ is a polynomial whose roots all have multiplicity less than n. In this case the finite branch places occur precisely at the places above the roots of F. We claim that the integrand cannot have a nonzero residue at any branch place. If p is a branch place centered above a root x_0 of F with branch index r, then the order of $G(x)$ must be a multiple of r and the order of dx is precisely $r - 1$. But the order of Y is some positive integer j less than r. In order for the integrand to have a simple pole at p, we must have $k * r + r - 1 + j = -1$. But this equation implies j is divisible by r contrary to our assumption. Thus with simple radicals we never have nonzero residues at branch places.

In the more general situation we have to change our model of the function field to achieve this condition. Branch places occur when the line of projection from our defining curve down to the x-axis is tangent to the curve. By a different choice of independent variable we change tangency points and thus arrive at a different set of branch points. For each pole of our integrand, there are only a finite number of projection directions that are tangent to the curve. If we replace X by $X = mY$ for some random integer m then for almost all choices of m, the resulting function field will not have branch places above any of the poles of our integrand. This can be checked by computing the discriminant of an integral basis for the new presentation of the function field. The branch places all lie above roots of this discriminant so we wish to choose m such that the integrand has zero residue at all places above zeros of the discriminant. If we let $D(x)$ be this new discriminant and let $g = f D(x)$ where $f dx$ is the integrand, then equation (7) computes the residues of the integrand over all roots of $D(x)$. We then check that $R(Z)$ has no non-zero roots, i.e. is a pure monomial in Z.

Chapter 6
Principal Divisors and Points of Finite Order

In this chapter we will present a decision procedure for computing the log term associated with each element of a **Q**-basis for the residues. In chapter five we showed how to construct a reduced divisor that described the pole zero ratios for the desired log term. We need to be able to test whether there exists some multiple of this divisor that corresponds to an actual function in $\mathbf{K}'(x, y)$.

We will first present an algorithm for determining whether the given divisor is principal, i.e. is the divisor of a function in $\mathbf{K}'(x, y)$. If this algorithm succeeds then we are done, but if not we need to try multiples of this divisor. If each of these tests fails then we need to know when we can stop, i.e. when we are guaranteed that there is no multiple of this divisor that is principal.

6.1 Principal Divisors

We start with a divisor description of the form $(h(x, y), g(x))$. This describes a divisor whose order at places over ∞ is 0, and whose order at all other places is the minimum of the orders of $h(x, y)$ and $g(x)$. We wish to determine if the divisor is principal, i.e. if there is a single function that has exactly the same orders as this divisor at all places in our function field. In particular such a function is a multiple of our divisor, and we will base our construction on the following proposition and corollary:

Proposition 6.1. *Let $h_1, \ldots, h_k$ be functions and D be a divisor such that $\overset{\min}{i} \operatorname{ord}_p h_i = \operatorname{ord}_p D$ at all finite places. Then the ideal generated by h_i over the ring of integral functions coincides with the multiples of D except at infinity.*

Proof. This is a restatement of the isomorphism between fractional ideals in an algebraic function field and multiples of divisors ignoring places at infinity. [22] $\quad\square$

Using the proposition we see that our desired principal generator will be an integral linear combination of $h(x, y)$ and $g(x)$. Every such linear combination will be a multiple of D except at infinity but most of them will have poles at infinity. If we can find one that has no poles at infinity then it will be a multiple of D everywhere. Since the degree of D is zero, any such function must have orders exactly equal to those specified by D and is thus uniquely determined up to a scaling by a constant. This proves the following:

Corollary 6.2. *If D is a divisor of degree 0 with no places at infinity, and $h_1, \ldots, h_k$ are as in the proposition, then D is principal if and only if the ideal generated by the h_i has an element with no poles at infinity.*

In chapter 2 we showed how to compute an integral basis $[w_1, \ldots, w_n]$ for our function field. Any integral multiple of $g(x, y)$ can thus be written as $\sum a_i w_i g$ where $a_i \in K[x]$. Thus we can rewrite our ideal as the $K[x]$-module generated by $(w_1 g, \ldots, w_n g, w_1 h, \ldots, w_n h)$. We need to determine whether this module contains a function that has no poles at infinity. If we have a $K[x]$-module basis for our ideal that is normal at infinity then a linear combination of basis elements is integral at infinity if and only if each summand is. Since each summand is a polynomial times a basis element and a polynomial can never have a zero at infinity, if the ideal contains a function with no poles at infinity, one of the basis elements must have no poles at infinity. Thus after computing a normal basis for our ideal, we only have to check whether any of the basis elements have no poles at infinity.

Theorem 6.3. *If D is a divisor of degree zero with no places at infinity, then D is principal if and only if a normal basis for the ideal of multiples of D except at infinity has an element that is regular at infinity.*

6.2 Good Reduction

The algorithm in the previous section will enable us to determine whether any given divisor is principal, but we need to know whether there is some power of the divisor that is principal. This problem that was thought to be insoluble in the early 1900s will be solved by the technique of "good reductions". Function fields in one variable whose constant fields are finite fields have the property that any divisor of degree zero has some power that is principal. Thus if we start testing the successive powers of a divisor, we are guaranteed that this process will terminate. Repeated applications of the algorithm in the previous section will enable us to determine the minimum power of any divisor that is principal. Thus we need to be able to reduce the constant field of our function field to a finite field in such a way that we can calculate the order of a divisor from the order of its image.

If we assume the defining polynomial for our function field $f(x, y)$ is monic in y and has integer coefficients, then its coefficients can be reduced module p, a prime integer. As long as the reduced polynomial remains irreducible, it defines a function field over the finite field Z/p. If the genus of the reduced field is the same as the original then we will say that our original field has "good reduction" modulo p. It can be shown [22] for any particular defining polynomial, $f(x, y)$, $Q(x, y)$ has good reduction at all but finitely many primes.

More generally if we have a discrete valuation of our coefficient field, we can choose a defining polynomial for our function field that is monic and such that all of its coefficients are contained in the valuation ring. Then we can apply the natural homomorphism to the residue class field of the valuation. If the resulting polynomial remains absolutely irreducible and the genus of the reduced field is unchanged, then we say we have "good reduction" at that valuation. Again we will have good reduction at "almost all" valuations. In the rest of this chapter when we say reduction modulo p, we mean reduction by a discrete valuation of the constant field, which extends the natural p-adic valuation of the rationals.

If we have good reduction then there is a homomorphism between the group of divisors of $\mathbf{K}(x, y)$ and their images. Since principal divisors map to principal divisors, we in fact have a homomorphism of the quotient group of divisors module principal divisors. Let D be a divisor of $\mathbf{K}(x, y)$ that has finite order n, i.e. such that D^n is principal. If n is relatively prime to the characteristic of the reduced field then a consequence of good reduction is that the order of the image of D under the reduction is also n. This is a consequence of the key theorem observed by Risch [47] and Dwork and Baldassarri [4] that makes "good reduction" useful.

Theorem 6.4. *The homomorphism between divisor class groups under good reduction is an isomorphism when restricted to divisors whose orders are relatively prime to the characteristic of the reduced function field.*

Let p be the characteristic of the reduced field,

Corollary 6.5. *If the divisor D has order $p^k n$ where $\gcd(n, p) = 1$, then the reduction of D must have order $p^j n$ for some $j \leq k$.*

Proof. Let the order of the reduction be $p^j m$. Since reduction is a group homomorphism, we must have $m|n$ and $j \leq k$. Since D^{p^k} has order exactly n, its reduction must have order exactly n. But the order of its reduction is a divisor of m and thus $n|m$ and so finally we have $n = m$. □

We have shown that good reduction preserves the "prime to p" part of the orders of divisors, but we need a way to recover the entire order of the divisor. The solution to this problem is to use two different reductions whose residue fields have different characteristics. Let p and q be the characteristics of the residue class fields of two different valuations each of which gives good reduction. If the order of a divisor D is $np^j q^k$ where n is relatively prime to both p and q, reducing D by the first valuation enables one to determine n and k while the second valuation provides a determination of j, thus completing the computation of the order of D. Note that if the two values of n obtained from the two reductions do not match, the D could not have finite order, and we have a simple test that will frequently yield an early termination of our computation.

6.3 Algorithms for Divisor Reduction

The previous section discussed the properties of good reduction stemming from any discrete valuation of our constant field. In general our constant field will be a finitely generated extension of $\mathbf{Q}$. We can assume this is presented as a sequence of transcendental extensions followed by a single algebraic extension. If n is the transcendence degree of our constant field over $\mathbf{Q}$, then in order to reduce our constant field to a finite field we need to perform $n+2$ reductions as outlined in the previous section. The first n reductions will replace parameters by integer values. After these substitutions the polynomial defining our constant field may no longer be irreducible. The factors correspond to different choices of discrete valuations we can make. So we might as well choose the least degree factor. At this point we have reduced our constant field to a finite algebraic extension of $\mathbf{Q}$, and if all the reductions were good all divisors of finite order will map to divisors of exactly the same order (since the characteristic is still zero). Finally we need to choose two prime integers, p and q, and perform reductions modulo p and then modulo q. Instead of testing whether each reduction step is "good" it is enough that the reduction from the original coefficient field to the finite fields are "good". This guarantees that all the intermediate reductions were safe.

We now need algorithms for reducing various objects of interest and guaranteeing that our reductions are "good". For simplicity we will choose all our reductions so that the denominators and leading coefficients of all polynomials of interest do not vanish, i.e. the defining polynomial for our function field and the components of our divisor models and integral basis. Again this will be true for "almost all" reductions. We can use the integral basis algorithm in Chapter 2 to test both that the defining polynomial remains absolutely irreducible and that the genus is unchanged. We merely need verify that our original integral basis for our function field is still an integral basis, and that there are no non-constant functions with no poles. The second condition guarantees absolute irreducibility and the first implies that the genus remains unchanged.

Given that our function field reduces well, we now ask whether our divisor descriptors reduce to locally principal models for the reduction of the desired divisor. Eichler [22] shows that the image of a principal divisor is the divisor associated with the reduction of the generator as long as the generator has a well-defined non-zero reduction. He also shows that this can be accomplished by a proper scaling of the generator. In our situation we have a divisor descriptor of the form $(g(x,y), p(x))$ where g provides a locally principal model at places over the roots of p and the divisor has order zero everywhere else. Thus p defines the projection of the support of the divisor. We assume that g and p have been appropriately scaled so they have well-defined non-zero reductions. We first require that our reduced divisor also have zero order at places over infinity. Since the reduction of p defines the projections of the support of the reduced divisor, this is equivalent to the non-vanishing of the leading coefficient of p. The remaining problem comes from the zeros and poles of g that did not lie above roots of p. It is possible that they may reduce to a place that is above a root of p. In order to ensure that the reduction of our divisor model does in fact reduce to a model of the desired divisor, we must guarantee this doesn't happen. We compute another polynomial $q(x)$ which is zero precisely at the finite zeros and poles of g that do not lie above roots of p. The support of the finite poles of g can be found by expressing g in terms of an integral basis and computing a common denominator $d(x)$ of the rational function coefficients. If we multiply g by the factors of d which are relatively prime to p, then we obtain a new g which has no extraneous finite poles, i.e. all of its finite poles lie above roots of p. We can similarly find the projections of the zeros of g by denominator of the representation of $1/g$ in terms of our integral basis. By definition p and q have no common zeros. Our divisor model reduces well if the reductions of p and q remain relatively prime. If this condition fails to hold, we can either choose a different reduction or choose a different g as a model for the desired divisor. The test that our divisor description reduces well is then simply that the leading coefficient of p doesn't vanish and that p and q remain relatively prime.

The condition for a divisor to reduce well can be re-expressed using the standard terminology from algebraic geometry [26]. A locally principal or Cartier divisor is usually specified by giving a collection of open sets U_i and function f_i such that the open sets cover the Riemann surface for our function field, and f_i/f_j has no zeros or poles on $U_i \cap U_j$. In our case it is convenient to define two closed sets V_1 and V_2 which are the complements of U_1 and U_2 respectively. V_1 is the set of places over the roots of q and at infinity. V_2 is the set of places over the roots of p. We define $f_1 = g$ and $f_2 = 1$. The condition that U_1 and U_2 are an open cover is equivalent to the statement that closed sets V_1 and V_2 have no points in common. By construction g has no zeros or poles on $U_1 \cap U_2$, i.e. the complement of $V_1 \cup V_2$. Thus our initial divisor description yields a well-defined Cartier divisor. We now check whether this remains true after "reduction modulo p." Again by construction $g = f_1/f_2$ has no zeros or poles except at places above roots of p or q or infinity. The condition that V_1 and V_2 have no points in common translates to p and q remaining relatively prime and no roots of p moving to infinity, i.e. the leading coefficient of p remaining non-zero. Thus we again arrive at the same conditions for good reduction of our divisor description.

Chapter 7
Conclusions and Future Research

This thesis has presented an algorithm for integrating algebraic functions that is "rational" in the sense that no algebraic extensions are made beyond those required to express the answer. This yields a more efficient and direct solution to the problem than the previous approach presented by Davenport and Risch. We have stressed the analogies between algebraic function integration and the well-known techniques for rational function integration. The main source of complications comes from the lack of unique factorization in an algebraic function field. We have restored unique factorization by working with ideals. The gcd of two elements is represented by the ideal they generate. After performing the necessary arithmetic on these ideals, we determine whether some power of the resulting ideal is principal giving rise to a logarithmic term in the answer. This ability to give an explicit presentation for an ideal or the associated divisor allows us to compute its order exactly instead of merely finding an upper bound. We perform more work in algebraic function fields over finite fields in order to perform less work over our original constant field where operations are much more costly.

Another central theme in our approach is the iterative reduction of singularities of the integrand. By developing a Hermite-like reduction we are able to reduce non-integrable problems to a simpler form. In fact we are able to prove as Hermite does that one can always reduce the integral of an algebraic function to one which has only simple poles at finite places.

The fundamental construction that we use is the integral basis. This is used to determine the actual singularities of the integrand, to find principal generators for ideals, and to test our function field for "good reduction." Two other useful applications of an integral basis are computing the genus of our function field and verifying whether our defining polynomial is absolutely irreducible. The existence of this basis allows us to reduce many problems to elementary row operations on matrices of rational functions. If a more efficient algorithm can be found for computing integral bases, then our entire algorithm will be similarly improved. The integral basis provides us with a global affine non-singular model for our function field and replaces the Puiseux series and Coates' algorithm used by Davenport. A "rational" algorithm to find the multiples of a divisor would yield a similar improvement in Davenport's implementation.

The natural extension of this work would involve handling mixed towers of algebraic, exponential and logarithmic extension, i.e. general elementary function fields. In the appendix I have given a partial algorithm for the case where you have a purely transcendental tower followed by an unnested radical extension. As Davenport observes in [17, appendix 4], Risch's original transcendental algorithm holds over any differential function field in which one can compute integrals and solve first order linear differential equations. Once you have computed bounds for the orders of the poles in the differential equations, both the algorithms presented here and those of Davenport would suffice for finding the solutions. Thus we can now also perform integrations over any elementary transcendental extension of an algebraic function field.

Extending this work for general elementary function fields may be simpler if first performed using local power series expansions as advocated by Risch in [46]. Once this local approach is well understood, then perhaps a global "rational" approach can be developed. The difficulties seem to involve dealing with places at infinity since one can no longer simply transform infinity to a finite place without creating a non-elementary function field tower. Also in attempting to solve a first order linear differential equation using "rational" techniques, one seems to generate a system of coupled differential equations over the function field one level down. This "reduction" seems to lead to an apparently more difficult problem. The development of algorithms for finding solutions to systems of first order linear differential equations over a given function field could yield a very elegant solution to the general problem of integration in "finite terms" [18].

Appendix A
Integration of Simple Radical Extensions

Although Risch has presented an outline of an algorithm for integrating mixed towers of algebraic and transcendental elementary functions in [46] and [47], unfortunately his algorithms require considerably more complex machinery than his earlier ones for purely transcendental functions [45]. Moses' implementation of the transcendental case [41] demonstrated its practicality, whereas there are as yet no implementations for Risch's more general algorithm [46].

This appendix will show how a combination of Risch's earlier techniques and the algorithms presented in this thesis can be generalized to begin to handle mixtures of transcendentals and unnested radicals. While this may seem a severe restriction, perusing an integral table such as [8] will show that fewer than 1% of the problems are excluded.

We will assume the reader is familiar with the terminology and results of [45]. We will use the term field to mean a differential field of characteristic 0. If F is a field with y algebraic over F of degree n such that $y^n \in F$, then we will call $F(y)$ a simple radical extension of F. Any element of $F(y)$ can be written as a polynomial in y of degree $n-1$ with coefficients in F.

A.1 Structure Theorems

Our first result will be a refining of Liouville's structure theorem for simple radical extensions. Let F be any differential field with K its field of constants and y radical over F. Risch's Strong Liouville Theorem states that $g \in F(y)$ is integrable if and only if there is a $v \in F(y)$, $c_i \in K(d)$, and w_i in $F(y, d)$ with d algebraic over K such that:

$$f = v' + \sum c_i \frac{w_i'}{w_i}. \tag{1}$$

We will call v the rational part of the integral and the rest the transcendental part of the integral. If we assume that F contains ω a primitive n^{th} root of unity then there is a unique differential automorphism of $F(y)$ over F such that $\sigma(y) = \omega y$. Define the operator

$$T_i = \frac{1}{n} \sum_{j=0}^{n-1} \frac{\sigma^j}{\omega^{ij}}.$$

Note that $T_i(y^j) = y^j$ if $i = j$ else 0. Thus letting $g = \sum g_i y^i$ and $v = \sum v_i y^i$, we have $T_i(g) = g_i y^i$. Also noting that T_i commutes with the derivations, $T_i(v') = (v_i y^i)'$. Since T_i sends the transcendental part of the integral to a sum of the same form, by successively applying T_j to equation (1) for $0 \le j \le n - 1$ we deduce that (1) g is integrable if and only if each $g_i y^i$ is and (2) the rational part of $\int g_i y^i$ is $v_i y^i$.

Now let $\mathbf{G}$ be a compositum of simple radical extensions, i.e. $\mathbf{G} = \mathbf{F}(y_1, \ldots, y_k)$ where $y_i^{e_i} \in \mathbf{F}$ and $[\mathbf{G} : \mathbf{F}] = \prod e_i$. Any $g \in \mathbf{G}$ can be written as a polynomial in the y_i's with coefficients in $\mathbf{F}$ where the degree of y_i in g is less than e_i. Then by repeating the previous argument for each y_i, one can show g is integrable if and only if each term is integrable. The subfield of $\mathbf{G}$ generated by a single such term over $\mathbf{F}$ is differentially isomorphic to a simple radical extension of $\mathbf{F}$ of degree at most the least common multiple of the e_i's. Thus integrals over compositums of simple radical extensions can be reduced to integrals over simple extensions frequently of much lower degree.

A.2 A Generalized Risch Algorithm

Let $\mathbf{F}$ be an arbitrary differential field and $\mathbf{E} = \mathbf{F}(\theta)$ where θ is transcendental over $\mathbf{F}$ and $\mathbf{F}$ and $\mathbf{E}$ have the same constant subfield. We will additionally assume exactly one of the following is true:

1. $\theta' = 1$.
2. $\theta' = v'/v$ for some $v \in \mathbf{F}$, i.e. $\theta = \log(v)$.
3. $\theta'/\theta = v'$ for some $v \in \mathbf{F}$, i.e. $\theta = \exp(v)$.

We will be making use of the fact that $\mathbf{F}(\theta)$ is a constructive Euclidean domain. Thus we can compute gcd's and hence square-free decompositions. We are interested in the case where $\mathbf{G} = \mathbf{E}(y)$ is a simple radical extension of degree n. Additionally we will require that y must depend on θ, i.e. y^n is not in $\mathbf{F}$. By the previous section we are reduced to considering integrands of the form Sy^i with $S \in \mathbf{E}$. We will find it convenient to rewrite this as R/y^{n-i} where $R = Sy^n$. By changing our choice of y we can assume an integrand of the form R/y. (Note this may involve changing our value of n.) Without loss of generality we can finally assume $y^n = P(\theta) \in \mathbf{F}[\theta]$ where P has no factors of multiplicity $\ge n$.

Let $R = A/B$ with $A, B \in \mathbf{F}[\theta]$ and B monic. After finding a square-free basis for P and B and performing a partial-fraction decomposition on A/B we can split our integrands into three cases $C/(V^k y)$, $C/(W^k y)$, and C/y where V is relatively prime to P but W is a square-free factor of P. Unlike the previous section, integrability of R/y does not guarantee integrability of each term in the partial fraction decomposition. However after the splitting we will be able to apply a variety of reduction formulae to these cases. There will be a strong similarity between our algorithms for reducing the integrands and Hermite's algorithm for rational function integration.

Since the case $\theta = \exp(v)$ has additional complications, we will treat it later. For the remaining cases a polynomial Q is square-free if and only if $\gcd(Q, Q') = 1$.

The first problem we encounter is that y' introduces new denominators, so we will choose a polynomial f such that $(f/y)' = g/y$ for some $g \in \mathbf{F}[\theta]$. If fact we will need an f of least possible degree. If $P = d \prod_i P_i^{e_i}$ is our square-free decomposition of P into monic factors with $d \in \mathbf{F}$, then define $f = \prod P_i$.

$$(f/y)' = f/y \left(\sum (1 - e_i/n) P_i'/P_i - d'/nd \right) = g/y.$$

Clearly $g \in \mathbf{F}[\theta]$

A.2.1 Case 1: $C/(V^k y)$ where $\gcd(V, f) = 1$

We want to find a function whose derivative when subtracted from the integrand will decrease k and not introduce any new denominators. We want a polynomial B such that

$$\left(\frac{Bf}{V^{k-1}y} \right)' - \frac{C}{V^k y} = \frac{D}{V^{k-1}y}$$

for some polynomial D. Since

$$\left(\frac{Bf}{V^{k-1}y} \right)' = \frac{(1-k)V'Bf}{V^k y} + \frac{B'f + Bg}{V^{k-1}y}.$$

Thus we must choose B such that $(1 - k)V'fB \equiv C \pmod{V}$. Since $\gcd(fV', V) = 1$ we can find B as long as $k > 1$. Thus we can continue this reduction process until $k = 1$.

A.2.2 Case 2: $C/(W^k y)$ where $W = P_j$

Since W divides f we must start with an apparent denominator of the form $W^k y$. Letting $h = f/W$ we have

$$\left(\frac{Bf}{W^k y} \right)' = \frac{Bg - kW'h}{W^k y} + \frac{B'h}{W^{k-1}y}.$$

Thus we want to choose B such that $Bg - kW'h \equiv C \pmod{W}$. $W = P_j$ implies $g \equiv (1 - e_j/n)W'h \pmod{W}$. Thus we have $B(1 - k - e_j/n)W'h \equiv C \pmod{W}$. Since $W'h$ is relatively prime to W and $e_j < n$, this equation is solvable for any k. Thus by repeated applications of this reduction step we can eliminate all factors of f from the denominators of our integrands.

A.2.3 Case 3: C/y

Here we will try to find a B such that the reduced integrand has lower degree. Let $B = b_j \theta^j$, with $b_j \in \mathbf{F}$.

$$(b_j \theta^j f/y)' = b_j' \theta^j f/y + j b_j \theta^{j-1} \theta' f/y + b_j \theta^j g/y. \tag{2}$$

Letting $m = \deg f$ we have two subcases depending on whether d is constant or not. If $d' = 0$ then $\mathrm{degree}(g) = m - 1$ else $\mathrm{degree}(g) = m$. Let $C = \sum c_i \theta^i$.

We will first assume that $d' = 0$. If $b'_j = 0$ then equation (2) has degree $j + m - 1$ else degree $j + m$. Thus equating formally highest degree terms we obtain:

$$c_{j+m} = (j+1)b_{j+1}\theta' + b_{j+1}\text{lcf}(g) + b'_j \tag{3}$$

where b_{j+1} is a constant. $\text{lcf}(g)$ is the leading coefficient of g

$$\text{lcf}(g) = \sum(1 - e_i/n)\text{lcf}(P'_i).$$

If $P_i = \theta^k + a_i\theta^{k-1} + \cdots$ then $\text{lcf}(P'_i) = k\theta' + a'_i$. If $\theta' = 1$ then equation (3) reduces to:

$$c_{j+m} = (j + 1 + \sum \deg(P_i)(1 - e_i/n))b_{j+1}. \tag{4}$$

If $j + 1 \geq 0$ then the coefficient of b_{j+1} will always be nonzero. Thus we can always reduce C until $\deg(C) < m - 1$.

If $\theta = \log(v)$ then equation (3) takes the form

$$c_{j+m} = b_{j+1}((j+1)\frac{v'}{v} + \sum(1 - e_i/n)(\deg(P_i)\frac{v'}{v} + a'_i)) + b'_j. \tag{5}$$

The coefficient of $b_{j+1}v'/v$ in equation (5) is precisely the coefficient of b_{j+1} in equation (4) and is therefore nonzero if $j + 1 \geq 0$. If the original problem is integrable then c_{j+m} must be integrable. Just as in [45] b_{j+1} is uniquely determined since θ is a monomial over $\mathbf{F}$ while b_j is determined up to an additive constant. In this case we can reduce C until $\deg(C) < m - 1$ only if either the original problem is integrable or at least the necessary set of coefficients are integrable.

We will now treat the subcase where d' is nonzero. Note that we must have $\theta = \log(v)$ for this to hold. Here we must first assume that there is no constant s such that sd has an nth root in $\mathbf{F}$. If there is such an s then we can pick a new generator $y(sd)^{-1/n}$ for G that puts us back in the previous subcase. Otherwise we have $\deg(g) = \deg(f)$ and after equating leading terms we have $c_{k+m} = b'_k - \frac{d'}{nd}b_k$. Our assumption about d guarantees that this equation can have at most one solution in $\mathbf{F}$. Thus we need to be able to solve first order linear differential equations for a solution in $\mathbf{F}$. If $\mathbf{F}$ is a tower of monomial extensions then [45] shows how to do this. If all the equations we set up have solutions we will reduce C so that $\deg(C) < m$.

A.2.4 Case 4: $\theta = \exp(v)$

The distinguishing characteristic of the exponential function is that it is a factor of its derivative. Thus we can no longer claim that a square-free polynomial must be relatively prime to its derivative. It will only be necessary to treat factors of the form θ^k specially. We begin by rewriting the square-free decomposition of P. $P = d\theta^j \prod P_i^{e_i}$ where no P_i is divisible by θ. We will again define $f = \prod P_i$, noting that now f is not divisible by θ. We next verify that $(f/y)'$ still is of the form g/y for some $g \in \mathbf{F}[\theta]$

$$\left(\frac{f}{y}\right)' = \frac{f}{y}\left(\sum(1 - e_i/n)\frac{P_i'}{P_i} - j/nv' - \frac{d'}{nd}\right) = \frac{g}{y}.$$

After performing a partial-fraction decomposition of the integrand, we can deal with all denominators other than θ just as in the previous cases. Thus we will now assume an integrand of the form $C/(\theta^k y)$, and we again write $C = \sum c_i \theta^i$. We are again trying to decrease k and letting B be an arbitrary polynomial we compute:

$$\left(\frac{Bf}{\theta^k y}\right)' = \frac{B'f + Bg - kv'Bf}{\theta^k y}.$$

Requiring the numerator to be congruent to C modulo θ is the same as equating constant terms

$$c_0 = b_0'f_0 + b_0 g_0 - kv'b_0 f_0.$$

f not divisible by θ implies f_0 is nonzero, thus we can divide through by f_0. Again θ a monomial will force this equation to have at most one solution in $\mathbf{F}$ for $k \geq 0$. As long as the equation continues to have solutions, we can reduce k to 0.

Finally we must deal with integrands of the form C/y in the exponential case. Here $\deg(g) = \deg(f)$ always, and we again assume a solution of the form $\frac{Bf}{y}$ and equate leading terms

$$c_{k+m} = b_k' + kv'b_k + b_k \mathrm{lcf}(g), \tag{6}$$

$$\mathrm{lcf}(g) = \frac{-j}{n}v' - \frac{d'}{nd} + \sum(1 - e_i/n)\deg(P_i)v'.$$

Equation (6) will have at most one solution as long as the coefficients of v' is nonzero. This coefficient is $\frac{k-j}{n} + \sum(1 - e_i/n)\deg(P_i)$. Since the third term is always positive, $k > 0$ or if $j = 0$ then $k \geq 0$ is sufficient to guarantee the v' is present in equation (6).

A.3 Summary and Conclusions

The reduction formulae in section A.2 have enabled us to find the rational part of our integral if it exists. If the original problem was integrable, all the remaining integrands must generate the transcendental portion of the integral. Note that cases 1 and 2 will always reduce any integrand whether it be integrable or not. In particular, for the case $\theta' = 1$ we see that any integral can be reduced to $\int A/(By)$ where B is square-free and $\deg(A) - \deg(B) < \deg(f) - 1$.

We have shown that the question of computing integrals in $\mathbf{F}(x, y)$ can be reduced to the problems of computing integrals in $\mathbf{F}$ and that of solving first order linear differential equations over $\mathbf{F}$. If $\mathbf{F}$ was constructed as a tower of monomial extensions, then [45] shows that these problems are solvable.

We claim that the algorithms presented here form a natural extension to those presented in [45]. By restricting ourselves to this special but very important case, we are able to generate the integral using nothing more than simple polynomial arithmetic. Although our formulas are somewhat complicated, we require no additional machinery than those necessary in Risch's original approach. [45] We have not worked out the details of the logarithmic part of the integral, but the techniques introduced in chapters 5 and 6 of this thesis should be generalizable to deal with this situation. The major problems involve dealing with poles at infinity which were finessed by a change of variables in the purely algebraic case. The restriction to unnested radicals guarantees no simple poles at branch places as in the purely algebraic case. Formulae for the residue at infinity similar to equation (7) of chapter 5 are needed. The fact that the principal parts of a function now only determines it up to a function in one fewer variables instead of up to a constant could cause some additional difficulty. We have presented what we hope are very usable practical algorithms, and we intend to implement them in the near future.

References

[1] Abhyankar, Sheeram, Historical Ramblings in Algebraic Geometry and Related Algebra, *Amer. Math. Monthly*, Vol. 83, 6 (1976) pp. 409–448

[2] Atiyah, M.F., and MacDonald, I.G., *Introduction to Commutative Algebra*, Addison Wesley Pub. Co., Reading, Massachusetts, (1969)

[3] Artin, Emil, *Algebraic Numbers and Algebraic Functions*, Gordon and Breach, N.Y., (1967)

[4] Baldassarri, F. and Dwork, B., On second order linear differential equations with algebraic solutions, *American Journal of Mathematics*, Vol. 101, 1 (1979) pp. 42–76.

[5] Barsotti, I., Varieta Abeliane su Corpi p-adici; Parte Prima, *Symposia Mathematica*, Vol. 1, (1968) pp. 109–173

[6] Berry, T.G., On Coates Algorithm, *Sigsam Bulletin*, Vol. 17, 2 (1983) pp. 12–17.

[7] Bliss, G.A., *Algebraic Functions*, Dover, (1966), first published as AMS Colloquium vol. XVI (1933)

[8] Bois, G. Petit, *Tables of Indefinite Integrals*, Dover, N.Y., (1961)

[9] Cassels, J.W.S., Diophantine Equations with Special Reference to Elliptic Curves, *Journal of the London Math. Society*, vol. 41, (1966) pp. 193–291

[10] Chevalley, Claude, *Algebraic Functions of One Variable*, Math. Surveys Number VI, American Math. Society, N.Y., (1951)

[11] Chou, Tsu-Wu Joseph, *Algorithms for the Solution of Systems of Linear Diophantine Equations*, Ph.D. thesis, University of Wisconsin, (1979)

[12] Chow, Wei-Liang, On the Principle of Degeneration in Algebraic Geometry, *Annals of Mathematics*, Vol. 66, No. 1, pp. 70–79, (1957)

[13] Chow, W.-L., Lang, S., On the Birational Equivalence of Curves under Specialization, *American Journal of Math.*, Vol. 79, pp. 649–652, (1957)

[14] Coates, J., Construction of rational functions on a curve, *Proc. Camb. Phil. Soc.*, Vol. 68, pp. 105–123, (1970)

[15] Cohn, Harvey, *A Classical Invitation to Algebraic Numbers and Class Fields*, Springer-Verlag, N.Y., (1979)

[16] Davenport, James, *On the Integration of Algebraic Functions*, Lecture Notes in Computer Science No 102, Springer-Verlag, N.Y., (1981)

[17] Davenport, James, *Integration Formelle*, (1983)

[18] Davenport, J. and Singer, M., Private communication, (1984)

[19] Deuring, Max, Reduktion algebraischer Funktionenkorper nach Primdivisoren des Konstantenkorpers, *Math. Zeit.* 47, pp. 643–654 (1941)

[20] Dieudonne, J., The Historical Development of Algebraic Geometry, *American Mathematical Monthly*, Vol. 79, 8 pp. 827–866 (1972)

[21] Duval, Dominique, *Une methode geometrique de factorization des polynomes en deux indetermines*, Institute Fourier, Grenoble, (1983).

[22] Eichler, Martin, *Introduction to the Theory of Algebraic Numbers and Functions*, Academic Press, N.Y., (1966)

[23] Ford, David James, *On the Computation of the Maximal Order in a Dedekind Domain*, Ph.D. thesis, Ohio State University, Depart. of Mathematics, (1978)

[24] Fulton, William, Hurwitz schemes and irreducibility of moduli of algebraic curves, *Annals of Mathematics*, vol 90, pp. 542–575 (1969)

[25] Hardy, G.H., *The Integration of Functions of a single variable (2nd ed.), Cambridge Tract 2*, Cambridge U. Press, (1916)

[26] Hartshorne, Robin, *Algebraic Geometry*, Springer-Verlag, (1977)

[27] Hasse, Helmut, *Number Theory*, Springer-Verlag (1980)

[28] Hermite, C., Sur l'intégration des fractions rationnelles, *Annales Scientifiques de l'É.N.S.*, 2 Ser., 1(1872) pp. 215–218

[29] Herstein, I.N., *Topics in Algebra* Blaisdell, Waltham, Mass., (1964)

[30] Horowitz, E., *Algorithms for Symbolic Computation of Rational Functions*, Ph.D. Thesis, U. of Wisconsin, (1970)

[31] Jacobson, Nathan, *Basic Algebra II*, W.H. Freeman and Co., San Francisco, (1980)

[32] Katz, Nicholas, Galois Properties of Torsion Points on Abelian Varieties, *Inventiones Mathematicae*, vol 62, pp. 481–502 (1981)

[33] Lang, Serge, *Algebra* Addison-Wesley, Reading, Mass., (1971)

[34] Lang, Serge, *Algebraic Number Theory*, Addison-Wesley, Reading Mass., (1970)

[35] Lang, Serge, *Introduction to Algebraic Geometry*, Addison-Wesley, Reading, Mass., (1958)

[36] Lang, S. and Tate, J., Principal Homogeneous Spaces over Abelian Varieties, *American Journal of Mathematics*, pp. 659–684 (1960)

[37] Liouville, J., Premier Memoire sur la Determination des Integrales dont la Valeur est Algebrique, *Jounal de l'Ecole Polytechnique*, Vol. 22, (1833) pp. 124–148

[38] Lichtenbaum, Stephen, Curves over Discrete Valuation Rings, *American Journal of Mathematics*, Vol. 90, pp. 380–405, (1968)

[39] Mack, D., *On Rational Integration*, Computer Science Dept. Utah Univ., UCP-38, 1975

[40] Matsuda, Michihiko, *First Order Algebraic Differential Equations*, Lecture Notes in Mathematics, No. 804, Springer-Verlag, (1980)

[41] Moses, Joel, Symbolic Integration, The Stormy Decade, *Communications of the ACM* Vol. 14, no. 8, pp. 548–560, (1971)

[42] Mumford, David, *Abelian Varieties*, Oxford University Press, (1970)

[43] Nering, Evar, Reduction of an Algebraic Function Field Modulo a Prime in the Constant Field, *Annals of Mathematics*, Vol 67, No. 3, (1958) pp. 590–606

[44] Newman, Morris, *Integral Matrices*, Academic Press, New York, (1972)

[45] Risch, R.H., The Problem of Integration in Finite Terms, *Trans. AMS*, Vol 139, (1969) pp. 167–189

[46] Risch, R.H., *On the Integration of Elementary Functions which are built up using Algebraic Operations*, Rep. SP-2801/002/00, System Development Corp., Santa Monica, Calif., (1968)

[47] Risch, R.H., The Solution of the Problem of Integration in Finite Terms, *Bulletin A.M.S*, Vol. 76, (1970) pp. 605–608

[48] Ritt, J.R., *Integration in Finite Terms*, Columbia U. Press, N.Y., (1948)

[49] Rosenlicht, Maxwell, Differential Extensions Fields of Exponential Types, *Pacific Journal of Mathematics*, Vol 57, 1 (1975)

[50] Rosenlicht, Maxwell, Integration in Finite Terms, *American Mathematical Monthly*, Vol 79, 9 pp. 963–972 (1972)

[51] Serre, J.-P., Tate, J., Good reduction of abelian varieties, *Annals of Math.*, pp. 492–517, (1968)

[52] Schmidt, Wolfgang M., *Equations over Finite Fields an Elementary Approach*, Lecture Notes in Mathematics, No. 536, Springer-Verlag, (1976)

[53] Shafarevich, I. R., *Basic Algebraic Geometry*, Die Grundlehren der mathematichen Wissenschaften, Band 213, Springer-Verlag, (1974)

[54] Shimura, G., Taniyama, Y., *Complex Multiplication of Abelian Varieties and its Applications to Number Theory*, Mathematical Society of Japan, (1961)

[55] Trager, B.M., *Algorithms for Manipulating Algebraic Functions*, S.M. thesis M.I.T., (1976)

[56] Trager, B.M., Algebraic Factoring and Rational Function Integration, *Proc. 1976 ACM Symposium on Symbolic and Algebraic Manipulation*, pp. 219–226, (1976)

[57] van der Waerden, B.L., *Modern Algebra*, vol 1, tr. Fred Blum, Frederick Ungar Publishing Co., New York, (1953)

[58] Walker, Robert J. *Algebraic Curves*, Dover, New York, (1962)

[59] Weil, Andre, *Courbes algebriques et varietes abeliennes*, Hermann, Paris, (1971), first published (1948) in *Actualites Scientifiques et Industrielles*, nos. 1041 and 1064.

[60] Weil, Andre, *Foundations of Algebraic Geometry*, Amer. Math. Society, N.Y., (1946)

[61] Zariski, Oscar, and Pierre Samuel, *Commutative Algebra*, Vol 1&2, D. Van Nostrand Co., Princeton, N.J., (1958)

[62] Zassenhaus, Hans, On Hensel Factorization II, *Symposia Mathematica* **15**, (1975) pp. 499–513

[63] Zassenhaus, Hans, On the Second Round of the Maximal Order Program, *Applications of Number Theory to Numerical Analysis*, Academic Press, New York, (1972) pp. 389–431

Comments on
Integration of Algebraic Functions

Barry M. Trager

1 Overview

We survey some more recent work on some of results presented in the thesis on Integration of Algebraic Functions [33]. The thesis was intended to provide practical algorithms for performing the integration of algebraic functions after the fundamental theoretical results of Risch in [29] and [30][1] proving that the problem was solvable. At roughly the same time, James Davenport developed and implemented his approach to the problem [16, 15]. Following the local approach suggested by Risch, Davenport developed implementations of Puiseux expansions and Coates' algorithm to construct multiples of a divisor [11]. Since Puiseux expansions required the use of algebraic numbers which are not required to express the final result, the thesis developed rational algorithms analogous to those used for the integration of rational functions [32]. Dominique Duval subsequently developed her elegant theory of Rational Puiseux Expansions [19] which reduced the use of algebraic numbers and presented conjugate expansions together. The thesis instead used an integral basis to express the integrand in a way which made its poles explicit. The integral basis algorithm presented was based on one originally developed by Zassenhaus and Ford for algebraic numbers [38, 21]. Mark van Hoeij later showed that with a clever Puiseux expansion implementation, one could use a variant of Coates' algorithm to give a fast algorithm

---

[1] These papers have been combined with other papers of Risch and now appear as *On the Integration of Elementary Functions which are Built Up Using Algebraic Operations* in this volume.

© Springer Nature Switzerland AG 2022
C. G. Raab und M. F. Singer (Hrsg.), *Integration in Finite Terms:
Fundamental Sources*, Texts & Monographs in Symbolic Computation,
https://doi.org/10.1007/978-3-030-98767-1_8

for computing an integral basis [35]. Subsequently Manuel Bronstein showed how the algorithms in the thesis for algebraic functions could be generalized to provide a complete decision procedure for the integration of elementary functions [4]. A very nice summary of the algebraic integration algorithm from the thesis is given by Schultz [31] along with additional techniques for improving the implementation.

2 Integral Closure

The elements of an integral basis generate the integral closure (or normalization) of the ring of functions on a plane curve. The primary sub-algorithms for the Ford and Zassenhaus construction compute the radicals of the discriminant ideal and the idealizer of a given ideal, i.e. the largest subring of a quotient field which sends a given ideal to itself. It was later noticed that the basic idea behind this algorithm could be generalized to compute the normalization of noetherian domains of any dimension, given the ability to compute ideal quotients, radicals, and an element in the conductor of the normalization [23]. At the same time Theo de Jong presented essentially the same algorithm based on some theorems of Grauert and Remmert [17] and a similar approach was also developed by Carlo Traverso [34]. Subsequently there has been a lot of effort to improve this general normalization algorithm and it has been implemented in the SINGULAR computer algebra system [24, 3].

3 Absolute Factorization

The next major algorithm presented in the thesis was absolute factorization, i.e. factoring over the algebraic closure of the coefficient field. In the thesis we took advantage of the property that an irreducible polynomial is absolutely irreducible if it remains irreducible over the exact coefficient field, i.e. the relative algebraic closure of the coefficient field in the function field defined by the polynomial. Since the field generated by the coordinates of any smooth point must contain the exact coefficient field, this gives a sufficient extension field over which to factor, as also observed by Traverso in [34]. Kaltofen in [27] shows that $2 \deg_x(f) \deg_y(f)$ terms in the series expansion of a solution $y = a_0 + a_1 x + \cdots$ at a smooth point can be used to decide the absolute irreducibility of f. It is enough to check whether the powers of this truncated series up to degree $\deg_y(f) - 1$ are linearly dependent, which can be done using linear algebra. In this way he produces a linear system which has a solution if and only if the polynomial is not absolutely irreducible. He also observes that no factorization is required to describe the algebraic number corresponding to the point of expansion. The entire algorithm works over the ring of polynomials modulo a square-free polynomial. This is similar in spirit to the dynamic evaluation "D5" principle of Della Dora et al. for efficiently working in algebraic extensions [18]. Another application of Kaltofen's construction is that it can be used to produce a bound which guarantees that for all primes p exceeding this bound, an absolutely irreducible polynomial will remain absolutely irreducible when reduced modulo p. A completely different approach to absolute factorization was proposed by Duval in [20]. She computes a basis for the exact coefficient field as the set of functions which are integral everywhere, including at infinity. She observes that the number of absolutely irreducible factors is the same as the dimension of the vector space $L(0)$ of multiples of the divisor (0), i.e. constants are functions with no poles anywhere. She uses an integral basis algorithm which describes functions with no finite poles and then performs a normalization at infinity to compute this basis of constants with no poles anywhere. To actually find an absolutely irreducible factor, letting $\wp$ be a smooth point on the curve, she computes a basis for the subvector space of functions which have no poles and a zero at $\wp$, $L(-\wp)$. This corresponds to functions which are zero on the unique component containing $\wp$. By computing the gcd of these generators and the original defining polynomial, she recovers the absolutely irreducible factor which vanishes on that component. In characteristic zero there is a more direct method to obtain a basis for the exact constant field since these are just the set of elements of the function field whose derivative with respect to x is zero. Any nontrivial constant (i.e. not an element of the coefficient field) will have a nontrivial gcd with f, and thus can be used to factor f. This approach is presented in [13] along with other properties which can be computed using the linear differential operator satisfied by f.

4 Hermite Reduction and Creative Telescoping

By Liouville's theorem, when the integral of an algebraic function can be expressed in closed form it is the sum of an algebraic function (from the same function field) and a linear combination of logs of such algebraic functions. To find the algebraic part of the integral, the thesis shows how to generalize Hermite's method for rational function integrals [25] to algebraic functions. After a rational change of coordinates we can assume our integrand has no poles or branch places at ∞. Taking care of the complication that the derivatives of our integral basis elements will in general have denominators, we show how to set up congruence equations to systematically reduce the denominator of the integrand until it has only simple poles. If the resulting integrand now has poles at ∞ then it is not integrable, otherwise we will attempt to express the remaining integral as a linear combination of log terms. The Hermite reduction procedure presented in the thesis requires that the integral be expressed using an integral basis. Bronstein [5] showed that Hermite reduction can be performed in a "lazy" manner avoiding the integral basis computation.

This Hermite reduction for algebraic functions has subsequently been shown to be useful in performing creative telescoping, which is a technique which can be used for constructing linear differential equations which are satisfied by parametric definite integrals. Given a function $f(x,t)$, creative telescoping is used to find a linear differential operator $L = \sum_i a_i(t)D_t^i$ and a function g, such that $L(f) = D_x g$. Given a definite integral $F(t) = \int_a^b f(x,t)\mathrm{d}x$, then $L(F) = g(b) - g(a)$. In [9] an algorithm is presented for finding telescopers for algebraic functions. They first construct a differential operator L_1 such that $L_1(f\mathrm{d}x)$ has zero residues everywhere. Then recalling that exact differentials are those of the form $\mathrm{d}g$, they observe that in characteristic zero, the space of differentials with zero residue modulo exact differentials has dimension exactly $2G$ where G is the genus of the algebraic function field. This shows the existence of a differential operator L_2 of degree at most $2G$ with coefficients which are constants with respect to D_x such that $L_2 L_1 f \, dx = dg$. They show how to use an algorithm given by Barkatou in [1] to compute L_2.

In [8] a reduction based approach is used for finding a telescoper for algebraic functions. A reduction is a linear map $[\cdot] : D \to D$ such that $f - [f] = D_x g$ for some g in the space of functions D. The Hermite reduction algorithm for algebraic functions is a reduction in this sense. As before, after a change of coordinates we can assume that the differential df has no poles at infinity and in this case we have that $[f] = 0$ if and only if f is integrable as an algebraic function, i.e. $f = D_x g$. This can be used to find a telescoper if it is known that one exists. Chen et al. then introduce a polynomial reduction for reducing the order at infinity and show that the image of the combined reduction is a finite-dimensional vector space over the space of constants with respect to D_x. This guarantees that the algorithm to find a telescoper will always terminate. Then in [10] they show that the same construction can be generalized from algebraic functions to Fuchsian D-finite functions. In [7] they use lazy Hermite reduction to design a reduction-based telescoping algorithm for algebraic functions in two variables.

5 Divisor Arithmetic

Risch showed that the residues of the integrand were the key to constructing the logarithmic part of the integrand. If we let $r_{\wp_1}, r_{\wp_2}, \ldots, r_{\wp_k}$ be the non-zero residues of the integrand at the places $\wp_j$, then we can compute $c_1, c_2, \ldots, c_m$, which form a basis for the $\mathbb{Z}$-module generated by the residues. Thus each residue can be expressed as $r_{\wp_j} = \sum_{i=0}^{m} n_{j,i} c_i$, with $n_{j,i} \in \mathbb{Z}$. This allows us to associate a divisor $D_i = \sum n_{j,i} \wp_j$ with each of our basis elements c_i. Each of these divisors has degree zero, i.e. $\sum_j n_{j,i} = 0$. If the divisor corresponds to the poles and zeros of an algebraic function, g, then we have found a term $\log g$. If instead some integer multiple s of the divisor corresponds to an algebraic function, then we would have the term $\frac{1}{s} \log g$. In the thesis we show how to construct the basic divisors $\wp_j$ from a residue formula. Since we have changed coordinates so that the integrand has no poles and thus no residues above infinity, each divisor is represented by a fractional ideal which represents the multiples of the divisor ignoring behavior above infinity. To multiply divisors, we just multiply the corresponding ideals. The problem of deciding whether the divisor is principal becomes the problem of deciding if the ideal contains an element with no poles at infinity. Using the concept of normal basis at infinity, the divisor is principal if and only if one of the generators of its normal basis has no poles at infinity. This is easy to check. Independently Hess [26] developed a similar approach for the general problem of computing the spaces of multiples of divisors in algebraic function fields. Since his divisors can have support above infinity, he works with pairs of ideals, one with respect to the ring of integral functions at finite places, and the other with respect to the ring of functions which are integral at infinity. To compute a basis for the vector space of functions which are multiples of a divisor, it is enough to compute the intersection of these two ideals.

The group of divisors of degree zero modulo principal divisors is known as the divisor class group and can be represented by a geometric object called the Jacobian of our curve. An algorithm for efficiently computing in the Jacobian of a hyperelliptic curve was presented by David Cantor [6]. If one raises ideals to large powers, the generators become very large, which makes the computation very expensive. For hyperelliptic curves with only one place at infinity, each divisor class can be represented by a reduced divisor of the form $\sum_{i=0}^{g} \wp_i - g\infty$, where g is the genus of the curve. He gives an algorithm for the efficient computation of these reduced representatives while computing in the Jacobian. An optimized algorithm for the integration of algebraic functions defined on a hyperelliptic curve was given by Bertrand [2] using Cantor's technique for computing in the Jacobian. Subsequently Combot [12] gave an improved algorithm for superelliptic integrals, i.e. those expressed using the nth root of a polynomial. Volcheck [36] also developed a polynomial time algorithm for computing in the Jacobian of general plane curves.

6 Points of Finite Order

Risch showed that finding the logarithmic component of the integral of an algebraic function can be reduced to the problem of deciding whether or not the divisors of degree zero determined from the residues of the integrand are torsion or not, i.e. whether there exists some multiple which becomes principal. Risch observed that if the curve is reduced modulo a prime $\mathfrak{p}$ of good reduction which divides a rational prime p, then the torsion subgroup whose order is relatively prime to p is injected under the reduction map. Using the property that divisor class groups for curves over finite fields are finite, this implies that one can bound the order of the torsion subgroup using two primes of good reduction and taking the product of orders of the resulting groups. In fact the Weil bound for abelian varieties over finite fields

$$\#A(F_q) \leq (1 + \sqrt{q})^{2g}$$

depends only on q, the order of the finite field, and g, the genus of the curve. This bound yields a decision procedure for algebraic function integration, i.e. gives a stopping condition. In the thesis, each divisor associated with a log term is reduced modulo two primes of good reduction containing distinct rational primes p and q. If the order of the divisor was $np^r q^s$ where $\gcd(n, pq) = 1$, then the reduction modulo the prime containing p would have order $np^u q^s$ with $u \leq r$ while the reduction modulo the prime containing q would have order $np^r q^v$, with $v \leq s$, i.e. the prime to p part is preserved by the first reduction and the prime to q part is preserved by the second reduction. If the u and v inequalities are not satisfied or the component n which is relatively prime to both p and q is not the same in both reductions, then the original divisor was not torsion. Using a more refined analysis, one can give sufficient conditions for the entire torsion subgroup to be preserved under reduction modulo p. As demonstrated in the appendix of [28], if $\mathfrak{p}$ is a prime of good reduction,

$p = \mathfrak{p}^e q$ with ramification $e < p - 1$, then the reduction map is injective on the torsion subgroup. In particular, if the curve is defined over the rationals, and $p > 2$ is a prime of good reduction, then this condition holds. In this case reduction modulo a single prime is enough to recover the order of the divisor if it was torsion. However, to exclude non-integrable cases efficiently, it might still be advisable to use two primes which satisfy the ramification condition, and check that the order is the same modulo both primes.

When integrals depend on parameters, Davenport [16] observes that torsion points must satisfy Picard–Fuchs operators. So this gives a necessary conditions for a torsion point. A more subtle question about integrals involving parameters is whether there are specific values for the parameter under which the specialized problem becomes integrable. Although Davenport claimed that this can happen for only a finite number of values, in [37] the issue is shown to be more subtle, and they give conditions under which Davenport's claim is true.

7 Good Reduction

To conclude the points of finite order problem, we need a way to test good reduction. The thesis defined good reduction by requiring that the defining polynomial for the algebraic function field remains absolutely irreducible and has the same genus when reduced modulo "p". This is equivalent to the usual requirement of the existence of a non-singular model for the curve whose reduction modulo p remains non-singular. While this can be tested using the Jacobian test for singularity, the construction of a projective non-singular model requires some work. One approach is based on the integral basis which provides an affine non-singular model, and one can test whether or not the ring remains integrally closed. A similar test can be applied to a second affine chart containing the ring of integral elements at infinity. In fact, these two charts can be patched together to make a projective non-singular model. A simpler approach can be developed using results from Fulton's paper [22]. He views a projective non-singular curve as a branched covering of the projective plane. Assuming the characteristic of the reduced model is bigger than the degree of the covering, he shows that a sufficient condition for good reduction is that distinct roots of the discriminant of the covering remains distinct under the reduction. It can be shown that the same criterion holds for a projective singular plane model, i.e. if the characteristic of the reduced curve is bigger than the degree and the number of distinct roots of a discriminant of a projective plane curve remain the same, then there exists a non-singular model whose reduction remains non-singular [14]. Given a projective plane model $F(x, y, z)$ which we can assume is monic with respect to z, we can compute its discriminant with respect to z, $D(x, y)$, which is also homogeneous. If the characteristic of the reduced field is bigger than the degree of F in z, then the shape of the square-free decomposition of the discriminant being preserved under reduction guarantees good reduction of our curve. This gives an easy test that both the genus and the number of absolutely irreducible components are preserved under reduction.

References

[1] Moulay A. Barkatou. On Rational Solutions of Systems of Linear Differential Equations. *J. Symb. Comput.*, 28(4):547–567, October 1999.

[2] Laurent Bertrand. Computing a Hyperelliptic Integral using Arithmetic in the Jacobian of the Curve. *Appl. Algebra Eng. Commun. Comput.*, 6(4-5):275–298, 1995.

[3] Janko Böhm, Wolfram Decker, Santiago Laplagne, Gerhard Pfister, Andreas Steenpass, and Stefan Steidel. Parallel algorithms for normalization. *J. Symb. Comput.*, 51:99–114, 2013. Effective Methods in Algebraic Geometry.

[4] Manuel Bronstein. Integration of Elementary Functions. *J. Symb. Comput.*, 9(2):117–173, February 1990.

[5] Manuel Bronstein. *The Lazy Hermite Reduction*. Technical report, Technical Report RR-3562, INRIA, November, 1998.

[6] David G. Cantor. Computing in the Jacobian of a Hyperelliptic Curve. *Math. Comput.*, 48(177):95–101, 1987.

[7] Shaoshi Chen, Lixin Du, and Manuel Kauers. Lazy Hermite Reduction and Creative Telescoping for Algebraic Functions. In *Proc. ISSAC '21*, pages 75–82, New York, NY, USA, 2021. ACM.

[8] Shaoshi Chen, Manuel Kauers, and Christoph Koutschan. Reduction-Based Creative Telescoping for Algebraic Functions. In *Proc. ISSAC '16*, pages 175–182, New York, NY, USA, 2016. ACM.

[9] Shaoshi Chen, Manuel Kauers, and Michael F. Singer. Telescopers for Rational and Algebraic Functions via Residues. In *Proc. ISSAC '12*, pages 130–137, New York, NY, USA, 2012. ACM.

[10] Shaoshi Chen, Mark van Hoeij, Manuel Kauers, and Christoph Koutschan. Reduction-based creative telescoping for Fuchsian D-finite functions. *J. Symb. Comput.*, 85:108–127, 2018. 41st International Symposium on Symbolic and Algebraic Computation (ISSAC '16).

[11] John Coates. Construction of rational functions on a curve. *Math. Proc. Cambridge Philos. Soc.*, 68(1):105–123, 1970.

[12] Thierry Combot. Elementary Integration of Superelliptic Integrals. In *Proc. ISSAC '21*, pages 99–106. ACM, 2021.

[13] Olivier Cormier, Michael F. Singer, Barry M. Trager, and Felix Ulmer. Linear Differential Operators for Polynomial Equations. *J. Symb. Comput.*, 34(5):355–398, 2002.

[14] Giorgio Dalzotto, Patrizia Gianni, and Barry M. Trager. Good Reduction of Plane Curves. Unpublished, N.D.

[15] James H. Davenport. Algorithms for the integration of algebraic functions. In *Proc. EUROSAM '79*, pages 415–425, London, UK, 1979. Springer-Verlag.

[16] James H. Davenport. *On the Integration of Algebraic Functions*, volume 102 of *Lect. Notes in Comput. Sci.* Springer, Berlin, 1981.

[17] Theo de Jong. An Algorithm for Computing the Integral Closure. *J. Symb. Comput.*, 26:273–277, 1998.

[18] Jean Della Dora, Claire Dicrescenzo, and Dominique Duval. About a New Method for Computing in Algebraic Number Fields. In *Research Contributions from the European Conference on Computer Algebra-Volume 2*, EUROCAL '85, pages 289–290, Berlin, Heidelberg, 1985. Springer-Verlag.

[19] Dominique Duval. Rational Puiseux expansions. *Compos. Math.*, 70(2):119–154, 1989.

[20] Dominique Duval. Absolute Factorization of Polynomials: A Geometric Approach. *SIAM J. Comput.*, 20(1):1–21, 1991.

[21] David J. Ford. *On the computation of the maximal order in a Dedekind domain.* PhD thesis, The Ohio State University, 1978.

[22] William Fulton. Hurwitz Schemes and Irreducibility of Moduli of Algebraic Curves. *Ann. Math.*, 90(3):542–575, 1969.

[23] Patrizia Gianni and Barry M. Trager. Integral Closure of Noetherian Rings. In *Proc. ISSAC '97*, pages 212–216, New York, NY, USA, 1997. ACM.

[24] Gert-Martin Greuel, Santiago Laplagne, and Frank Seelisch. Normalization of rings. *J. Symb. Comput.*, 45(9):887–901, 2010.

[25] Charles Hermite. Sur l'intégration des fractions rationnelles. *Annales Scientifiques de l'É.N.S.*, 1:215–218, 1872.

[26] Florian Hess. Computing Riemann–Roch Spaces in Algebraic Function Fields and Related Topics. *J. Symb. Comput.*, 33(4):425–445, 2002.

[27] Erich Kaltofen. Fast Parallel Absolute Irreducibility Testing. *J. Symb. Comput.*, 1(1):57–67, 1985.

[28] Nicholas M. Katz. Galois properties of torsion points on abelian varieties. *Invent. Math.*, 62(3):481–502, Oct 1980.

[29] Robert H. Risch. *On the integration of elementary functions which are built up using algebraic operations.* Technical report, System Development Corp Santa Monica Calif, 1968.

[30] Robert H. Risch. The solution of the problem of integration in finite terms. *Bull. Amer. Math. Soc.*, 76(3):605–608, 1970.

[31] Daniel Schultz. Trager's Algorithm for Integration of Algebraic Functions Revisited. https://sites.psu.edu/dpsmath/files/2016/12/IntegrationOnCurves-2hhuby8.pdf.

[32] Barry M. Trager. Algebraic factoring and rational function integration. In *Proc. ISSAC '76*, pages 219–226. ACM, 1976.

[33] Barry M. Trager. *Integration of Algebraic Functions.* PhD thesis, Massachusetts Institute of Technology, 1984.

[34] Carlo Traverso. A study on algebraic algorithms: the normalization. *Rend. Sem. Mat. Univ. Politec. Torino*, 44:111–130, 1986.

[35] Mark van Hoeij. An Algorithm for Computing an Integral Basis in an Algebraic Function Field. *J. Symb. Comput.*, 18(4):353–363, 1994.

[36] Emil J. Volcheck. Computing in the Jacobian of a plane algebraic curve. In Leonard M. Adleman and Ming-Deh Huang, editors, *Algorithmic Number Theory*, pages 221–233, Berlin, Heidelberg, 1994. Springer Berlin Heidelberg.

[37] Umberto Zannier. Elementary integration of differentials in families and conjectures of Pink. In *Proceedings of the International Congress of Mathematicians—Seoul 2014. Vol. II*, pages 531–555. Kyung Moon Sa, Seoul, 2014.

[38] Hans Zassenhaus. On the Second Round of the Maximal Order Program. In *Applications of Number Theory to Numerical Analysis*, pages 389–431. Elsevier, 1972.